Prosser James

The Therapeutics of the Respiratory Passages

Prosser James

The Therapeutics of the Respiratory Passages

ISBN/EAN: 9783337191412

Printed in Europe, USA, Canada, Australia, Japan

Cover: Foto ©berggeist007 / pixelio.de

More available books at **www.hansebooks.com**

THE THERAPEUTICS

OF THE

RESPIRATORY PASSAGES

BY

PROSSER JAMES, M.D.,

LECTURER ON MATERIA MEDICA AND THERAPEUTICS AT THE LONDON HOSPITAL MEDICAL COLLEGE;
PHYSICIAN TO THE HOSPITAL FOR DISEASES OF THE THROAT AND CHEST; LATE PHYSICIAN TO
THE NORTH LONDON CONSUMPTION HOSPITAL; CONSULTING PHYSICIAN TO THE CHILDRENS'
HOME INFIRMARY, VICTORIA PARK; CORRESPONDING MEMBER OF THE ACADEMIES OF
MEDICINE OF LYONS, MADRID, AND BARCELONA, ETC., ETC., ETC.

NEW YORK
WILLIAM WOOD & COMPANY
56 & 58 LAFAYETTE PLACE
1884

PREFACE.

This book is not intended as a manual for students—it is hoped, indeed, that senior students as well as practitioners may peruse it with profit—but it is mainly intended for those who have left the schools and entered upon the responsibilities of practice. Such readers, having no longer the fear of examiners before them, may be interested in points upon which they would not otherwise dare to spend their time.

No doubt many faults will be found by critical readers, but not more than those of which the writer is sufficiently conscious. Some portions may be found too diffuse, others too condensed, and those who only refer to a question here and there may fail to see the connection between them, or perhaps to find that which they seek. Those who read the book through will not fail to discover the thread on which I have strung my beads—observe, I do not call them pearls. But as few may do this, it may not be amiss to say that there is a distinct purpose in the work, and it is with "malice aforethought" that some portions have been so extended and others so contracted.

Such as it is I commit this production to the candid consideration of my professional brethren, who have heretofore received my writings with so much favor.

Prosser James, M.D.

3 Dean Street, Park Lane, London.
October, 1884.

CONTENTS.

CHAPTER I.
INTRODUCTORY.. 1

CHAPTER II.
NUTRITION IN RELATION TO THERAPEUTICS............................. 3

CHAPTER III.
RESPIRATION.. 14

CHAPTER IV.
FOOD AND DIET—THE PROXIMATE PRINCIPLES OF FOODS—FOOD-STUFFS.... 21

    I.—Nitrogenous, Plastic or Albuminous Foods, 22 ; II.—Non-Nitrogenous, 23 ; III.—Inorganic Substances, 24 ; Nutritive Value of Foods, 26.

CHAPTER V.
PREPARATION OF THE FOOD-STUFFS—DIGESTIVE FLUIDS.................. 32

    Saliva, 33 ; Gastric Juice, 34 ; Pancreatic Juice, 36 ; Bile, 37 ; Succus Entericus, 38.

CHAPTER VI.
VARIATIONS IN THE DIGESTIVE PROCESS.............................. 39

    A.—The Body, 40 ; B.—Aliment, 40.

CHAPTER VII.
ALIMENTS AS REMEDIES—NUTRIENTS AND ANALEPTICS.................... 44

    Promoters of Construction, 45.

CHAPTER VIII.

Iron.. 56

CHAPTER IX.

Phosphorus and its Compounds 64
  Free Phosphorus, 68 ; Hypophosphites, 72.

CHAPTER X.

Aids to Digestion.. 74
  Stimulants to Digestion, 78.

CHAPTER XI.

Transfusion.. 79
  Injections into Serous Cavities, 82 ; Hypodermic Injections of Blood and Food, 83.

CHAPTER XII.

Water—Diluents—Beverages... 85
  Beverages, 89.

CHAPTER XIII.

Exercise and Rest.. 94

CHAPTER XIV.

Alcohol.. 97

CHAPTER XV.

Denutrients... 107
  Antiphlogistics, 107 ; Bleeding, 109 ; Counter-irritation, 117 ; Evacuants, 119 ; Mercury, 121 ; Diaphoretics, 133.

CHAPTER XVI.

Antipyretics.. 142
  Cold, 145 ; Quinine, 157 ; The Salicyl Compounds : Salicin, Salicylic Acid, and the Salicylates, 166 ; Kairin, 171 ; Chinolin or Quinolin, 174 ; Resorcin, 174 ; Veratria, 174 ; Digitalis, 175 ; Aconite, 177.

CHAPTER XVII.

NEUROTICS... 187
    Narcotics, 187.

CHAPTER XVIII.

PNEUMATICS.. 208
    Expectorants, 213 ; General Expectorants, 222; Antiseptic and Disinfectant Pneumatics, 249 ; Antispasmodic Pneumatics, 255 ; Sedative and Anodyne Pneumatics, 262 ; Contra-Expectorants—Central Pneumatics, 263.

CHAPTER XIX.

TOPICAL PNEUMATICS 276
    Methods of Inhaling, 282 ; Uses of Inhalations and other Topical Pneumatics, 294 ; Solids, 303.

INDEX 309

THERAPEUTICS

OF THE

RESPIRATORY PASSAGES.

CHAPTER I.

INTRODUCTORY.

The word therapeutics is often used in much too restricted a sense. It is derived from θεραπεύω, which is commonly translated, I cure, but which is susceptible of a far wider signification. It may be fairly rendered, I take care of, or render service to (the sick), and thus includes whatever relief to suffering or help to restoration may be conferred. Thus the therapeutist is the medical attendant, not the mere prescriber of more or less potent medicines. His materia medica includes all those materials which may be pressed into medical service. Thus it is as much his business to remove an accumulation of fluid from the pleura by means of the aspirator as to promote its absorption by the lymphatics, and the instrument is as much within his use as a remedy as the administration of the often inefficient absorbents. To apply a bandage or a splint, nay, to fashion one from any material at hand is as much a therapeutical proceeding as to order a poultice or prescribe a narcotic. Nor is this a novel doctrine, for during the thirteen dark centuries which followed Galen every one admitted the importance of the "non-naturals," as they were called by the Peripatetics, and we know that long before him these subjects were ably treated by Hippocrates, to whom, indeed, Galen looked up with reverence, and whose method he did much to revive, avowedly taking him as his guide, although it must be admitted that he was sometimes led away by the unsubstantial speculations of the dogmatists who repudiated for empty hypothesis the solid result of the patient and attentive study of nature inculcated by the sage of Cos. These non-naturals are commonly defined as the principal things which do not enter into the composition of the body, but are nevertheless necessary to its existence. Galen called them the procatarctic causes, προκαταρκτικος, principal, from προ, and κατα and ἄρχομαι, I begin. He considered that when well used and properly disposed they contributed to the health, but otherwise to the derangement of the system. They are air, meat and drink, exercise and rest, sleep and waking, retention and excretion, and the passions and affections of the mind. But long before the Hippocratic writings laid such broad foundations for Greek medicine the importance of such subjects had attracted the attention of the sages of earlier civilizations. Thus we find diet and regimen treated of in the oldest records of Hindoo medicine. These are written in the most ancient form of Sanscrit, and are believed to extend from the third to the tenth century before the Christian era. The Buddhists, too, have records of early thoughts on these subjects, on which even ancient Chinese literature is not destitute of advice, overlaid though it undoubtedly is with accumulations of the strangest and most preposterous fancies. But it is not worth while to pursue this subject, for it is obvious that the effects of food, air, and other agencies would be observed by every race quite as early as the influence of medicinal plants, and we know that the observation of such influences dates from the remotest times, while scarcely a savage tribe is known but possesses, either as a treasured secret or as a common belief, a conviction of the value of some kind of herb or other substance for the relief of disease.

CHAPTER II.

NUTRITION IN RELATION TO THERAPEUTICS.

THE body may be regarded as an apparatus for the manifestation of energy, and it possesses the power of self-repair and self-adjustment. Growth, repair, and the production of energy constitute the work of the animal machine. All substances by means of which the body can maintain its nutrition or either of the processes named may be called food or aliments (*alimentum*, food, nutriment, from *alo*, I nourish). Without a due supply of food nutrition fails, and the body can neither repair its waste nor develop energy. The action of the body on food may be said to be broadly one of oxidation. On the other hand, the result of vegetable life is deoxidation or reduction. The vegetable is an agent for transmuting carbonic acid, water, and ammonia into other compounds, and, further, light and heat into chemical affinity. The work of the animal is to carry the vegetable material into higher structures, then to destroy these compounds and change their affinities into other manifestations of energy—ultimately into heat. Thus matter and force are alike ever moving in a circle. From the mineral kingdom the vegetable prepares compounds out of which the animal can construct its tissues, which having served their purpose in the body are oxidized and then returned to the inorganic world.

As the production of energy is mainly, if not entirely dependent on oxidation, the atmospheric oxygen might perhaps be included in the above wide definition of food, but it is more conveniently considered further on, and moreover such a use of the term does not well accord with the idea universally entertained of food, which rather corresponds with the word aliment. Water, however, as the one necessary beverage, may very well be included. Medicines do not maintain nutrition; sometimes they interfere with it, and are often said to be substances which act on unhealthy nutrition. Nevertheless no hard and fast line need be drawn. Thus we have analeptics (from αναλαμβανω, I restore), which restore, and sometimes we speak of a class of nutrients. Poisons may either induce unhealthy or prevent healthy nutrition. But here again there is no sharp line between medicines and poisons. Nature, indeed, does not rejoice in hard and fast lines, so that, useful as we find them in classification, it is well to remember that this is founded on artificial distinctions, and consists of groupings to aid the memory. There is philosophy in the popular proverb in its twofold reading: "What is one man's food is another man's physic," or "poison," as some people phrase it.

Further, we may safely assert that food is, in numerous cases, the one medicine of prime value, as will appear if we bestow a little closer attention on nutrition. This process must consist of both waste and repair, and may therefore be looked at in this twofold aspect. It is necessary, first, for the aliment to be presented to the tissues in a condition in which it can be utilized, and then it must become incorporated with or assimilated by these tissues, while at the same time the worn out particles must be removed. In lowly organisms like the amœba the process seems tolerably simple: the jelly-like body simply flows around and encloses such food as comes in its way; from this is dissolved apparently the soluble portion which thus unites with the jelly-like mass, and this in its turn extrudes the insoluble residue in a manner somewhat the reverse of that by which it surrounded the whole. But the higher we rise in the scale of animal life the more complex becomes the process, and the more its several stages are differentiated, and so we have to consider the nutrition of our own bodies through the process of digestion, absorption, assimilation, nutrition proper, or the renewal and waste of tissue, and elimination and excretion; that is to say, it is essential for our food to undergo certain preliminary changes before it can be presented to the tissues in such a form that they can act upon it or be acted upon by it.

Some substances may be taken which are at once rejected by the digestive organs as useless or inconvenient, but with them others may be ingested which are of great value, and are therefore at once turned to account. The nutritious particles, then, are taken up, changed by digestion, absorbed by the alimentary canal, poured into the blood, of which they thus become a part, and are carried by the circulation to every portion of the body. From the blood are derived nutrient fluids which escape from the minutest vessels and continually bathe every particle of every tissue, and while thus bathed it would seem that each particle yields up to this bath whatever is useless or worn out, while at the same time all such effete material is replaced by the fresh contained in the bath. A double process is thus continually going on, waste and repair, and both are attended by the liberation of energy. On the one hand we have wear and tear, the vital machine, as we may say, is being constantly consumed in doing its own work: this is a sort of oxidation, a slow combustion in which the tissues are being burnt, and the residue of such combustion, the ashes, so to say, must be got rid of, and accordingly are thrown into the surrounding stream and returned to the blood, which thus becomes a kind of sewer, and therefore in its turn has to be purified in order to fulfil its double office. Provision is accordingly made for removing these useless or worn-out products of combustion. On the other hand, to counteract this continual destruction, or wear and tear, we have an equally constant process of construction, or repair, uninterruptedly going on, so that as any used up product is thrown into the blood new material is taken from the same fluid to supply its place; all such new material, as we have seen, must be derived from the food. It is true we cannot observe the intimate molecular movements of nutrition as we watch the reactions of chemicals in the laboratory, but we can trace the aliment into the blood; this fluid finds its way to every part; we next trace the disappearance of certain nutrient ingredients, and then we find, not them, indeed, but the products of their combustion among the constituents of the excretions. And these continual changes, these marvellous metamorphoses are necessarily accompanied with the liberation of energy which ultimately manifests itself as heat. So long as all these changes continue we have life; the arrest of them all is death. In health they all go on harmoniously, there is a constant balance, so to say, between destruction and construction. Let this balance be disturbed and disease in some form will result. That some disturbance may occur within certain limits without much injury is clear, especially if only for a short period, but a prolonged disturbance, even if in so slight a degree as to be imperceptible, will in time inevitably make itself manifest. To compensate for slight disturbance in one direction we have a sort of reserve: the body can lay by a store of material and force upon which it may draw when the aliment is insufficient, and to prevent the effects of temporary deprivation. Hence nutrition is often compared to banking. There is a balance, which may vary from time to time in amount, but we must pay in as well as draw out, or a time will come when our checks are valueless, and that time will depend on the amount of the balance and the demands made upon it.

So, as we have said, the body is a sort of machine, an apparatus for the manifestation and employment of energy, every such manifestation being, of course, attended by wear and tear, but it possesses the power of self-repair, and the material for this purpose is obtained from the aliment. Life, then, may be said to consist in a series of perpetual changes in the tissues of the body, which in accordance with a physical law necessarily give rise to the liberation of energy, and these changes are of two kinds, destructive and constructive. If this be so, the greater or the more rapid the changes the more actual the life, and hence it has been plausibly argued that the more rapid the destruction of tissue, provided aliment be supplied for the construction of new, the greater will be the degree of health. With due limitations such a statement may be allowed to pass, but must not be accepted as certainly true to any extent and in all circumstances. We may, indeed, admit that work necessarily being attended by waste, increase of work should lead to more rapid waste, as it undoubtedly does, and this again to more rapid repair, more active life, greater health. This is so within certain limits, exercise is enjoyable, the waste caused by it is repaired quickly, energy is freely liberated, and bounding health with full joyous life results. But there is a point when fatigue comes on and we find that each individual machine, though all are constructed on the same lines, is specially adapted for its own work and its own rate of performing that work, however true it may be, that as a self-repairing, self-growing, self-adjusting, and even self-improving machine, it may within certain limits attain toward the perfection of others.

There is one more point. Strange as it may appear, when we view life from this standpoint it is closely related to death. We speak of the stoppage of the machine, but the analogy fails. Construction, repair ceases, but destruction goes on, or rather, after a pause a new mode of destruction begins. We have seen that all the waste is indeed only a sort of combustion or oxidation of our tissues. Well, after death, as we call it, these tissues decay, and what is that but oxidation? We consign the no longer animate clay of our beloved ones to the grave, where it slowly moulders back to mother earth. The result would be the same if we adopted the plan of cremation, and burnt, that is, rapidly oxidized, their mortal remains. In either case the products of the oxidation of the body are the same—chiefly water, carbonic acid, ammonia, and salts; and these are the substances which during life result from the oxidation of the tissues and food which, as no longer serviceable, are excreted. So that as in life organic material is continually being oxidized and returned as waste to the inorganic kingdom, so in death the same result happens to all the body at once. The stages, indeed, through which the particles pass may vary, the rate of travelling may differ, the stopping-places may not be the same, but the ultimate goal, the end of the journey is the same in death as in life. "So in the midst of life we are in death," as Luther said, whose phrase has been copied into our burial service. Had Watts a poetic prescience of this when he wrote:

> "The moment when our lives begin
> We all begin to die?"

To return. For our present purpose we may regard healthy vital action, life, if you will, as a process of continual destruction on the one hand, counterbalanced by compensatory repair and construction on the other. But this normal condition may vary, and then we have disease. Always during life destruction is going on; always, therefore, its products are accumulating and always must be cast off. Now it is easy to see how these actions may vary in rate or be otherwise imperfectly carried on, and that such interruptions or imperfections may occur in any of them.

The rate of the destructive changes may be too slow or too fast, and in either case mischief will ensue. The work may be hindered, or even arrested at any point. The removal of the waste products may be sluggish, and then accumulation may take place within the tissues, at the starting-point, or they may be poured into the blood and accumulate there from sluggishness in the organs of excretion. These products are for the most part highly injurious; when carbonic acid is prevented from being thrown out by the lungs death soon occurs, as it also does, though less quickly, when either the skin or the kidneys cease to work. Then again the tissues may be imperfectly oxidized, in which case, instead of products precisely adapted for rapid removal, other intermediate compounds may be thrown into the blood—cinders, as it were, instead of ashes—and these may set up disease by their directly poisonous qualities, or by the difficulty experienced in removing them. Uric acid is such an intermediate compound, taking the place of the much more soluble urea, and the relation of gouty affections to the presence of that acid in the blood has been carefully investigated. The whole machinery of the body is extremely delicate, precisely adapted to accomplish its ends, and when any part fails to efficiently perform its work, or when a foreign substance ill adapted to be utilized is present, we need not wonder that such a deviation from the conditions designed for its operations is so injurious when we see that a grain of sand in the oil with which a watch is lubricated may bring its movement to a standstill.

Further, even when elimination is equal to the task, it is easy to see that destruction may vary in rapidity. Such variation within moderate limits may occur in health, but if prolonged or excessive must give rise to disease. The body cannot work with tissues that need renewing, and therefore a certain rapidity of destruction and repair, or perhaps we should rather say renewal, is essential. Indeed, the more rapid the changes within a certain degree the more vigorous is the health, so that slowness of the process may leave imperfection in the organs. On the other hand, excessive rapidity of destruction is also dangerous. The power of repair is limited, and if destruction goes on so rapidly that repair cannot keep pace with it emaciation will take place. The body will indeed feed on itself, as we say, for a time, use up its stores which have been laid by, but when they come to an end it will then consume itself. This is not the ideal of health; rather it constitutes one form of disease. The body wastes, though certain functions are going on actively, and unless in some way the balance can be redressed, the end of such life is inevitable.

Now let us turn to the other side—construction or repair. This, like its opposite, may vary in rate or may be impeded or arrested, and that at any point. If the rate of construction more than keeps pace with wear and tear some of the superfluous supply is rejected, cast out of the body as useless, but some of it may be utilized, stored up in the body, and thus we may have general increase of weight instead of emaciation, and this in some circumstances may go on to corpulency or obesity. The necessity for abundant supplies to produce increase of weight is well understood by feeders of stock. We have seen that repair is effected from aliment carefully prepared within the body itself—elaborated from the food-supply. Now this food-supply may be cut off from the outside, the body starved, as we say. So the processes of elaboration may be impaired or arrested, or the aliment, when duly prepared, may be prevented from reaching the tissues. In either of these cases the same result ensues—death by starvation. So, also, instead of absolute want of aliment there may be deficiency, when nutrition will flag; or again, instead of insufficiency of nutrient material it may be of inferior quality, in which case other consequences are observed—those spoken of as malnutrition. The results of deficiency of nutrient material resemble those of excessive waste, for wear and tear is always going on in the living body, even in the most complete repose; energy has to be employed and its liberation is attended by waste, oxidation, consumption of material. Merely to keep up the animal heat a large consumption of fuel is necessary—indeed, most of the food of animals may be regarded as such fuel—so that insufficiency of this fuel leads to loss of power as well as to decrease of weight. It is in this way that absolute deprivation of food soon brings about death.

Starvation has been carefully studied by Chossat.[1] From his experiments it would appear that death was greatly accelerated by a low temperature and delayed by warmth. Some of the animals, moreover, on approaching death were revived by the application of warmth and, being then supplied with food, recovered. On the other hand, a reduction of the external air hastened death, just as we know that exposure to cold in conjunction with deprivation of food is so much more rapidly fatal. In all cases there was a great loss of heat. At first the fall was slight and gradual, but later it became rapid, and when it reached twenty-nine or thirty degrees Fahrenheit below normal death always took place. The variations of temperature were, however, not confined to a regular progressive fall; on the contrary, there were frequent rises and falls, both of them four times greater than the diurnal variations of health, presenting the phenomena of an extreme degree of hectic. Besides the temperature, the degree of emaciation served as an index to the approach of death, which always took place as soon as the loss of weight amounted to four-tenths of the normal, from which, perhaps, we may learn to measure the danger of progressive wasting. It seems as if the arrest of nutriment was at first partly compensated for by slower waste; for example,

---

[1] Chossat: Recherches experimentales sur l'inanition. Paris, 1843.

Chossat found frequently that constipation occurred, and this we should anticipate, but he reports that it was afterward sometimes replaced by a peculiar diarrhœa. Perhaps, however, we ought scarcely to denominate as diarrhœa the evacuation of the grass-green dry fæces, or even the liquid saline matters he observed, especially as in most cases of persons who have been enclosed in mines, or otherwise accidentally deprived of food, the constipation seems frequently to have been nearly complete.

So somnolency or stupidity may for a time mark the mental state, though later on convulsions or delirium may ensue, in either of which life may fail, or it may terminate in gradually increasing torpidity. It is interesting to note that nervous tissue—the highest, as we regard it—seems the last to fail, and after death was often found to have preserved its full weight, as if the highest function of animal life must be supplied at whatever cost to the members.

We see, then, how the consequences of starvation naturally resemble those produced by the interrupted supply of food brought about by disease. Stricture of the œsophagus, cancer of the stomach, pressure of a tumor on the thoracic duct alike interrupt the supply of aliment to the tissues, if more gradually yet no less certainly, and the results are the same, though the symptoms may vary. The emaciation would alone be fatal, even if other complications did not arise to hasten death. In other cases where food is insufficient, where it is not properly digested, where assimilation in any of its stages fails, we have progressive emaciation, which will be fatal unless the obstacle to the flow of nutriment from the mouth to the tissues can be removed in time. Again, the aliment may be abundant but its quality unsuitable, in consequence of which the digestive organs, although active, may be unable to extract what is wanted for the system. Or the fault may be in some point of the digestion, or elaboration of the aliment, so that only unsuitable material is presented to the tissues, which are thus starved, or half-starved, or ill-fed. If all the material accessible were to be unfit for use in a given tissue, and therefore rejected, that tissue would be starved; if the supply were too limited it would be half-starved or deprived of nutriment in proportion to the limitation; if unsuitable aliment were taken to make the best of, then the tissue would deteriorate. Such products of imperfect elaboration as are rejected must be carried away by the blood in order to be got rid of by excretion. If they were to be retained in the blood its quality would be deteriorated; and, indeed, they might act as powerful poisons, as we know that many used-up products do; or again, they might set up disease in the organs on which was thrown the work of excreting them.

What we have said of nutrition has been for the most part with reference to the body as a whole, but very similar statements might be made as to each of its parts, and we may study the process as a local one with equal profit. Interrupt the blood-supply of a part and it perishes; prevent the blood from leaving it and the effects are also disastrous; changes in the rate of circulation in parts are common enough in disease. But it is not necessary here to dwell further on interferences with local nutrition.

With regard to the function of aliment, though we have hitherto chiefly kept in view the maintenance of tissues and the repair of their wear and tear, we must not forget its work as a liberator of energy. This is manifested as movement, either muscular or molecular, as chemical or electrical action, and ultimately as heat. Indeed, the larger part of our food is required for keeping up the animal heat; this explains why a fall of the temperature of the body so constantly attends starvation; why when function is active there is increase of heat, and vice versa.

Here it may be asked, Must, then, all food be converted into tissue before it can generate action; may not energy also arise from the transmutations of nutrient material dissolved in the blood, which may itself be regarded in many respects as a fluid tissue? This question has given rise to much debate among chemists and physiologists, some holding one view and some the other. Liebig maintained that muscular force is always obtained from the nitrogenous constituents of muscle. At one time, he regarded the action as a simple combustion, but later he modified his view, supposing that the oxygen might not necessarily cause though it must take part in the action, and that no oxidation of nitrogenous material could occur until it had become organized tissue, so that all the energy of the animal machine must be derived from its own combustion. Others, however, hold that energy is liberated by the combustion of alimentary material, generally non-nitrogenous, in the blood, and it has been shown that for a short period great labor may be undergone on a non-nitrogenous diet, and that while there is always a relation between the quantity of nitrogen ingested and excreted, nevertheless this last amount is not so related to the work of the body. Still, no manifestation of energy takes place without the participation in some manner of nitrogenous structures, which, if they do not carry on, at any rate appear to initiate the action. In this action the tissue is worn out. It has been shown by Pettenkofer and Voit that the tissue-changes determine the absorption of oxygen, its conversion into ozone, and its use in combustion; so that its appropriation and use appears to be entirely dependent on the action of these tissues.

Hitherto nothing has been said about the control exercised over nutrition and other processes by the nervous system, because the points to be principally enforced are presented more simply by the omission; and our readers as medical men will be able to supplement the sketch given. Whether great believers or not in the functions ascribed to trophic nerves, they will know well enough that every action of the organism is dominated by the nervous system; and we may defer further consideration of its influence over nutrition in its several stages with the remark that such influence may also be varied somewhat in health, much more in disease, and that the variations may appear as increase, decrease, or perversion of the normal action.

Besides, we cannot undertake to give a complete dissertation on the physiology of nutrition; but only to set forth what may seem necessary for the purpose in hand. Some readers, indeed, may already be impatiently asking, What has all this to do with therapeutics and more particularly the therapeutics of respiratory diseases? To them I reply, Much every way. In the first place, as already stated, food *is* physic, and what has been said as to nutrition goes far to show how and why. The balance between destructive and constructive tissue-changes being disturbed, it may often be restored by the proper use of aliments or nutrients, or, as we call them in such case, *analeptics*. So, too, it may happen that in opposite conditions the restriction of nutriment may suffice. In a third case, the impaired or perverted nutrition may be remedied by regulation of the diet. In all such cases alimentary substances constitute our materia medica and the principles of dietetics our therapeutics. It may be added that in few diseases are questions relating to nutrition of greater importance than in those that affect the respiratory organs.

In the next place, articles which have little or no claim to be regarded as food, but which are commonly spoken of as medicines, are taken into the system by the same channels as food, and follow the same route to the inner arcana of the economy, and having produced their influence there are again carried out of the body by the same way, and are therefore liable to be acted on by the same influences throughout that journey. The gastric juice, the liver, the pancreatic and intestinal juices may all modify our medicines or be modified by them. So, too, the blood may act upon them or be acted upon by them before they can reach the organs or tissues on which their influences are exerted. So, again, the several parts with which they come into contact—stomach, alimentary canal, etc.—before entering the system may be influenced by their local action. It is true that some of them are at once taken into the blood without suffering any change in the digestive organs; others we deliberately introduce into the cellular tissue, and so avoid the route by the alimentary canal; others, again, we introduce into the lungs by inhalation, obtaining in this way, in some cases, invaluable local action, and in others exceedingly rapid effects on the whole system in consequence of their speedy absorption; others, again, we direct immediately into the blood-stream through the veins. But the statement remains, that, whatever the channel, it may or may not be locally affected by the medicine. In the former case the action may be reflected, so to say, and give rise to an indirect effect upon the system. From the air-passages, when absorbed, medicines are taken up unchanged by the blood, but in the alimentary canal they are liable to be acted upon by the digestive fluids. It is clear enough that the gastric juice may be reinforced by acids or neutralized by alkalies, but other changes are also important, though too generally neglected.

Passing to the next stage, the effects of our remedies may be principally manifested in the blood, which may be impoverished or enriched or otherwise changed by their action, as in the case of hæmatics (αἷμα, blood), hæmatinics on the one hand or spanæmics (σπανός, poor, and αἷμα, blood) on the other. Having thus reached the circulation the medicines, either in their original form or changed by the blood, are carried by it to the various tissues, and may produce their principal effects on them, as for example neurotics (from νεῦρον a nerve), which act upon the nervous system. Or, again, they may influence the function of organs, such as the heart, on which cardiac stimulants or cardiac depressants produce their effects, and of course through that the whole vascular system may in some cases be affected. Still later, having done certain work or liberated energy—either hastening, retarding, or otherwise changing the nutrition of the body or of some part of it, or of special tissues, or perhaps having influenced some special organ—such medicines, or "poisons," may be either deposited in the tissues or the viscera, or, as is much more frequently the case, may be rapidly eliminated. In this last stage of their journey in and out of the system, during the very process of excretion they may affect considerably any parts through which they pass—again hastening, retarding, or interfering with the work of the excretory organs, most frequently the first, as we see in the extensive group of evacuants.

The medicines we thus administer are very various in their nature and properties. Some are simple salts on which the body often appears to have little influence and which are quickly restored to the mineral kingdom. Others exercise much greater influence, often chiefly or entirely of a chemical kind, and after some changes are excreted in such compounds as might be anticipated. Others linger still longer in the system—if they do not become tissues, at least form some close union with them, and consequently have some claim to be regarded as supplementary food or a partially necessary aliment. Iron, for instance, must be supplied to the blood, which requires it for the production of hæmoglobin. The amount required is usually obtained from the food, but sometimes—perhaps from excessive waste, perhaps from deficient supply, perhaps in consequence of loss of blood or of some disturbance of the balance not so easily traced—it is not present in sufficient quantity, and then the blood thus impoverished, so far as concerns one of its ingredients, fails to perform its work perfectly, and anæmia (from a, privative, αἷμα, blood) is said to be present. This may often be removed by a suitable diet, because the iron which the system lacks is present in appropriate food. So it may by fresh air, exercise, and other influences which stimulate oxidation, and in other cases by chalybeates (from Chalybs, χάλυψ, a name for the metal derived from the Chalybes, a people in Pontus, who used to work it), because they put within the system the missing metal, which may thus far be considered an aliment. And so with regard to some other medicines. But further, not only may nutrition be modified, but all organs, all tissues, all fluids, perhaps all modes of energy developed within the body, may be influenced in some way by agents introduced or brought to bear upon it, and which therefore constitute our materia medica.

CHAPTER III.

RESPIRATION.

We have seen how nutrition consists of a perpetual balance between destruction and construction. Tissue waste is incessantly going on, and new material is as constantly brought to take the place of the old, while the débris must as regularly be removed. The process is analogous to that of respiration. We inspire fresh air, we expire carbonic acid and some other material; and this work, too, may be impeded or may be arrested, and that, also, in any part of its progress. The oxygen passes into the lungs, is taken up by the blood, is carried by the circulation to the tissues, where its work is done, is then returned as carbonic acid in the venous stream, brought back to the lung and expired into the atmosphere. If we cut off the supply of fresh air asphyxia is speedy, if we limit its quantity the process is slower, and it matters not whether the access of air be prevented by something external or within the body. Closure of the larynx or trachea by the intrusion of a foreign body is as fatal as hanging or enclosing an animal in a vessel containing no oxygen. Obstruction of the bronchi or pulmonary cells by the products of disease may also asphyxiate. So, also, the atmosphere may be so contaminated as to be unfit for the purposes of respiration. Again, to prevent the exhalation of the carbonic acid is just as fatal as to arrest the entrance of air, and such prevention may occur at any part of the journey. How often in respiratory diseases is the patient poisoned by the carbonic acid which he is unable to get rid of, asphyxiated by the product of his own function which the lungs fail to excrete. So, too, the rate of respiration, which varies somewhat in health, may in disease present great modifications.

But we must examine the function of respiration a little more in detail. The first point is, that oxygen, being continually used in the tissues, must constantly be brought to them. This is accomplished by the circulation, which also takes away the carbonic acid and other refuse. The oxygen passes into the blood mainly by diffusion, by which process also carbonic acid passes from the tissues to the blood and from that to the air. Although a portion of the interchange between the body and the atmosphere takes place elsewhere the lungs are the chief organs of this process. Through the alternate inspiration and expiration a regular ebb and flow of tidal air is established and the stationary air is renewed by diffusion from this. The ordinary amount of air in the lungs of an adult after a full inspiration averages 330 cubic inches. Of this the tidal air, or that which is inspired and expired in every respiration, is estimated at only 20 cubic inches; the complemental air, or that which can be inspired after an ordinary inspiration, 110 cubic inches; the reserve, or that which can be expired after an ordinary expiration, and the residual, or that which remains after the fullest possible expiration, are each estimated at 100 cubic inches. What we call the vital capacity is the sum of the tidal, complemental, and reserve, that is to say, the total amount of air which can be given out by expiration after the fullest forcible inspiration. This in a man 5 feet 8 inches high, averages 230 cubic inches, and this is what we measure by the spirometer. Dr. John Hutchinson showed ("Med. Chir. Trans.," 1858) that the vital capacity differs with the height, posture, weight, and age of the patient, but is much more modified by disease. Every additional inch of stature from five to six feet enables the patient to breathe eight cubic inches additional. As the movements of the thoracic walls are more free in the erect posture the vital capacity should be taken in that position. It increases from fifteen to thirty-five years of age and then diminishes to sixty-five. Usually it increases with the weight of the person. There is a great diminution in almost all thoracic diseases, not only in tumors, abscesses, and effusions, which may be said to displace the air, but even in the early stage of phthisis, in which the respiration is evidently weakened. Instead of estimating the air contained in the chest we may measure the movements of its walls. Sibson's chest-measurer registers the changes in the horizontal posture in the antero-posterior diameter of the chest; Ransome's stethometer measures the movements of the walls in several diameters; Riegel's, Marey's, and Burdon-Sanderson's ("Hbk. Phys.," p. 291) furnish graphic records.

In ordinary inspiration the two principal means of enlarging the chest are the descent of the diaphragm and the elevation of the ribs; expiration is mainly accomplished by elastic reaction. In labored inspiration all the muscles which can elevate the ribs, or which can contribute to the support of those which can, are brought into play; and in labored expiration all those which can depress the ribs or press on the abdominal viscera or afford a fixed support to those which act in this way, are thrown into powerful action. The associated respiratory movements, both facial and glottic, may be observed even in tranquil respiration, but are greatly exaggerated in dyspnœa.

In respiration the air loses four or five per cent. of oxygen but gains four per cent. of carbonic acid and becomes saturated with moisture. Hæmoglobin is the purveyor of oxygen. This body, without any further change, takes up in the lungs oxygen, holding it loosely in combination and may then be called oxyhæmoglobin, which is carried by the circulation to the tissues. These rob it of its surplus oxygen, and so it becomes again reduced to hæmoglobin and must go back to the lungs to obtain a further supply of oxygen. The amount of change varies in different tissues, and, indeed, in the same tissues at different times, *e.g.*, whether they are at work or at rest. It will be observed that we treat oxidation as taking place in the tissues, though it has sometimes been thought that part at least of the process was carried on in the blood, as if certain oxidizable substances were taken up into the blood and there oxidized. If such were the case, the chief oxidation would be in the blood, but all available evidence points to the conclusion that the locality of oxidation is the tissues. It is true that certain reducible substances may be detected in the blood, but only in too small quantity to be of any account; moreover, shed blood has no action on various substances which are unquestionably oxidized in the living body. Grape-sugar, for example, when added to blood is not oxidized, even though the mixture be kept at the temperature of the body; and further, a slight excess of sugar in the system instead of being oxidized, is discharged unchanged. So too, pyrogallic acid passes through the system without change, although it is a very oxidizable body. Citric and other acids are scarcely at all oxidized in the body, and even when given in combination with alkaline bases are only partially oxidized. We conclude, then, that it is in the tissues that the change takes place, and this explains why certain diffusible substances which the tissues refuse to take up are removed by the secreting organs, whereas it would seem that they must be oxidized if that process took place to any considerable extent in the blood.

The blood, then, by virtue of its corpuscles, these by virtue of their hæmoglobin, takes up oxygen in the lungs, becomes, as we say, arterial, and goes as such to the tissues where the oxygen-tension is low. They therefore receive oxygen from the blood and store it up in some stable combination, leaving the hæmoglobin reduced, that is, more or less of it, according to the activity of the tissue. As carbonic acid is continually produced in the tissues, its tension is always higher there than in the blood, and therefore it passes into the stream. Accordingly, the venous blood has not only its hæmoglobin reduced but its carbonic acid tension increased, so that on reaching the lungs the gas streams through the vascular and alveolar walls till the tension without is equal to that within. Then the stationary air, having lost oxygen and gained carbonic acid, has a lower oxygen tension and a higher carbonic acid tension than the tidal air, in consequence of which rapid diffusion between these two takes place, so that at last the air issues from the body poor in oxygen but rich in carbonic acid. It is calculated that it requires from six to ten respirations to completely renew the air in the lungs.

Variations in the atmospheric pressure necessarily produce important effects. Living at the bottom of an atmospheric ocean some fifty miles deep, the normal pressure of nearly fifteen pounds to the square inch gives us thirty thousand to forty thousand pounds total pressure on the surface, according to the size of the person. This is at the sea-level. On lofty mountains it is considerably less and it diminishes rapidly in a balloon ascent. To gradual changes we can accommodate ourselves, but in mountaineering they are perceptible and, of course, aggravated by the necessary exertion. In ballooning the ascent is so sudden that very unpleasant symptoms are often produced from the diminution of the pressure. Thus the congestion of the capillaries of the skin and free mucous surface leads to sweating, mucous discharges, and sometimes hemorrhages. The capillary resistance being lessened, the heart beats more frequently, and some dyspnœa is felt, the respirations are deeper but somewhat irregular. Muscular weakness and exhaustion are also complained of, with either dyspnœa or oppression and constriction of the chest. Perhaps these symptoms are due to insufficient oxygen and imperfect elimination of carbonic acid. Vomiting is sometimes produced, which may, perhaps, be attributed to stimulation of the vagal centres in consequence of deficient oxygen. As the blood is drawn to the surface the brain has a less supply and this often leads to faintness. Diminished secretion of urine may be explained in the same manner.

The effects of increased pressure may be seen in descending below the sea level, as in mines and in works carried on in an atmosphere of compressed air, as in the diving-bells or caissons for laying the foundations of bridges. In this case the skin becomes pale, perspiration is diminished, and the respirations are from two to four less in the minute; inspiration is easier, but expiration is prolonged, and there is a distinct pause between the movements; the vital capacity increases, so does the secretion of the kidneys; muscular efforts are made with more energy and activity; the heart, meeting with more resistance from the contraction of the cutaneous capillaries, beats slower, and the pulse-curve is lower; there is a subjective sensation of warmth. Persons should not return suddenly to the normal pressure or the blood will be drawn to the surface, as if the body had been placed in a vacuum, and there will be hemorrhage from the nose, mouth, ears, etc., while the sudden removal of blood from the centres may bring on paralysis.

The oxygenation of the blood may be impeded or totally arrested by various circumstances, as by want of oxygen when an animal is placed in confined air, in a vacuum, in an atmosphere of nitrogen, or under water; in the fœtus from the separation of the placenta or pressure on the cord; by interruption of the cutaneous respiration by varnishing the skin. We may also have closure of the air-passages, either partial or complete, temporary or permanent, as in strangulation, spasm of the glottis, pressure from without by tumors, accumulation within, as of mucus in the bronchi or even hemorrhage. Then, again, we may have collapse of the lung from the entrance of air or liquid into the pleura, as in pneumo-thorax and pleuritic exudation. So, too, partial destruction of the lung, as in phthisis, or cessation of respiratory movement, or embolism of the pulmonary artery will bring about the same condition.

In a chamber of rarefied air an animal will die before the oxygen is exhausted, a result attributed by Hoppe-Seyler to the evolution of gas in the blood, which disturbs the circulation. In compressed air, the death is believed by Paul Bert to be due to the elimination of carbonic acid being hindered.

In the complete absence of oxygen, or when the gas is entirely prevented from reaching the alveoli, death speedily ensues. As the oxygen diminishes the carbonic acid increases, the accessory muscles of respiration are then called into action and the movements are deeper but slower. This degree of dyspnœa is a compensatory act, bringing in more oxygen when it can be obtained and ceasing on its arrival. A further want of oxygen sets up clonic convulsions—that is, spasms of all the muscles of the body. These seem to be due to stagnation of blood in the brain—venous blood, poor in oxygen, rich in carbonic acid—for they occur when arterial blood is prevented from going to the brain, as by ligature of the carotids and vertebrals, as well as when venous blood is prevented from returning, and also after hemorrhage. A further want of oxygen causes the centres to lose their irritability and asphyxia results.

When there is a long-continued want of oxygen, but only of a moderate degree, there is an adjustment between the requirements and the supply, the functions associated with oxidation diminish, the temperature is lower, the muscles are flaccid, the respiration is more rapid, the smaller arteries and capillaries relax, and the darker color of the imperfectly arterialized blood is seen through the lips, mucous membranes, and portions of the skin. This is cyanosis.

The supplementary respiration performed by the skin resembles in character that carried on in the lungs. In both we see a rich plexus of capillaries separated from the atmosphere by epithelial structure, but in the skin the latter is many times thicker than in the lungs, as we find the carbonic acid excreted through the cutaneous surface in much smaller amount than that removed by the pulmonary surface—the proportion being as one to thirty-eight—but the water removed by the skin is about two pounds per day, double the amount which escapes in the form of vapor from the lungs.

As respiration involves the continual co-ordinated acts of many muscles the nervous control of these movements demands attention. The harmony of their action is complete, and that in labored as well as in gentle breathing. Observation and experiment teach us that the co-ordinating impulses proceed from the medulla, and from a small point in the medulla which we call the respiratory centre. The respiration goes on after the removal of the brain above the medulla and on section of the cord just below it, though the thoracic movements cease; the centre continues in action, for the facial and glottic movements continue, but these cease when the recurrent laryngeal nerves are divided. When that small portion of the medulla which Flourens called the *nœud vital* is destroyed, breathing ceases forever, though every other part of the nervous system be uninjured, and as the inhibitory vagus centre is generally stimulated at the same time, the heart's beat is arrested and the animal killed. This *nœud vital*, then, we call the respiratory centre, and its rhythmic action is not merely reflex but automatic. Respiratory impulses start *de novo* from this centre, though their character may be altered by afferent impulses arriving at the moment when they are generated. As the centre communicates with the lungs by the vagi no doubt afferent impulses started by any stimulus to their peripheral endings continually ascend and produce modifications, which changes have reference chiefly to the distribution in time of the efferent impulses. When the vagi are divided respiration becomes much slower, the pauses between inspiration and expiration being lengthened, but each respiration is fuller and deeper, so that the oxygen consumed and the carbonic acid lost in a given time are about the same, the loss in rate being made up in extent. Stimulation of the divided vagi by a gentle interrupted current quickens the respiration, and by careful management the natural rhythm may be restored, but a too powerful current accelerates the breathing too much, and may throw the diaphragm into a tetanic condition and bring about a standstill of respiration in the inspiratory phase. The respiration is also rendered slower by stimulating the central end of the superior laryngeal branch of the vagus, and that whether the trunk of the vagus (below the origin of that branch) be divided or not, and this stimulus may bring about a standstill in the phase of rest, that is in the normal expiratory phase. We conclude, therefore, that the superior laryngeal nerve contains fibres the stimulation of which inhibits the respiratory centre, while the main trunk of the vagus contains fibres the stimulation of which augments the action of that centre; but as in some cases there is a retardation, we conclude further that the main trunk contains some inhibitory fibres, though the accelerating ones are much more numerous, and these accelerating fibres appear from the result of dividing them to be always at work. The rhythmic impulses of the respiratory centre seem to be excited by the blood, or rather its condition: the more venous its character, the greater is the stimulus. The breathing becomes quicker and deeper whenever the arterialization is defective. Greater activity in the tissues—loading up carbonic acid and using oxygen, as in muscular exertion—makes the respiration more active, as does any hindrance to air entering the lungs. In obstruction to respiration, as the blood becomes more venous the nervous discharges of the centre are more vehement, and, so to say, overflow and set other muscles in action, and this may go on until the centre is exhausted. Instead of being too venous, the blood may be made less venous—as by breathing oxygen, or by forcible voluntary breathing, or by artificial respiration vigorously carried on. Then, the blood having obtained all the oxygen it needs, the centre is no longer stimulated by a want of it, and respiration ceases until tissue-changes render the blood again venous. In such case the centre is at rest, and we have what many physiologists call "apnœa," following in this Rosenthal, who thus employed the word in 1864, but it must be remembered it had long before been in use for a different purpose, and is still applied by medical writers in its original sense. This physiological "apnœa," then, is the converse of dyspnœa, the centre resting when the blood is more arterialized, and being stimulated when it is more venous. The stimulus is the want of oxygen, not excess of carbonic acid, because dyspnœa is produced in an atmosphere of nitrogen in which there is no impediment to the exit of carbonic acid, and that gas does not accumulate in the blood. So, too, breathing an atmosphere containing much carbonic acid, but at the same time more than the normal percentage of oxygen, does not produce true dyspnœa, but only narcotic poisoning. It may be admitted that the state of the blood may affect to some extent the peripheral ends of afferent nerves, and that impulses thus started—as, indeed, impulses started in any way—may contribute indirectly to the result; but the effect is produced mainly by the direct action of the blood on the respiratory centre, for when the cord is cut below the medulla, and both vagi also divided, defective aëration still causes increased activity of the centre, as shown by the facial respiratory movements. Again, if we cut off the blood-supply from the medulla by tying the vessels dyspnœa is produced, though the blood is not altered generally, but there is a deficiency of oxygen in that region with accumulation of carbonic acid. So, if the blood in the carotids be warmed the same effect follows, viz.: dyspnœa, the heated blood apparently hurrying on the activity of the cells, so that the normal supply does not suffice.

We have spoken of the respiratory centre as a fact, but, though generally admitted, its existence is strenuously denied by Brown-Séquard, who says he has seen respiration continue after removal of the supposed centres, and, indeed, of the whole medulla, and his statements have been corroborated by P. de Rokitansky, Langendorff, Nitschmann, and others. Not long ago Langendorff, as the result of a lengthy research, expressed himself as disposed to completely decentralize the respiratory motor impulses, but the most recent experimenter, Dr. Léon Fredericq (*Archiv. für Physiologie*, 1883), finds himself compelled to admit the existence of a central organ for respiratory impulses in the medulla oblongata.

CHAPTER IV.

FOOD AND DIET—THE PROXIMATE PRINCIPLES OF FOODS—FOOD-STUFFS.

As the one thing which is necessary to be supplied frequently, food should take a first place in therapeutics. There is no question about our power to regulate or to arrest the supply; we can cut off all food for a time or we can diminish it. On the other hand, we can supply it more freely, but in this case we may not be able to assist its assimilation; often appetite fails, and then we may feel that we are in the position of the man who took his horse to the brook but could not make him drink. The arrest of supply, a fast, is a most powerful therapeutic agent. Further, we can regulate diet in reference to its various ingredients; we can exclude some constituents and introduce others. The substances we use as food are so numerous that to simplify their grouping we employ their proximate principles, which are either organic or inorganic—the latter class comprising water and salts.

The organic elements of the food were divided by Liebig into two classes:

I. Plastic, or proteinaceous, comprising nitrogenous substances, which he held were alone concerned in repair, growth, nervous and muscular energy.

II. Elements of respiration, afterward better styled calorifacient principles, comprising the organic non-nitrogenous substances which he considered were employed in heat production.

No doubt a very direct relation exists between nitrogenous food and muscular work, but that it must be converted into tissue before it can liberate force has not yet been proved. It has, in fact, been shown that heavy labor can be undergone for a short period on a non-nitrogenous diet.

Prout took milk as the type of a perfect food, and no one will deny that it is so for the young, who both subsist and grow upon it alone, and flourish best when nothing else is added to it. The ingredients of milk are: (a) nitrogenous; (b) fatty; (c) saccharine; (d) mineral—and we cannot do much better than follow this classification with a little enlargement.

I.—Nitrogenous, Plastic, or Albuminous Foods.

These are of two varieties:

A. Substances containing nitrogen in the same proportion as albumen, viz.: two of nitrogen to seven of carbon, nearly. These substances are sometimes called *albuminates,* and this group includes albumen, fibrin, syntonin, myosin, globulin, and casein, from the animal kingdom, with glutin and legumin from the vegetable. Their work is tissue repair, regulation of oxygen, absorption, and utilization; under special circumstances they may form fat and liberate energy.

B. Substances containing nitrogen in the proportion of two to five and one-half of carbon, or four to eleven. This class comprises gelatin, chondrin, keratin, ossein, etc. These perform similar functions under particular circumstances, but far less perfectly. Some have doubted whether they possess any nutritive value at all. The experiments of Pettenkofer and Voit, of Plosz and Györgyai, Maly, Latschenberger, and others appear to indicate that the gelatinoids cannot replace the albuminoids in the formation of tissues, although they may be perhaps substituted for metabolism.

Nitrogen is present in every tissue which can liberate any form of energy, and seems to be necessary for its manifestation, even if it only serves as an instrument. Nerves, muscles, gland-cells, the cells suspended in liquids, all are nitrogenous. So, too, are the digestive liquids; the ferments contained in these liquids, ptyalin, pepsin, pancreatin, are nitrogenous—so are biliary compounds. Cut off the nitrogen from the food and function languishes, not at once, perhaps, but as soon as the store in the body is exhausted. We may add that the nitrogenous food-stuffs are not completely exhausted of their energy in the body; the chief product of their combustion, urea, carrying away about one-seventh part of the potential energy. Voit's experiments seem to show that albumen is perhaps accumulated in the general fluids of the body, from which it is taken for the repair of the tissues as they may require it, and that this dissolved albumen is more easily metabolized than that which has become tissue. A full supply, perhaps an excess, of such dissolved albumen is essential to healthy functional activity; but further excess is at once metabolized. Thus the excretion of nitrogen depends on the amount in the food and not on muscular exertion, as Liebig thought. In support of the more recent view, it may be remarked that the metabolism of albuminoids is never entirely suspended, nor is it increased by exercise, which, however, as previously proved by Dr. Edward Smith, enormously increases the excretion of urea.

We pass on now to the next class.

II.—Non-Nitrogenous.

These consist of hydrocarbons or fats and carbo-hydrates of starches and sugars, to which we may add vegetable acids and pectous substances.

A. *Fats* or *hydrocarbons* are composed of carbon, hydrogen, and oxygen; the oxygen being insufficient to combine with all the hydrogen in order to form water. The unoxidized hydrogen is in the proportion of about one to seven of the carbon. The fats are more energetic as calorifacients than the amyloids, about two and a half times as effectual, and they are also of value as nutrients, especially of the nervous system, and no doubt help in the production of energy. Perhaps, also, they assist disintegration. Fat is always present in the nerve-tubules and in the ganglionic centres. It is also of some use in digestion, perhaps aiding the solution of proteids; further, it seems to assist in their conversion into tissue. Even artificial digestion is hastened by the presence of fat, and it has been proved experimentally that albuminous food deprived of all fat lies longer in the stomach. The power of the pancreatic fluid is largely expended in emulsifying fats, and chymification and absorption into the blood are thus secured. It would appear that the white corpuscles get their first impulse from fat, which thus directly helps to make blood, and it seems to have something to do in cell development. If, then, fat promotes digestion, chymification, the absorption of nutrient fluids, and their transmutation into blood, or rather corpuscles; and, further, seems to enter cells and take a part in the process by which the nutrient juices derived from the blood become tissues, and afterward assists the disintegration and removal of used-up material, it is not surprising that a deficiency of this aliment should prove disastrous, and that in some diseases characterized by malnutrition, the administration of easily digestible fats should prove our most valuable therapeutical resource. Cod-liver oil is thus an important aliment as well as a medicine.

B. *Carbo-hydrates* or *amyloids* may fairly be called hydrates of carbon, as their oxygen and hydrogen are in the proportion to form water. The proportion of water to carbon is about three to two. These substances are the starches and sugars. They are converted into fats by deoxidation and, as we have seen, they also give rise to energy and, of course, supply animal heat. Dextrin is included in this group, though the proportion of oxygen and hydrogen is not precisely the same. Cellulose, from its form and aggregation, is only adapted for animals with an appropriate digestive apparatus. As the amyloids are so closely related to the fats, it is natural to inquire whether the former can altogether replace the latter. Certainly not conveniently, and as a matter of taste alone the human race has always preferred to eat both. After long discussions it is now admitted that fat may be derived from carbo-hydrates. A man's living on fat, free diet, proteids, amyloids, salts, and water being supplied, would seem to be a question of digestion, though a full experiment of this kind has not been made. He can exist for a time on proteids, fat, salts, and water—but health is not sustained on such diet. Parkes found in some experiments that when the health failed under certain dried foods, it was at once improved by the addition of fats and amyloids without nitrogen. It would seem that some fat may be obtained from albuminates, and in diabetes sugar seems sometimes also derived from them, but they cannot replace either sugar or starch, as the system would be injured by the excess of nitrogen introduced. We conclude, then, that a due proportion of all classes is necessary for health.

C. *Vegetable Acids.*—With these we may perhaps associate pectin and its allies. In this group the oxygen is more than enough to combine with all the hydrogen as water, except in acetic and lactic acids, in which there is no excess. They furnish but little energy in oxidation, but being converted into carbonates assist in maintaining the alkalinity of the blood.

III.—Inorganic Substances.

A. *Salts* are essential to health, they seem to be largely concerned in nutrition, facilitating the removal of effete matter as well as the entrance of nutrients. Perhaps they promote the transmutation of liquid colloid into solid tissues, as well as the converse solution of these. They seem to be concerned in digestion, absorption, assimilation, sanguinification, and tissue building; later their presence promotes disintegration, secretion, and excretion. Proteids are incapable of osmosis, but the salts and diffusible acids aid in their conversion into peptones. An opposite effect may be produced in the alkaline blood and this again reversed in the acid tissues. As to disintegration, acid compounds are produced by oxidation, and they appear to give to the salts a solvent power on the débris. Salts have also been regarded as regulators of energy as well as nutrition, and some of them give solidity to bone, etc. A few words may be added respecting the several salts.

Chloride of sodium is universally recognized as of great importance. The instinct of animals as well as of the human race shows its necessity. Feeding his victims on food deprived of salt was once one of "man's inhumanities to man." Almost everywhere, in all ages, among savages and civilized alike, salt has been regarded as an emblem of wisdom, riches, comfort, and even immortality. It forms about half the weight of the salines in the blood, but it does not enter into the tissues, appearing to act rather as a medium. We cannot much change the amount in the blood. Dilute solutions run off by the kidneys, concentrated ones by the bowels; thus if we drink sea-water it is not absorbed. Most unorganized and waste products of the body form crystallizable compounds with salt; possibly this is why it seems to be concerned in both absorption and secretion.

Phosphates.—Alkaline or basic phosphate of soda is always present in the blood, but acid phosphate of potash appears in the flesh; their functions must be different, and in some degree opposed. The alkalinity of the blood keeps the fibrin and albumen in the liquid colloidal state. The basic phosphate of soda also absorbs carbonic acid, and is the chief agent for its removal, acting like an alkaline carbonate which is capable of replacing it in case of its insufficiency. In the herbivora the proportion of these two salts is the reverse of that found in the carnivora. In fact, we may say that whenever sodium phosphate is insufficient in quantity to carry away the carbonic acid it is supplemented by the alkaline carbonate, and where phosphoric acid is deficient it is replaced by carbonic acid. The acid phosphate of potash in muscular substance reverses the alkaline action of the blood, promoting the transudation of nutrient material as well as the solution of effete matter. Earthy phosphates, especially the lime-salt, are present in all hard tissues, not only in the bones and teeth, but also in the flesh. Possibly their presence directly promotes solidification; they appear to be agents in cell-growth. Lime is present in morbid growths and in rapidly growing cells in considerable amount. Weiske found that when calcium phosphate was excluded from the food of a goat its bones did not lose their lime; but perhaps it was drawn from other parts, as nutrition was evidently interfered with, the animal becoming weak and dull. Phosphorus, like sulphur, is thought to enter the system with the albuminoids.

Iron and *manganese* are both present in the blood. Iron in some form is further necessary for the coloring matter of muscle, and in some degree is found in almost all structures, just as it is in almost all foods. Manganese is the chief mineral of the corpuscles in white-blooded animals as iron is in red-blooded. In hæmoglobin and myochrome the iron is combined with albumen, or rather globulin. They absorb oxygen from the atmosphere, and give it up in the presence of reducing agents. The functions of the iron compound are therefore respiratory, they are purveyors of oxygen, taking it from the air in the lungs, carrying it thence over the whole body to be employed in oxidizing tissue.

Dr. Edward Smith calculates that an adult man needs daily from 32 to 79 grains of phosphoric acid, from 51 to 175 grains of chlorine = 85 to 291 of table-salt, from 80 to 171 of soda, from 27 to 107 of potash, 2.3 to 6.3 of lime, from 2.5 to 3 of magnesia, which last is also an essential constituent of some tissues. It would seem that only a small proportion of these salts is stored up in the system, for Mr. Lawes found in fattening pigs that only twelve ounces out of eleven pounds present in the food were retained, and this amount was chiefly earthy phosphates. Generally there is sufficient saline in our ordinary food and drink, except of sodium chloride, and to supply this table-salt must be added.

B. *Water* is necessary in large quantities. It is present in every tissue, even the most solid, as the teeth; it forms seventy-five per cent. of muscle, and about seventy per cent. of the entire weight of the body; it constitutes 79.5 per cent. of the blood. Not less than about thirty pounds pass in and out of the alimentary canal by secretion and absorption. Bidder estimates that 28.6 pounds of chyle and lymph pass daily through the thoracic duct, a quantity equal to nearly one-fifth of the body weight. Barail reckoned that for a healthy adult the average amount introduced per day in the food and drink was two thousand grammes. Some water also appears to be formed in the body, but not more than perhaps three hundred grammes, many experiments and observations showing an excess of two hundred to five hundred in the amount removed over that introduced. After forming part of the solids and fluids, it is discharged, according to the researches of Valentin, Laboisier, Seguin, and others, by four channels, viz., twenty per cent. through the lungs, thirty per cent. through the skin, forty-six per cent. by the kidneys, and four per cent. through the bowels. The uses of water are obvious: it dissolves the food, carries it into the circulation, and distributes it through the body; it combines with all the tissues and lubricates every part; it dissolves the waste and carries that away; thus we see it is an active agent in absorption and secretion, in construction and destruction. Besides this, it regulates the temperature as it evaporates from the skin and lungs. Its quality is important; it carries into the system minerals, as do the foods. It would not be advisable to abandon our ordinary drinking-water in favor of distilled; in fact, as Letheby observed, "the water of a country may determine the diet of its inhabitants." "The soft waters of the lakes of Scotland, for example, may have had something to do with the choice of brown meal, which contains so much saline matter; and but for the calcareous waters of Ireland the potato could not have become a national food." Nevertheless there are times when we may be glad to place patients on distilled water, in order that it may carry off other salts from the body, as, for example, where there is a tendency to deposit calculi. So drinking largely on the one hand or a so-called dry diet on the other become important therapeutical measures.

Nutritive Value of Foods.

Animal and Vegetable Diet.—The amount of food required by the body varies with the work it has to do and other circumstances. Further, the proportions of the several classes best adapted varies with the kind of work required.

The following table gives the amounts required:

	Subsistence.	Rest.	Standard.	Hard work.
	Ounces.	Ounces.	Ounces.	Ounces.
Proteids	2.0	2.5	4.31	6.0 to 7.0
Fats	0.5	1.0	3.53	3.5 to 4.5
Amyloids	12.0	12.0	11.71	16.0 to 18.0
Salt	0.5	0.5	1.70	1.2 to 1.5
Total water-free food...	15.0	16.0	21.25	26.7 to 31.0

The amount required for bare subsistence, that is for the maintenance of the internal work of the body only in a state of complete repose, is given in the first column. But as a man could probably not exist on that amount without loss of weight, we give in the next column the requirements of ordinary rest, which presupposes that only gentle exertion should be undertaken, and is the minimum for an adult male weighing 150 pounds. This calculation is for water-free food, that is each constituent is reckoned theoretically as water-free, but in what we term ordinary solid food there is from one hundred to one hundred and fifty per cent. water, so that thirty-two to forty ounces of such food would be required. For ordinary work a larger supply is needed and standard daily diets have been constructed, partly by calculation and partly by experiments. We give the mean of several such standard diets in our third column, but perhaps this should be a little more liberal. If we allow twenty-three ounces of water-free food for a man of 150 pounds, this gives 0.15 ounce for each pound, or nearly one-hundredth part of the weight of the body. This is dry food, and would correspond with fifty to sixty ounces of ordinary solid food, to which may be added the fifty to eighty ounces of water or other liquid taken, which would make a total supply of seventy to ninety ounces of fluid = 0.5 ounce for each pound of the body-weight. When working hard, men take much more food whenever they can get it, and for hard labor diet the greatest increase should be in the proteids and fat. Of the diets given, those most adapted for our patients would naturally be those for rest and ordinary work, but as some patients imagine they ought to consume large quantities, we have added a fourth column to our table, suitable for a person undergoing very laborious work; it gives the figures calculated by the late Dr. Parkes as sufficient for a soldier on service in the field. Of course in acute cases the nature of the diet may have to be greatly changed—here we are only speaking of that which is required in health, as a basis for the consideration of what may be desirable in disease. The average proportion of the several classes, taking the mean of the figures given by Moleschott, Ranke, and Pettenkofer and Voit, is for 100 of proteids, 82 fats, 272 amyloids, and 23 salts. As nutrition depends so largely on the chemical interchanges of nitrogen and carbon, many calculations and experiments have been made as to the nutritive value of different foods. In the best diets the nitrogen should be in the proportion of one to fifteen of carbon. Tables based on the nitrogen alone lead to very unsatisfactory conclusions; for instance, ham and red-herring would be, according to the amount of nitrogen contained, more than nine times as nutritious as human milk, both being calculated in the dry state. In framing, therefore, a table of alimentary equivalents we must consider all the constituents. The late Dr. Letheby constructed the following table showing the percentage, proportion of nitrogenous and carbonaceous matters, the latter being calculated as starch.

The habits and instincts of the human race are in favor of a comparatively rich nitrogenous diet, and the easiest to digest appears to be one containing meat; for though the chemical composition of animal and vegetable albumen is nearly the same, and they serve similar purposes, the facility of digestion is not to be overlooked. A judicious admixture of meat and vegetables is therefore generally recommended, though it is not to be denied that men can live in perfect health on vegetable food; there are, indeed, races who do this. Some advocates of animal food, however, consider that the highest kind of work cannot be accomplished on such a diet, and Dr. Carpenter must perhaps be enumerated among these, for he says: "And while, on the one hand, it may be freely conceded to the advocates of 'vegetarianism' that a well-selected vegetable diet is capable of producing (in the greater number of individuals) the highest *physical* development of which they are capable, it may, on the other hand, be affirmed with equal certainty that the substitution of a moderate proportion of animal flesh is in no way injurious, while so far as our evidence at present extends, this seems rather to favor the highest *mental* development." Dr. Graily Hewitt, in his address at the British Medical Association, assigned to a defective meat diet a large amount of "weakness" which he found prevalent and regarded as an antecedent of distinct disease, and the *Lancet* recently stated that three-fourths of a pound of meat fairly represents the daily quantity which suffices to maintain an adult in health. Certainly many exceed this, taking meat two or three times a day, even when leading sedentary lives, but, as the same writer points out, sooner or later they pay the penalty. On the other hand, many—especially women—reduce their meat to a minumum, scarcely taking an average day's allowance in a week, and it may easily be understood that they do not possess the same vigor as those who take more; but the question naturally arises whether these persons are not underfed in other articles also, and that their weakness is due not so much to want of meat as to want of food.

Table of the Nutritive Values of Food (DR. LETHEBY).

	Water.	Albumen, etc.	Starch, etc.	Sugar.	Fat.	Salts.	Nitrogenous.	Carbonaceous, as starch.	TOTAL PER CENT. Carbonaceous to one nitrogenous.	Nitrogen.	TOTAL PER CENT. Available carbon.[1]
Bread	37	8.1	47.4	3.6	1.6	2.3	8.1	55.00	6.8	1.25	28.21
Wheat flour	15	10.8	66.3	4.2	2.0	1.7	10.8	75.50	7.0	1.66	38.57
Barley meal	15	6.3	69.4	4.9	2.4	2.0	6.3	80.30	12.8	0.97	36.61
Oat meal	15	12.6	58.4	5.4	5.6	3.0	12.6	77.80	6.2	1.94	40.44
Rye meal	15	8.0	69.5	3.7	2.0	1.8	8.0	78.20	9.8	1.23	38.48
Indian meal	..	11.1	64.7	0 4	8.1	1.7	11.1	85.35	7.7	1.71	43.09
Rice	13	6.3	79.1	0.4	0.7	0.5	6.3	81.25	12.9	0.97	39.03
Peas	10	23.0	55.4	2.0	2.1	2.5	23.0	62.65	2.7	3.54	38.55
Arrowroot	18	..	82.0	..	..	..	..	82.00	..	..	36.44
Potatoes	75	2.1	18.8	3.2	0.2	0.7	2.1	22.50	10.7	0 31	10.98
Carrots	83	1.3	8.4	6.1	0.2	1.0	1.3	15.00	11.5	0.20	7.28
Parsnips	82	1.1	9.6	5.8	0.5	1.0	1.1	16.65	15.1	0.17	7.91
Turnips	91	1.2	5.1	2.1	..	0.6	1.2	7.20	6.0	0.19	3.76
Sugar	5	..	..	95.0	..	..	..	95.00	..	..	42.22
Treacle	23	..	..	77.0	..	..	..	77.00	..	..	34.22
New milk	86	4.1	..	5.2	3.9	0.8	4.1	14.95	3.6	0.63	8.55
Cream	66	2.7	..	2.8	26.7	1.8	2.7	69.55	25.7	0.42	32.17
Skim milk	88	4.0	..	5.4	1.8	0.8	4.0	9.90	2.5	0.62	6.26
Butter milk	88	4.1	..	6.4	0.7	0.8	4.1	8.15	2.0	0.63	5.53
Cheddar cheese	36	28.4	..	..	31.1	4.5	28.4	77.75	2.7	4.37	47.77
Skim cheese	44	44.8	..	..	6.3	4.9	44.8	15.75	0.3	6.90	27.82
Lean beef	72	19.3	..	..	3.6	5.1	19.3	9.00	0.5	2.97	12 98
Fat beef	51	14 8	..	..	29.8	4.4	14.8	74.50	5.0	2.28	39.99
Lean mutton	72	18.3	..	..	4.9	4.8	18.3	12.25	0.7	2.82	13.95
Fat mutton	53	12.4	..	..	31.1	3.5	12.4	77.75	6.3	1.91	40.33
Veal	63	16.5	..	..	15.8	4.7	16.5	39.50	2.4	2.54	25.22
Fat pork	39	9.8	..	..	48.9	2.3	9.8	122.25	12.5	1.51	58.89
Green bacon	24	7.1	..	..	56.8	2.1	7.1	167.00	23.5	1.09	77.52
Dried bacon	15	8.8	..	..	73.3	2.9	8.8	183.25	20.8	1.36	85.53
Ox liver	74	18.9	..	..	4.1	18.9	10.25	0.6	2.91	13.34	
Tripe	68	13.2	..	..	16.4	2.4	13.2	41.00	3.1	2.04	24.36
Poultry	74	21.0	..	..	3.8	1.2	21.0	9.50	0.4	3.23	13 99
White fish	78	18.1	..	..	2.9	1.0	18.1	7.25	0.4	2.79	11.64
Eels	75	19.9	..	..	13.8	1.3	9.9	34.50	3.5	1.53	19.93
Salmon	77	16.1	..	..	5.5	1.4	16.1	13.75	0.8	2.48	13.60
Entire egg	74	14 0	..	..	10.5	1.5	14.0	26.25	1.9	2.16	18.18
White of egg	78	20.4	..	..	..	1.6	20.4	..	..	3.14	9.49
Yolk of egg	52	16.0	..	..	30.7	1.3	16.0	76.75	4.8	2.46	41.55
Butter and fats	15	..	..	..	83.0	2.0	..	207.50	..	..	92.22
Beer and porter	91	0.1	..	8.7	..	0.2	0.1	8.70	87.0	0.02	3.92

[1] The available carbon consists of all the carbon of the carbonaceous constituents of the food, and of the carbon of the nitrogenous after deducting the carbon of the urea which is excreted, 100 of dry nitrogenous matter yielding 31.23 urea.

The advocates of meat diet say that the carnivora are more active, or at any rate that they can respond to sudden calls for exertion more easily than the herbivora. But we must not be content to contrast the sudden spring of the tiger with the slow steady move of the domesticated cow. The antelope will start with amazing rapidity and keep up its run for a great distance. The high-bred hunter and the plodding donkey are both vegetable-feeders. The carnivora appear to be fiercer in their nature, but that would be almost necessary to enable them to seize their prey. It is said that the acorn-eating bears of India and America are mild and tractable, while the flesh-eating polar-bears are savage and untamable. Experiments have been made by feeding bears on different diet, and these point to the same conclusion, but allowance should be made for animals kept in captivity. Men also seem to differ somewhat in their disposition and powers of endurance according as they live on vegetable diet or partake of meat. Hindoo navvies employed in the boring of tunnels have felt driven to forsake their vegetable diet and live like their English fellow-laborers. Liebig attributed the Englishman's energy to the nature of his food, and says: "Compare the English statesman, who, in expounding his views or maintaining a debate in Parliament, delivers a speech lasting five hours or more, who at sixty years of age retains the capability of taking part in field sports, with the German philosopher, who at the same age keeps up with difficulty the remains of his power in order to be capable of work, while he becomes fatigued by a walk of a few hours." It is a flattering picture, and in the present day we may see a foremost statesman with physical energy enough to amuse a leisure hour by felling trees, but it is to be feared that the average Englishman will not be able by any diet to display an equal amount of physical and intellectual energy. Again, it may be remarked that not only sustained work but sudden putting forth of energy is seen in men by no means remarkable for their meat-eating. Some recent statistics seem to show that the English, after all, are not such large meat-eaters in comparison with other nations as has been generally thought. Even if they are, they are certainly not more warlike nor more implacable than their neighbors; indeed, not a few are disposed to grumble at John Bull as having grown slow to take offence. It may, then, be doubted whether the deductions that have been made can be sustained. Certainly, the contrast with an ill-fed rice-eater should not be made, but with a well-fed vegetarian, who takes full supplies of corn and lentils as well as rice. The argument, from the degree of complication of the alimentary canal, seems rather to belong to the digestibility of the food. Further, it seems certain that different nitrogenous foods have varying nutrient values. Fibrin and gelatin are not meat and cannot supply its place. Majendie found that dogs fed on raw meat alone for one hundred and twenty days retained their health, while three times the amount of isolated fibrin was insufficient to preserve life, though much gelatin and albumen were given at the same time. The glutinous materials of wheat and barley, though chemically nearly alike, are not of equal nutritive power. With regard to the non-nitrogenous foods, their varying value for nutrition depends largely on the digestibility. Starches seem to differ much in this, though all of them have to become sugar. Yet sugar will not completely replace starches in our food ; though perhaps this is partly a matter of taste, or it may very well be that cane-sugar may be less assimilable than that produced by the action of ptyalin on starch. As to hydrocarbons, animal fats are more easily digested than vegetable ; some with great flavor offend the palate or disagree with the stomach. Excess of any of the classes of food may pass through the alimentery canal undigested.

CHAPTER V.

PREPARATION OF THE FOOD-STUFFS—DIGESTIVE FLUIDS.

The digestibility of articles of food is closely related to their physical properties and these we may considerably modify by cooking. Dr. Beaumont's experiments on Alexis St. Martin will be remembered by our readers, and they certainly give some useful indications, but his tables have been so often quoted that it is unnecessary to repeat figures which may be so easily referred to. Moreover the variations in cooking introduce difficulties. It would be well if the light of science could be reflected in the kitchen. Cooking may be compared to some of the operations of pharmacy, and it is as important that food should be well cooked as that medicines should be accurately dispensed.

It is obvious that the food should bear a certain relation to the processes by which its nutritious elements are to be extracted. We cannot feed on the proximate principles but must prepare them in our bodies. Milk is indeed a perfect food for the young, but adults seem to want something else. It may therefore be left to the edentulous race which some whimsical sophists anticipate will succeed us to prepare a fluid aliment for adults, as the present age has made artificial foods for those of its infants who are unhappily deprived of their mother's milk.

The world seems content to employ its own digestive organs, and even those who are no gourmands seem to have no desire to renounce the pleasures of the table. Suitable food being then obtained, to what processes is it subjected? Berzelius compared digestion to rinsing. The food is taken into and passed along the alimentary canal, being detained at intervals while various digestive liquids are poured upon it. By these the nutritious elements are dissolved out and carried into the system as the solvents thus loaded are reabsorbed. The amount of these digestive liquids is often unappreciated. According to the researches of Bernard, Bidder, and Schmidt no less than three gallons are secreted into and reabsorbed from the alimentary canal every twenty-four hours. The following are the daily proportions of the several digestive fluids and of their chief constituents:

Amounts of Digestive Fluids Secreted Daily, and the Proportions of their Chief Constituents.

	lbs.	Solid matters. grs.	Active principles. grs.
Saliva	3.53	231	116 of ptyalin.
Gastric juice	14.11	2,963	1,482 of pepsin.
Pancreatic juice	0.44	309	39 of pancreatin.
Bile	3.53	1,234	1,058 of organic ferment.
Succus intestinalis	0.44	46	28 of organic ferment.
Total	22.05	4,783	2,723 of special solvents.

These liquids are not merely solvents, but each possesses a power of changing the food by virtue of the peculiar ferment it contains. Each not only dissolves nutrient principles but transforms them.

SALIVA.

This secretion is slightly alkaline. Except when fasting it contains only about one per cent. of solid matter, but half of that consists of the nitrogenous ferment called *ptyalin*, possessing similar properties to diastase, for it converts starch into glucose. Mialhe called it animal diastase, and calculated that it could transmute eight thousand times its weight of starch into a soluble glucose. It is, then, as a first stage in the digestion of starches that ptyalin is of value, for it has no action on proteids or fats. Rodents and animals which feed on woody matters have large salivary glands and prolong the contact of the food with the saliva. Dogs swallow their meat without mastication, and their viscid saliva has little use except as a lubricant. In horses, oxen, sheep, and cats there is not enough ptyalin to appreciably convert the starch. In human infants there is no ptyalin, showing that they ought not to have starchy food. In the adult the action is indubitable; it is an easy experiment to show that glucose appears in a solution of starch after it has been held a moment in the mouth. Some dextrin is mostly found also, showing that the ptyalin either transforms the starch into dextrin, and then that into glucose, or else that it first splits the starch into glucose and dextrin, and then transmutes the latter also. It would seem that other intermediate bodies may also be formed as an erythro- and achroo-dextrin. The close analogy between ptyalin and diastase has led to the use of the latter as an *amylolytic;* for example, in extract of malt we have a substance, if well made, rich in diastase, therefore possessing the property which characterizes ptyalin. This is one reason for the use of that extract as a medicine, but it is also a nutrient on account of its other ingredients. When administered with a view to its amylolytic powers it should, of course, be given with starchy food.

Gastric Juice.

As soon as the food reaches the stomach it is subjected to very different changes—not that it is necessarily plunged into a flood of gastric juice, but that acid liquid is being rapidly secreted, indeed, has begun when the food was in the mouth and perhaps before; for we are informed that the strongest juice is obtained from pigs which have been kept hungry and killed just after being excited by savory food which they have not been allowed to eat. As the churning movements of the stomach proceed, each part of the food is brought into contact with the acid-secreting mucous membrane. Thus the contents of the stomach increase in acidity and in the presence of acids it is found that the action of ptyalin on starch is suspended. The most important constituent of the gastric juice is another ferment—

Pepsin, the presence of which was first indicated by Schwann, who may therefore be considered its discoverer, though Wasmann first separated it in a comparatively pure state. In the *presence of an acid*, pepsin converts proteids into a soluble form of albumen, the albuminose of Mialhe or the peptone of Lehmann. As the product seems to differ somewhat according to the body from which it is derived, we may speak of peptones in the plural as we do of proteids. The action of pepsin, like that of ptyalin, is evidently that of a ferment; it facilitates or sets up a change which may be produced in its absence under special circumstances and it is not exhausted in the act of digestion. Its action is arrested when the liquid becomes saturated with peptones, but then we have only to dilute the fluid and the process immediately recommences. Its power is enormous; Wasmann found an acid liquid calculated to contain one part of pepsin in sixty thousand would dissolve meat, and Lehmann found one hundred parts of dog's gastric juice would digest five of coagulated albumen. The acidity of natural gastric juice is equal to .02 per cent. of hydrochloric acid. It would seem desirable that experiments in artificial digestion should be made with the same strength, though it is not uncommon to employ a much more acid liquid. Lehmann held that lactic acid was the natural one and that the best proportion was such as would saturate 1.27 of potash. Other acids—lactic, butyric, phosphoric, or at least acid phosphates—may be found in fresh juice, and may be made to serve in artificial digestion, but not so effectually as hydrochloric, and the degree of acidity which answers best varies with the acid. The evidence points to hydrochloric as the chief natural acid and the proportion is so constant that some have considered that a kind of compound is formed which they speak of as pepto-hydrochloric acid. This is very different from ptyalin, which only needs a faint alkaline liquid.

The action of pepsin is, then, on proteids, which it converts into peptones. As, moreover, fats and starches are enveloped in proteids, they are loosened, the starch-granules and the oil-globules being thus set free, but for further changes they have to wait. The chyme is from time to time squeezed through the pylorus and may even be accompanied by imperfectly digested lumps of solids. That which most resists the gastric juice usually tarries longest in the stomach.

But there is another point here: peptones and sugars are remarkably diffusible, they easily dialyze; they may therefore pass at once into the gastric capillaries. Chyme taken from a stomach in full digestion gives evidence of the presence of some parapeptone, but not of peptone; and food has been found to disappear from the stomach of an animal after ligature of the pylorus. We conclude, therefore, that the peptones are absorbed at once from the stomach.

Rennet.—Gastric juice acts on milk, but there seems some difference. It precipitates the casein, but this effect is not due to acidity, for it will do so after it has been neutralized. It is scarcely due to the pepsin, for Brücke's preparation will not act, though a glycerine extract of the mucous membrane will. This looks as if another ferment might be present in the latter, and which would be the active principle of rennet. Dr. Roberts calls this "curdling ferment."

Mucus-ferment.—It would seem, too, as if there were another ferment located in the mucus of the stomach, for in its presence the gastric juice is able to convert cane-sugar into grape, and the ingestion of cane-sugar seems to provoke secretion of mucus. This mucus-ferment may be analogous to ptyalin but is not identical.

Gas.—A certain amount of air is carried into the stomach with the food and saliva. It is returned having mostly been deprived of its oxygen, for analysis of eructated gas give only nitrogen and carbonic acid with occasionally a trace of organic material. A small quantity of carbonic acid may come from the tissues of the stomach, but most must diffuse from the blood, while the nitrogen is derived from the air of which the oxygen has been used. In flatulency the enormous quantity of gas disengaged is chiefly carbonic acid, and where can it come from except the blood? It may be admitted that some persons have acquired a trick of swallowing air like a cribbing horse, but in such cases the amount would not be very great. Fermentative decomposition of saccharine food or of changed amyloids may undoubtedly set up flatulence and prove a source of some of the gas. But in many cases the stomach is again and again distended, and that very rapidly, after the removal of large quantities of gas. I have seen, when the epigastric region has been tense as a drum, highly tympanitic, and the patient suffering agonies, instantaneous relief follow rapid removal of the gas, so complete that percussion elicited no evidence of its presence, and yet two or three minutes afterward the distention has returned, to be again relieved in the same way, and this course of events repeated for hours together, the patient suffering alternations of agony and ease.

Pancreatic Juice.

After passing the pylorus the chyme has to mix with pancreatic juice and bile, both of which being alkaline tend to neutralize it, but it remains somewhat acid a little longer and has been found so much lower down, but perhaps it might have been reacidified. The conversion of amyloids, which was retarded or arrested in the stomach, is carried on much more actively in the small intestines, chiefly by the pancreatic juice, which contains a ferment resembling ptyalin in its effect on starch but much more energetic and with a much wider area of action. This ferment has not been absolutely isolated, but a very active substance, pancreatin, is obtained by methods such as are employed to prepare pepsin. Pancreatic juice not only acts on amyloids, which is its most characteristic property, but it dissolves proteids and converts them into peptones, and moreover it emulsifies fats. One essential distinction between peptic and pancreatic digestion is that the former is an acid and the latter an alkaline process; reversing this condition will in neither case arrest the action, and further, as peptic requires a particular degree of acidity for perfection so does pancreatic of alkalinity. The .02 per cent. of hydrochloric acid required for perfect peptic digestion is represented by one per cent. of sodium carbonate for pancreatic. In each process peptones are produced very similar, if not alike, but the by-products differ. Instead of parapeptone, which is an acid albumen, we get a sort of alkali albumen, and it seems not unlikely that the first action of pancreatic juice is to convert the proteid into an intermediate body between ordinary and alkali albumen.

In the pancreatic digestion of proteids two crystalline bodies, *leucin and tyrosin*, appear, and the amount of peptones which can be recovered shows a great loss—a loss increasing with the time of digestion, and far exceeding that in peptic digestion; thus a considerable amount of proteid is so completely broken up as to be a proteid no longer. Such a complete change does not occur in peptic digestion. We do not yet know in what proportion proteids are thus hurried into crystallines, nor even how far pancreatic digestion goes, nor whether peptones are formed and rapidly absorbed, but inasmuch as leucin and tyrosin are always present, we conclude that they are formed in normal digestion, and it has been conjectured that excess of proteid food is hurried on to this later stage. *Indol* also appears, but is probably due to decomposition induced by the presence of organisms, for when experimenting with fresh juice its odor is very marked and the mixture swarms with bacteria, but when well prepared pancreatin is employed and precautions taken to exclude atmospheric germs there is no odor, although carbonic acid and nitrogen are disengaged.[1] Kuhne found no indol was formed in the presence of salicylic acid. Fermentative changes may be set up in the small intestines, and not improbably a little sugar is converted into lactic acid. Perhaps when excess of proteids is eaten these fermentative changes may give rise to disorders of the late digestion, especially to flatulence. The possibility of butyric acid fermentation at this stage is also interesting, as suggesting one method by which amyloids may possibly become fat. Pancreatic juice does not act on the gelatins, in which it contrasts with gastric juice.

On fats the emulsifying action is familiar to all, but besides this the pancreatic juice also splits some neutral fats into their respective acids and glycerine; when an alkali is present these fatty acids would form corresponding soaps, and this is no doubt one of their uses. It is only when emulsified that fats to any extent enter the lacteals, and when fat is undigested we naturally look to the pancreas. It is true subsidiary structures may assist or replace it, as the duodenal or intestinal juice, but the most important aid is

Bile.

This fluid has no action on proteids: with free fatty acids it forms soaps. It has a slight emulsifying power and in some animals will convert starch into sugar. Bile, or even a solution of the biliary salts throws down a precipitate from chyme, which is a parapeptone and carries down with it mechanically the pepsin, so that the supernatant liquid has lost its peptic power even if reacidified. Though peptic acid is arrested by bile it promotes the pancreatic digestion of proteids and it helps the digestion of fats, for both experiment and observation show that the passage of fat undigested through the alimentary canal may be due to defect of bile as in obstruction of the biliary duct as well as to deficiency of pancreatin. The secretion of bile is singularly irregular, in which it is distinguished from the other digestive fluids. When the chyme passes by the biliary orifice it seems to cause gushes of the secretion, which precipitate parapeptone, and this forms a lining to the membrane which may possibly prevent a too rapid passage of undigested matter. Recent investigations confirm the idea that waste materials of different tissues are removed in the bile. It has been doubted whether biliary matters absorbed by the blood again pass into the bile, but Schiff and more recently Baldi (*Lo Sperimentale*, 1883) have shown that the liver rather than the kidneys

---

[1] Vide Brownen: Action of Digestive Ferments on Food and Drugs. Trans. Brit. Pharm. Conf.

remove biliary matters which they had injected into the stomach and also into the blood direct. This secretion, however, presents many analogies with that of the kidneys; both essentially depend on the collective waste of the system, the liver having a special excretory faculty for the biliary materials and the kidneys for others. The irregularity of the flow of bile is also interesting in reference to the use of so-called cholagogues. The late Dr. Hughes Bennett's experiments led him to conclude that we have no agents except food for directly stimulating the hepatic secretion; Dr. Rutherford, however, does not quite confirm this. Neither Schiff nor Baldi found either food or medicine excite the flow, but nevertheless clinical observers are well aware that medicines may at least hurry the secretion along the intestine.

Succus Entericus.

It is said to come from the glands of Lieberkuhn, but much uncertainty exists as to its source as well as to its action. It might almost be compared to laboratory washings; it is said to convert proteids into peptones and to emulsify fats, but both these actions have been denied. It is also said to change cane into grape-sugar, and also into lactic acid, and this last into butyric acid, disengaging carbonic acid and hydrogen. Its most characteristic action appears to be reinforcing pancreatine. By the time the contents of the alimentary canal arrive at the ileocæcal valve they have been largely, if not entirely deprived of their nutritive principles; they are about as fluid as in the duodenum, secretion and absorption having been thus far about equal. After passing the valve they become acid, and this must be from fermentative changes, as the secretion of the walls continues alkaline. In the large intestine absorption of water and some soluble constituents goes on and no doubt the secretion has still some influence. Although digestion may be said to be pretty well finished at the ileocæcal valve it is still capable of continuing ; the contents are not merely residues, but to some extent products of the processes previously carried on. They contain a ferment similar to pepsin, and another resembling ptyalin or trypsin. This is very important in relation to rectal alimentation. We have often kept patients for long periods by this plan, but of late years we have been able to improve the method by using the artificially prepared ferments. Still it has been experimentally shown that not only peptones and sugar but albuminoids and amyloids, such as white of eggs, casein, and starch, have been absorbed when introduced through a fistulous wound, and we know clinically that nutrient injections will support life ; we know, too, that artificially formed peptones are nutritious, for dogs fed on them and non-nitrogenous food put on flesh.

CHAPTER VI.

VARIATIONS IN THE DIGESTIVE PROCESS.

REGARDING digestion, then, as the process of extracting the nutritious elements from the food by means of fluids of various qualities, we observe that these are not merely solvents but change insoluble substances into soluble. Albuminoids become peptones, amyloids turn to sugar, and these two readily diffuse into the circulation. Fats are emulsified or split up into their respective acids and glycerine and a portion of these saponified; the emulsified fats pass into the lacteals. Most of the materials for nutrition thus enter the system by the small intestine, some by the lacteals, and some by the portal vessels. Most of the fat passes through the lacteals, but some soaps have been detected in the portal blood and in the thoracic duct after a meal, but the quantity is small and though some fat is introduced in the form of soap into the circulation, this is altogether a subsidiary process, for the digestion of fats invariably leads to their presence in portal blood and still more in the lacteals.

The sugar and peptones should naturally take the direct way of the portal blood or the indirect route of the lymphatics; perhaps some of these may accompany the fats into the lacteals, and it has been conjectured that possibly less diffusible proteids may also find access in this way, for example casein and parapeptone. There is this difficulty about peptones —they cannot be detected in the general circulation nor in the portal blood, nor yet in the chyle during digestion. They vanish! It looks as if they must be taken up as fast as they are formed into the circulation, and there immediately reconverted into albuminates. As to sugar it seems to go both ways, for some, though not much, is found in chyle as well as in portal blood; but we cannot say which way the bulk goes, nor do we know whether amyloids all become sugar or whether a little may not be transformed by the fermentative process previously described.

But the elaboration of the aliment does not always go on so smoothly; disturbances in the process may take place at any point in the journey of the food from the mouth to the tissues. And the disturbing circumstances may concern either the body or the aliment.

A.—THE BODY.

Disturbances occur at various points, for example in the stomach the peptic digestion may be impaired; in the small intestines the pancreatic. We may have indigestion or difficult digestion (*a*) of animal food; (*b*) of fat, in which case meat is generally omitted from the diet; (*c*) of amyloids and of sugar; (*d*) of water—this last being much more frequent than some have supposed. In the next stage absorption may be interfered with at its commencement, or now and then the thoracic duct may be obstructed. Besides the chemical changes, the mechanical movements by which digestion is promoted may be interfered with, as may the nervous influences which regulate them as well as the secretions. To secure the richness and purity of the blood it is necessary for all these functions to be carried on regularly; and in case the circulating stream becomes deteriorated it can only supply inferior digestive fluids, and the whole system will suffer from its impoverishment, while the newly made chyle is in its turn deteriorated, and thus we have established a vicious circle in which one part of the organism reacts injuriously upon another, and is for that reason in its turn itself injured.

B.—ALIMENT.

This may vary in, I., quantity; or in, II., quality. As to

I. QUANTITY, there may be, 1, excess; 2, deficiency. As to excess it may be (*a*) occasional; (*b*) habitual.

1. *Excess.*—(*a*) An excessive meal may not be digested, it is heavy, but some articles of close texture will give a sensation of weight to delicate stomachs even in small quantity. Thus an ounce of plum-pudding may seem heavier than four or five times as much meat or custard. Sometimes after a heavy meal the stomach strikes, as it were—the gastric juice fails, the food remains undigested, there is *embarras gastrique*, the food decomposes, and gases are set free, chiefly carbonic acid with some carburetted hydrogen and sometimes a little sulphide, or some organic material which pollutes the breath and adds to the distress. Unless the semi-putrid mass be got rid of by vomiting or purging feverish symptoms are set up, the blood is poisoned, and jaundice or other symptoms occur. Certain Roman gluttons well understood how an emetic would prevent such consequences of too heavy a meal. Sometimes the food passes through the alimentary canal undigested or partially digested, but with less irritation, for in habitual heavy feeders meat fibre, vegetable fibre, starches and fats are all found in the fæces. Although there may be no excess in the food actually taken in twenty-four hours, it may happen that at some period of the day the amount may be too great for the powers of the stomach, because the meals are either too close together or, on the

other hand, too far apart; in the latter case the appetite and capacity of the stomach being subjected to long periods of repose interrupted by distention.

(b) Habitual excess may occur from too much food or from the quality being too nutritious. Sometimes imperfectly digested material may pass into the blood. The risks of too much nutriment vary with the quality of the food. Excess of proteids gives rise to plethora, congestion, and perhaps enlargement of the liver. If much exercise be taken it may be burnt off, but with too little exercise oxidation is deficient, as if enough oxygen for the proteids was not absorbed, and then some products not fully changed remain in the system and cause disturbance, or else in passing out trouble the organs by which they are removed. Gout is one of these evils, but it is no doubt partly due to taking liquids which retard metabolism as well as indigestible articles, for large meat eaters often escape gout so long as they are engaged in laborious employments. Experiment has shown that excessive proteids without other food produces in a few days fever, diarrhœa, and albuminuria with great depression, the last symptom being probably the effect of the salts in the meat. In habitual excess of fats and starch the results differ, attacks of acidity and flatulence are common. In consequence of the retarded metabolisms of nitrogen corpulence may ensue and muscular debility, upon which the heart may suffer and saccharine urine may be present. Lastly, the meals may all be small, but if they are too numerous there is excess.

2. *Deficiency.*—Acute starvation is best studied in famines, chronic in the underfed poor. The effects of deficiency of the various principles has not received so much attention as it deserves. Dr. Parkes kept a man for five days on fat and starch, and attempts have been made to feed patients with rheumatic fever in this way, but the removal of all nitrogen soon brings about great depression and the patients need to be carefully watched or exhaustion will follow. When the deprivation of proteids is incomplete but yet considerable, there will ensue, although later, debility and anæmia. Amyloids may be withheld for a considerable time if fat be given, but if both fat and starch are withheld a few days will bring about illness in spite of a full supply of proteids. Deficiency of fat gives rise to malnutrition in various forms, and total deprivation is dangerous. Very often the addition of fat to the diet or the introduction of oil as a medicine is a most important measure. Deficiency of salts is also attended with ill results.

II. QUALITY.—Improper food may produce the effects of deficiency in so far that it takes the place of more nutritious substances; but it may prove more injurious than this, or even than excess. Sometimes it may be positively poisonous. Then good food may become bad by keeping or other circumstances, or it may be improperly cooked or insufficiently masticated; these or other circumstances may render it indigestible.

Meal-times.—The time of taking food and its distribution in appropriate meals deserves a word. We have seen that the stomach may be overloaded at one time and starving at another. Savages will take enormous quantities at once, when they can procure it, and starve at other times. In civilized life we regulate our diet with advantage. Taste has much to do with this. Three meals a day is perhaps the most simple plan and is usually adopted in hospitals and public institutions.

According to Dr. Edward Smith, the daily distribution of the food, supposing a physiological diet of 4,300 grains carbon with 200 grains of nitrogen to be taken, should be somewhat in this manner:

Relative Proportions of Food at Different Meals.

	Carbon. grs.	Nitrogen. grs.	Equal to	
			Carbonaceous. ozs.	Nitrogenous. ozs.
For breakfast	1,500	70	6.62	1.04
For dinner	1,800	90	7.85	1.34
For supper	1,000	40	4.52	0.59
Total in the day	4,300	200	18.99	2.97

The continental plan of two meals daily agrees with many, and affords the advantage to some people of having all their work between them, but the interval is too long for weakly people and they should therefore take a light luncheon between and be careful to avoid overloading the stomach at their regular meals. The times of the meals, as well as the articles of the diet, may often be adapted to the age and constitution of the individual, as well as to the work he is called upon to perform.

In invalids who are walking about we may vary and arrange the diet of health, keeping to regular meals and taking care to provide particular proportions of the different classes of alimentary principles. The digestibility of the food is in such cases of the first importance, but very often when we are desirous of improving nutrition we are defeated by failure of appetite, and even when this is not the case we may have to supplement ordinary diet with articles which become imperative in acute diseases. In these last meals are abandoned, often no solids can be given, the patient is in bed, rest is imposed, and we return to the simplest food —liquids. Milk, supplemented perhaps by beef-tea, broths or soups, or in other cases gruel, so lauded by Hippocrates and employed ever since, or barley-water, or other forms of light nourishment, afford what is required. Where there is any difficulty about the digestion of these they should be peptonized. Milk must be curdled in digestion, but when considerable quantities of curds are thrown up it shows that the casein has not been merely coagulated but that hard masses have become conglomcrated. Mixing with lime-water or soda, or peptonizing, will obviate this inconvenience. A considerable quantity of liquid food may be taken in the twenty-four hours when it is given in very small quantities but very frequently. Two or even three ounces per hour is often easily taken, but every two hours is mostly enough to attempt, and thus it is easy to administer a pint and a half daily. I have known a teaspoonful of liquid aliment taken every five minutes for days together with the very best results.

CHAPTER VII.

ALIMENTS AS REMEDIES—NUTRIENTS AND ANALEPTICS.

FROM what has preceded we see that nutrition demands the maintenance of a continual balance between the food-supply, the power of the system and its work. Another balance has also to be maintained between the several classes of alimentary principles or food-stuffs, and these are so related to the work that we may arrange dietaries by varying them according to the work done. Suppose, then, the food to be sufficient, a due proportion of the food-stuffs to be included, further, that digestion in every stage is easy, absorption unimpeded, and the blood continually renewed with wholesome material: this implies the integrity of the presiding nervous influence, vigor of the circulation, perfection of mechanical movements—a chain, any link of which may be weak. But if all these actions are normal the blood is renewed by the fresh chyle, reinforced by the lymph and continually poured into the stream through the thoracic duct. Up to this point all has been, as it were, absorption from the outside, but the blood now passes through every tissue and here begins, so to say, a pouring out of the nutritious material into the intercapillary spaces, an impoverishing process so far as the blood is concerned (and the same holds good of secretion), but a nutritive one as to the tissues; these have to select or appropriate from the nutritious plasma each what it wants. A kind of elective affinity may be said to exist by which this is accomplished; thus the tissues leave a residual liquid around them, which is next returned by the absorbent system to the blood, as it is only defective by the ingredients appropriated. That which is not absorbed by the lymphatics, together with the effete matter cast off by the tissues, the results of denutrition or oxidation, constitute the final residue, which passes into the venous capillaries by diffusion and gives to the venous blood its positive characters. The products of denutrition and disintegration, the results of the destructive changes, are carried away by the blood to be eliminated by the organs set apart for this purpose, and so maintain the purity of the blood. Thus is maintained the everchanging balance between repair and wear and tear, or rather renewal and waste. These continual changes, appropriation or construction on the one hand, rejection, destruction, and removal on the other hand, must keep pace with each other. Too rapid removal leads to decrease, too much renewal to increase of weight, and such deviations affecting parts of the body give rise to atrophy, hypertrophy, and other allied conditions.

Seeing the importance of these changes, can we influence them? Can we hasten, retard, or alter them? Unquestionably yes. Deviations, except within narrow limits, mean disease. We see, therefore, diseases of quickened or of retarded tissue-change, or other forms of malnutrition. Our therapeutics must be directed to prevent such deviations or to obviate their results. Both in health and in disease we may promote or hinder, we may hasten or retard these changes, and that from either side. Thus we may stimulate either construction or destruction, supply or removal, and we may do this from either side. Let us begin with the

Promoters of Construction.

We have dwelt so long on aliment and digestion because the promotion or prevention of nutrition is fraught with such important consequences to our patient, and furnishes us with our most potent remedial measures. To prevent or retard construction we have only to intercept or limit the food supply. When no aliment is taken death soon occurs, but when only little is taken this event is delayed. A compensatory process occurs in the lessening of destruction, for under the influence of scanty supply the body relaxes its work, a retardation of tissue changes ensues, and so for a time the balance may be redressed. But prevention, if easy, is not so often important as promotion. How can this be accomplished? Simply by supplying nutrients, for the natural stimulus to nutrition is the presence of the needed material, just as healthy blood calls forth the action of the heart. So abundance of plastic matter promotes nitrogenous transformation, and excess of amyloids and fats may lead to obesity. We vary, then, the food and go beyond the usual range of ordinary diet; we administer, as we say, nutrients, which are the supporters of metabolism, and when imperfect health exists they are true analeptics—restorers. There is, indeed, another way of promoting construction—an indirect way—and that is by hastening destruction, for if the organism is in good working order it is obvious that inasmuch as construction and destruction go on *pari passu*, increase of the latter will create a demand, for renewal and repair will be effected from the fresh supply. In health this is what takes place when work is increased but the diet not restricted. Work, then, exercise, within the limits adapted to the capacity of the individual, is a promoter of construction, because it hastens destruction, and so we can prescribe exercise, gymnastics, and so forth. Further, we can hasten removal through the various excretory organs by stimulating them to increased action: many of our evacuants may thus be made indirect promoters of secretion by exciting increased elimination. Once more, not only may we hasten secretions and remove them, but we may even take away a portion of the nutrient fluid which has been prepared and will of course have to be replaced, this direct depletion, or blood-letting, within certain limits and under certain circumstances, becoming thus a direct promoter of construction. But of this indirect method hereafter, we have first to consider the direct promoters of construction.

NUTRIENTS.—*a. Milk*, as itself a complete food, may be first considered. We have already seen its importance, and sometimes even in chronic cases an absolute milk diet is employed. Milk cures are very common in Switzerland, but they are by no means modern, being at least as old as the time of Galen, who sent his scrofulous patients to Stabia to undergo them. In early phthisis, or rather in the pre-tubercular stage, a milk cure in a mountain resort has often been beneficial, but some influence must be assigned to the fresh air and suitable climate. It may be necessary to employ alkalies, as already mentioned, to facilitate the digestion of milk, but a simple method, and often completely effectual, is to dilute it with water, hot or cold, according to taste and season. Aërated waters are also of great use for this purpose. The most certain mode, however, of securing its digestion is to peptonize it. Whey is sometimes employed; it contains the sugars and the salts, but the fat and casein have been removed, consequently it is a very fluid but only slightly nutritious diet. Buttermilk is in some places a favorite; in it the sugar has been converted into lactic acid, and the butter as far as possible removed. Koumiss is a pleasant fluid in great favor on the Steppes of Russia, prepared by fermenting mare's milk. It therefore contains a little alcohol and carbonic acid besides butter, casein, milk salts, and lactic acid. A similar fluid is now prepared from cow's milk; it is very easily digested, makes a nutritious liquid which is agreeable to feverish patients, and is almost always easily digested. Perhaps some of the fame of the koumiss cure, as carried out in Russia, may be due to the outdoor life in the singularly pure atmosphere of the Steppes.

b. Beef-tea contains all the salts with a little albumen and gelatin, and some other nitrogenized material. It stimulates the nervous system, but it is not so important a nutrient as is generally supposed, especially when, as is too often the case, the fibre is not powdered and added to it; given alone it stimulates and often provokes fever, quickening tissue waste but not supplying sufficient for repair. Mixed with arrowroot or other starchy food, or taken alternately with milk, it is much more valuable. Broths and soups, to be really useful, ought to be made in the same way. A patient ought not to be fed on proteids alone, or even on proteids and fats, but some farinaceous food should be added to them. The salts in beef-tea would tend to hasten waste. It will be seen that the various extracts of beef, valuable as they may be as stimulants, can scarcely be considered nutrients. Eggs afford the means of adding highly nitrogenous material to the diet and can generally be digested in convalescence.

c. Fatty Nutrients.—In respiratory diseases the digestion is sometimes good and at other times it fails, and our greatest difficulty is to maintain nutrition. In acute inflammation a fever diet may be necessary, but generally more nutrients should be given than in fevers. In chronic diseases, especially phthisis, there is often a state of malassimilation, in which indigestion of fat and even repugnance to it is prominent, and this may occur before the phthisis is pronounced and be a sort of premonition of it. The organs apparently cannot easily utilize it, and the question is whether we can induce them to do so. Digestion of proteids, perhaps, goes on well enough, and this supplies a blood rich in albuminates, consequently full of material for repairing albuminous structure—like the lungs. It is quite conceivable that under such circumstances excess of proteids may be taken, and that, perhaps, under the mistaken notion that they will improve the nutrition and arrest emaciation. Is it not rather possible that the presence of such excess may irritate the lungs and excite rapid cell-proliferation? It seems sometimes as if the more such patients needed fat, the more they disliked it. Some may be taken in the form of milk, more in cream, clotted cream, or butter; but to digest these fresh air, out-of-door exercise will perhaps be necessary, and when these as well as fat meat are eliminated from the diet, from an apparently unconquerable repugnance, our difficulties increase. What is the use of ordering codliver oil for a patient who is sickened by the thought of butter? Can we present oil or fat in a form which can be taken? Many attempts have been made to disguise the taste and appearance. We have seen how fat is digested, can we not commence the process? Will emulsifying it answer? Very often yes. Pancreatic emulsion, prepared from the best beef-fat by means of pancreatin, was introduced by Dr. Dobell a few years ago and has been found a most valuable nutrient.

d. Farinaceous and Saccharine Nutrients.—It is not necessary to mention the various articles of this kind suited to invalids. The weakest of them—rice-water, barley-water, gruel, as well as vegetable juices and fruits, all have their places and are utilized in turn. Ripe fruits contain from ten to fifteen per cent. of solid matter, consisting of sugar, free acid, nitrogenous substances, and salines. Dried fruits, being richer in sugar, are more nutritious. An attempt has been made to feed consumptive and other patients exclusively on ripe grapes. This grapecure can of course only be carried out where the grapes flourish, and the influence of fresh air and climate is to be considered. There is not enough nitrogen in the fruit to supply the wants of the healthy system, consequently a certain amount of bread is permitted to most patients, and certainly all predisposed to phthisis or other wasting diseases should not be placed on too rigid a diet even for a short time, though no doubt other patients may be benefited by such a restriction. The usual plan adopted in the grape-cure is to allow as much of the ripe fruit as the patient feels able to take, beginning with about a pound and increasing to three, five, seven, and sometimes eight pounds. The first meal is taken in the early morning, and eaten in the vineyard; not so the others, one of which would correspond with our breakfast, another with evening dinner, and another at bedtime. In the middle of the day, after the morning walk, a meal of bread is taken and water to drink. Diet of this kind continued from four to six weeks must have a considerable effect and cannot be without risk for tuberculous constitutions. At first the grape-cure sets up purgation, it also stimulates the kidneys and sometimes proves in this way too debilitating. It might therefore be considered to act by elimination, bringing about destructive changes. No doubt this is so, and to this is due the good results in lithiasis, portal congestion, plethora, etc. Here, too, is the risk in other diseases, but inasmuch as some of these have been improved it would seem that the removal of the waste has brought about increased repair, and thus the grape-cure has some claim to be considered an indirect promoter of construction.

e. Salts.—The saline substances which enter into the tissues and the blood are in a certain sense nutrients, and one or other of them has sometimes been spoken of as food. It is to be observed, however, that the salts are only one class of the necessary ingredients of our food, and, though exceedingly important, have perhaps the least power of sustaining life. When administered in extra quantities they possess other actions more medicinal than nutritive, and will therefore be considered more conveniently farther on. We have seen how they promote metamorphoses and perhaps they may be called more naturally restorers than promoters of nutrition. Such a distinction may be drawn between nutrients proper and analeptics or restorers, but it is not always observed, the words being frequently interchanged. We have, however, as a matter of convenience placed these nutrients together, and now pass on to analeptics.

ANALEPTICS.—*Cod-liver oil* has been known for more than a century, Dr. T. Percival having been acquainted with it as a remedy for chronic rheumatism about 1771. He sent a communication to the Medical Society, October 7, 1782, with a letter from Dr. Robert Darbey, giving an account of its accidental introduction some ten years previously to the Manchester Infirmary (*Lond. Med. Jour.*, 1783). A number of articles or pamphlets upon its use were published before it was recommended by Dr. Bardsley in 1807, who tells us it was in much favor in Lancashire (*Med. Rep.*), and after him by Dr. Hughes Bennett, whose work appeared in 1841. The last writer demonstrated its great value in phthisis, since which it has become generally employed in that disease as well as in others in which fatty nutrients and analeptics are required. In 1849, in the first report of the Hospital for Consumptives, the physicians of the institution, as the result of its extended use, stated that "no other conclusion can be drawn than that cod-liver oil possesses the property of controlling the symptoms of pulmonary consumption, if not of arresting the disease to a greater extent than any other agent hitherto tried." In 1853 the Paris Academy of Medicine offered a prize for the best essay on the therapeutical uses of cod-liver oil. The prize was awarded to Dr. E. Taufllieb, but Dr. F. Dubois also received very high encomium. Both essays were published, and as may be expected are full of information. Dr. de Jongh published a monograph in 1853, containing many analyses and much other information. From this time interest in the subject has not abated and numerous communications tending to confirm our estimate of its value have been published. The oil contains about seventy to eighty per cent. of oleine and about fifteen to twenty-five per cent. of margarine, with certain biliary matters derived from the livers. It seems very complex, and analyses give us oleic, palmitic, and stearic acids with glycerine, acetic, butyric, phosphoric, and sulphuric acids, peculiar substances soluble in alcohol or ether, traces of iodine and bromine, phosphorus, lime, magnesia, and soda. A trace of iron is found, but this is probably accidental; it has been thought that some of the other inorganic ingredients may also be accidental. A substance named gaduin was extracted by De Jongh, but is not very important. By a reaction with ammonia in distillation propylamine is obtained. The biliary principles are felninic, bili-felninic, and cholic acids and bilifulvin. The iodine has been thought to account for some of the properties, but it is very small in amount; De Jongh was unable to find more than the four-hundredth part of a grain in one hundred grains of oil. The sulphuric acid test by no means proves the purity of the specimen, it only shows the presence of biliary matter, and so far indicates that it is derived from liver. No doubt the livers of various species of gadus are continually employed, and that other fish have also been used, either alone or to mix with genuine cod-liver oil. Perhaps the smell and the taste are the best indications of quality, a distinct odor of ordinary fish-oil is suspicious. Great improvement has taken place of late years in the preparation, and the best specimens are not nearly so disagreeable to the palate as formerly.

All that we have stated respecting fatty aliments may be applied to this oil, which has the advantage of being pre-eminently easy of digestion. It therefore presents us with an easy means of introducing a fatty nutrient and analeptic. Moreover, the experiments of Naumann, Manz, Simon, Dugald Campbell, and others, tend to show that this oil facilitates absorption and also oxidation, and general experience not only confirms this, but shows that under its influence the blood improves in quality. It is therefore both a supporter and promoter of nutrition. At the same time it does not disturb any of the functions of the body, unless sometimes digestion, for in excess it is apt to give rise to dyspepsia and perhaps some diarrhœa. We should, however, remember that the same may be said of all oils.

There can be no question that as animal fats are so much easier of digestion than those of vegetable origin, so this fish oil is more digestible than the former. Small quantities undoubtedly promote digestion, as we have seen other fats do. Dr. Pollock found one or two ounces a day given to pigs, one ounce to sheep, and from three to nine ounces to bullocks helped to fatten them, but larger quantities deranged digestion.

Many attempts have been made to trace the therapeutical properties of the oil to one or other of its constituents, but without success, and unless all its value is to be ascribed to the peculiarly digestible oleine, we must consider its virtues as dependent upon this, but perhaps reinforced by other ingredients. A peculiarly digestible fatty food should be able, upon the principles we have laid down, to accomplish most of the results we observe to follow a course of the oil. Clearly, then, the first indication for its administration is defective nutrition, as soon as some degree of emaciation and loss of strength, with perhaps anæmia is observed, this medicine should be administered and will often suffice to restore health, this is particularly noticeable in children and young persons, and such a condition often precedes wasting diseases, which thus taken in time may be arrested. In scrofula, besides a similar form of debility, we get more decided evidences of malnutrition, among which may be named glandular affections, as well as a tendency to frequent catarrhal attacks of the respiratory mucous membrane, and even pneumonia. It was in chronic rheumatism that the oil first obtained its reputation, and in some of these cases it will be successful—generally, I am inclined to think, in patients of a strumous disposition.

But the disease in which it has obtained the greatest vogue is phthisis, in which it is consequently prescribed almost as a matter of routine; and we may perhaps be permitted to speak of this in its several forms as one disease without entering minutely into its pathology, for it cannot be denied that in all forms, whether tubercular or fibroid, scrofulous or catarrhal, the most important indication is to improve nutrition. In the so-called pre-tubercular stage we have no more valuable remedy. In the advanced stages of chronic phthisis it has less chance of restoring nutrition, though it sometimes arrests the emaciation and produces other good effects, such as relieving the cough, retarding the progress of lung mischief, and so prolonging life. Unfortunately the oil is sometimes prescribed without reference to the state of the digestive organs, and sometimes too large a dose is given to begin with. It is not well to give it during febrile attacks, and therefore it is a good plan to suspend it for a time when, as so often happens, a little febrile excitement occurs. It will often be found that the pulmonary symptoms subside under the influence of the oil, which has therefore been called by some persons an expectorant, but the true reason of the improvement is rather the promotion of nutrition. Of such cardinal importance is this, that it is generally better to put aside the favorite expectorants and narcotics, which have often done so much mischief, and to attend entirely to the appetite, and direct all our efforts to the maintenance of the digestive and nutritive functions. If lung symptoms require relief, it should be given without disturbing the stomach; inhalations will often suffice for this, and are among our most useful remedies, since they scarcely ever interfere with nutrition. When the oil cannot be digested, attention to the *primæ viæ* will often suffice to produce toleration of very small doses. Alkalies, bismuth, bitters, or even iron may be called for, but not so often as is generally supposed, for just as very small quantities of other nutrients will produce an appetite, *e.g.*, a teaspoonful of milk or beef-tea, so will small doses of cod-liver oil; and patients who at first feel a great repugnance for the remedy, often come to take it without dislike. Emulsions are more easily taken by many than the simple oil, and often more easily digested, but I do not recommend them to be made with potash. Much better emulsions, without the drawback of the alkali, and with additional nutritive qualities, are made with extract of malt; these, however, are not easily prepared on a small scale. A very nutritious emulsion may be made with yolk of egg and gum arabic, to which sugar may also be added, and this should be freshly prepared. A jelly can be made with isinglass. The glyceritum-vitelli may also be utilized for this purpose. It is common to flavor such emulsions, but most persons will prefer them more simple, and to me the well-made emulsions with extract of malt seem far the best, and if taken pure and cold may almost be called palatable, though if mixed with warm fluids, as some have thoughtlessly directed, the flavor of the oil is intensified.

When the oil is taken pure, attempts are made to disguise it by the vehicle—the various bitter and aromatic infusions may usually be taken for this purpose; when iron and quinine are given, many patients prefer to float their oil on their mixture. I must, however, condemn the too common recommendation to employ whiskey or brandy for this purpose; from one to three tablespoonfuls of these spirits is often recommended with a dose of the oil, a method of using alcohol which has nothing in its favor. If, as it is sometimes said, the stimulus promotes the digestion of the oil, the dose for such a purpose ought to be small, and ether is very much better than alcohol. Porter and ale are less objectionable. Chewing orange- or lemon-peel, or any other strong flavoring substance, even a pinch of salt, are innocent methods; or Pavesi's plan of deodorizing with charcoal and flavoring with coffee is not objectionable. The addition of more active medicines, such as phosphorus, iodide of iron, etc., whatever may be said as to the utility of the drugs, renders the oil generally less easy to take. It is said that the pure oleine obtained from cod-liver oil sometimes agrees when the oil itself cannot be tolerated.

The dose must bear a due proportion to the digestive power. Few people can digest more than an ounce a day, and the experiments already mentioned show that the limit is soon reached. It is usually better to begin with one or two teaspoonfuls once a day, and as the taste becomes less objectionable, twice a day. With delicate stomachs it is well to begin at bedtime, as the oil is digested during sleep, and thus disagreeable eructations are not perceived. As soon as it is easily tolerated the best time for administration would seem to be soon after meals, as that would be the period at which it would be acted upon by the pancreatic juice, and pass with the other products of digestion into the lacteals, and exercise such influence as it may possess in the process of sanguification. When the taste is not at all objected to, it may sometimes be given with advantage between the meals, especially if the patient can take other kind of nutrients, say milk, egg, etc., at the same time, and really stands in need of additional food. It is, however, unfortunately too often found that where nutriment is so much needed there exists the greatest repugnance toward it, and then the best way is to endeavor to create a desire for food by the application of the natural stimulus—that is to say, by giving it in minute quantities. In such cases cod-liver oil in very small doses will sometimes effect the purpose, especially in the form of malt-emulsion. We may also endeavor to assist its assimilation by the addition of other substances as well as by administering medicines to improve the digestive organs. The most important of these is

Ether.—We have seen how fat, and therefore how cod-liver oil is digested—that we may assist the process by emulsifying it, or by the administration of pancreatin or malt extract. Instead of supplying artificially the digestive fluid, or a substitute for it, can we stimulate the pancreas and duodenal glands to increased secretion by varying the diet or by the administration of medicines? How to attempt this by diet may be easily deduced from what has preceded. With regard to medicine, ether deserves considerable attention. The object of its administration is not to disguise the oil or to coax the stomach into receiving it, but to help its digestion; not, indeed, directly, though it may have some influence in this way, but by a direct action upon the pancreas. Dr. Balthazar Foster proposed to utilize this action at the Oxford meeting of the British Medical Association in 1868, and he has employed ether in conjunction with oil as well as a substitute for it. He was led to this practice by the experiments of Claude Bernard, who found that when ether was introduced into the stomach a considerable flow of pancreatic juice soon afterward followed. Not only so, but there was always a degree of vascular congestion excited in the digestive tract, such as is caused by the normal excitement, food. The pancreas also became red and turgescent, as it does during digestion, and Bernard accordingly administered ether in order to increase his stock of pancreatic juice. For he tells us that this fluid issues by drops, more or less frequently, from the tube of an animal with a pancreatic fistula; but when ether is given it flows much more freely, and the liquid has not lost any of its characteristics. Still further, he found that when two rabbits were killed soon after they had received fat and ether into the stomach, one of them having had the pancreatic duct tied, its lacteals contained only transparent lymph, while those of the other were filled with milky fluid. From his investigations it seems that the digestion of fats should be increased by ether. If it stimulates at once the pancreas and glands of the duodenum to pour out their secretions more freely, these must necessarily act upon the cod-liver oil presented at the same time, and if it not only emulsifies within the system the oil, but further assists its absorption, it presents us with a most valuable remedy. Dr. Foster proceeded to put this conclusion to the test of clinical experience, and he found the results highly satisfactory. Under the influence of ether he often observed a return of the power of taking oil and fatty food which had been previously distasteful or had even excited sickness, increased appetite, improved general nutrition and increase of weight, diminution of cough and expectoration, and cessation of night-sweats. "In a further more systematic enquiry on 50 dispensary cases, in which only those were considered improved in whom physical exploration of the chest and decided increase in weight confirmed the patient's impressions, we find that of 16 cases in the first stage 7 improved, 5 remained stationary, and only 4 became worse. Of 19 cases in the second stage 6 improved, 6 remained stationary, 7 became worse. Of 15 cases in the third stage 7 improved, 5 remained stationary, and 3 became worse. The increase of weight in the first stage averaged seven and one-half pounds, in the second eight pounds, in the third five pounds. Some of the cases returned to their ordinary avocations and were able to earn their livelihood. In the cases that did best the power of taking fats, which had been impaired, returned, and the gain in weight was maintained for many months. It may be added that in all the cases there had been marked wasting before the ether was given, although cod-liver oil had been taken in at least half the cases.

Dr. E. L. Fox, of Clifton, and Dr. Mapother, of Dublin, confirmed Dr. Foster's opinion of the advantages of etherized oil, which is further sustained by some observations of Dr. Ramskill on the use of olive oil as a remedy. A committee of the New York Therapeutical Society (Dr. Andrew H. Smith, chairman) reported (1879) that ether often enabled the oil to be taken when it had previously disagreed, but they had no cases bearing on the ultimate effect of the ether on nutrition. I may add in conclusion, that etherized oil gave satisfactory results at the North London Consumption Hospital when I was attached to it.

The ether may be given in the oil, *æther purus fortior* being employed, in the proportion of ten minims to two drachms of oil to begin with, afterward increasing to fifteen and twenty. It has the advantage of masking the unpleasant flavor of the oil. Some patients prefer an ether mixture taken before or after the oil, or it may be floated on such a mixture, from fifteen to thirty minims or more of spirits of ether in the necessary quantity of water.

The general effects of ether somewhat resemble those of alcohol, but are much more evanescent. The liquid is rapidly absorbed and as rapidly eliminated, for the most part through the lungs. The odor is perceived in the breath very quickly after a dose is taken, even when enclosed in capsules. Generally, too, a portion is returned by eructation almost immediately. Even when injected into the peritoneum Majendie found the odor rapidly perceptible in the breath of the animal. And in a case in which half an ounce had been taken shortly before death, at the post-mortem examination the odor was distinctly perceived on opening the skull. Very large doses produce poisoning somewhat in the same manner as alcohol, but very considerable quantities may be taken when the system becomes habituated to its use. Christison mentions a man, who for several years got through sixteen ounces every eight or ten days. Bucquet, the chemist, is said to have taken a pint a day to assuage the agonies of schirrus of the colon, of which he died.

In small doses its exciting effect is perceived, increasing the power and the frequency of the heart's beat, bringing on flushing of the face, warmth of the surface, sometimes perspiration; there is also some mental excitement, which soon subsides into calmness, or even torpor. So transient, however, is the effect and so rapid the elimination by the lungs, that generally a full medicinal dose will have disappeared in an hour, from which it may be argued that moderate doses at short intervals are most likely to be of service. It is often given in spasmodic and nervous diseases where there is a good deal of pain or distress without inflammation, *e.g.*, in colic, in gastralgia, flatulency, nervous palpitation, etc. With regard to respiratory diseases it is recommended in spasmodic asthma, pertussis, various forms of dyspnœa, some forms of chronic catarrh, and in phthisis. With regard to this class of diseases it will be observed that it is chiefly as an antispasmodic that it is employed. And it often affords rapid relief, if only temporary, in all forms of spasmodic neurosis. In catarrhal cases it is to be tried with caution, and often inhalation with steam will be the best mode of administering it. It may be advantageously combined with conium for this purpose. An ethereal tincture has often been employed, but a much more effectual remedy is the vapor-conii, with the addition of ether.

The ordinary dose of ether is from twenty to sixty minims. The spirit is more convenient, as it mixes freely with water, thirty to one hundred and twenty minims may be given. The compound spirit—Hoffmann's anodyne—contains some ethereal oil and seems more anti-spasmodic. Capsules or perles of Dr. Clertan are convenient, especially when the taste is objected to, but they must be washed down quickly lest they burst *en route* and cause local irritation.

CHAPTER VIII.

IRON.

This metal is found in the tissues, but to a much greater extent in the blood, of which it is one of the most important constituents. Some have denied that medicinal doses are absorbed; but it must be derived from the food, is in fact a necessary constituent of our diet, and when from any cause we have reason to believe it is deficient in quantity, it seems rational to give it as we should other proximate principals. Moreover, many experiments have shown that after its administration larger quantities have been found in the blood as well as in the secretions, besides which absorption has been proved to take place from the cellular tissue. Those salts which are insoluble in water may easily be dissolved or acted upon by the gastric juice. The pure metal, minutely divided as in reduced iron, is perhaps first oxidized by the help of water, hydrogen being set free, and this accounts for disagreeable eructations of sulphuretted hydrogen occurring when any sulphur is present and this preparation is employed. Formerly it was said that all protosalts quickly became persalts in the stomach, but though such a change rapidly occurs outside the body it is not necessarily so within. In fact, some experiments seem to indicate that the reverse change may occur. Stenhouse found that persalts were readily reduced by organic substances, and Claude Bernard, after injecting into the jugular vein a persalt, found only a protosalt in the renal secretion. If the persalts obtain access to the system it seems probable that they can only exist as such a very short time. As to the salts of the organic acids, it seems they may be absorbed at once when the acid is oxidized and the nascent metal left free to form other combinations, a carbonate being, according to Rabuteau, most likely formed in the same way that similar salts of the alkalies are decomposed and their carbonates formed. Protoxide and carbonate would be easily converted into protochloride by the gastric juice. So, perhaps, would sesquioxide passing first through the state of perchloride, and it should be observed that protochloride does not precipitate albumen and could therefore be absorbed. Even the salts of the mineral acids, if given sufficiently diluted, need not necessarily coagulate albumen or act as astringents on the mucous membrane, and might therefore be absorbed, but we certainly find that these preparations are more apt to interfere with digestion, which seems to show that ordinary medicinal doses are astringent. Probably iron is generally absorbed as an albuminate, and it is certainly for the most part eliminated in albuminous secretions. To this last fact Gubler points as explanatory of the remote astringent action of iron, supposing the metal to be separated by the secreting surface from the albumen and then to exercise its natural astringency. A large proportion of iron is eliminated from the bile, and some have thought, therefore, that all, or nearly all, passes to the portal circulation without entering the general circulation. But it has also been supposed that much of the iron in the bile is derived from effete blood-corpuscles being broken up in the liver. Still we must admit some may be absorbed and eliminated through the bile or the intestinal secretion, some may be taken up with the fats and products of late digestion, while yet another portion may pass on undigested. These results may vary with the preparation, the dose, the degree of dilution, the time of administration, and the state of the digestive organs, and most of these determining conditions are under the control of the physician, and it is desirable so to regulate them as to obtain all the benefits of the remedy without any inconvenience. We have said that iron is in the tissues as well as the blood, and this is why, as an essential ingredient, we regard it as one of the nutrients in the sense in which we have already discussed them. In the blood it is in combination with the hæmoglobin, forming about 0.42 per cent., its chief use being as a purveyor of oxygen ; in the lungs the hæmoglobin takes up oxygen and thereby becomes oxyhæmoglobin, the result of this being that the purple venous blood becomes the scarlet arterial. This is carried in the vessels to the tissues, to which it gives up its oxygen and so is reduced again. Thus it is by virtue of their hæmoglobin that the red corpuscles are carriers of oxygen to the tissues, and it is curious to observe how easily the oxygen is associated and dissociated without disturbing much their molecular state. Other gases may also be taken up, *e.g.*, carbon-monoxide, but this is less easily dissociated, hence its poisonous qualities, for this gas is useless in respiration and the person is asphyxiated though the blood is found red.

The iron is in intimate union with the red corpuscles, and this union would seem to be not merely chemical. Some have conjectured that it may be present in the metallic state ; but Liebig, Mialhe, and others consider that it is present as a peroxide, which in contact with moist tissues might become protocarbonate, and this on exposure to oxygen in the lungs would give up its carbonic acid and reassume its former state. But there are difficulties in accepting this theory, which would account for the properties of the corpuscles by the purely chemical action of the contained iron. We must recognize, however, that these corpuscles are the purveyors of oxygen to the tissues, and that the iron is essential to the hæmatin, and no doubt concerned in its function. Iron is also present in myochrome. Now, if the iron brings the oxygen to the tissue and also takes some part in the oxidation of tissue, it may be argued that this is a disintegrating process, and therefore that the metal should not be considered as a promoter of construction. We have, however, already shown that it is an essential ingredient of the body, that it is usually supplied in the food and eliminated in the secretions. It is therefore entitled to be considered, among the proximate principles of aliment, the most important in its class, perhaps; moreover, deficiency of its supply is followed by characteristic symptoms, which are removed by its administration, apart from which clinical observation has universally pronounced that it possesses tonic properties. Further, even if it should be shown to hasten destructive tissue changes, we have seen that this process within certain limits indirectly promotes construction; and, finally, it may be said that the presentation of the oxygen to the tissue, though it stimulates oxidation, being a necessary constant condition of nutrition, may be regarded much in the same light as the presentation of the other nutrient principles, which, as we have seen, stimulates construction.

If the iron increase oxidation it should raise the temperature, as it should also if it excited more active constructive changes. Pokrowsky believed that he did detect a rise in the temperature after the exhibition of iron, and that in some cases within five hours after the first dose, so that it could not be due to an increase in the number of the red corpuscles, as that could only take place slowly; the elimination of urea was at the same time increased. It is unfortunate that the iodide of iron was employed, as the iodine could not be without its effect. Picard has shown that a definite constant relation exists between the amount of iron in the blood and the amount of oxygen. Recent investigations by Rabuteau tend to show that iron increases oxidation, and many regard it as the active agent in the red corpuscles for condensing oxygen into ozone, though they are not prepared to admit, with Sasse, that the metal can supply the place of red corpuscles as an ozonizing agent; if it could, we should be able to accomplish more by its use.

Somewhat contradictory results have been reported as to the effect of iron on the pulse. It is mostly believed to give tone to the circulation and to make the pulse fuller and more forcible, though often slower; but others state that it increases the frequency of the pulse, and it is a common observation that it appears to increase congestions. So far as it favors oxidation, it tends to increase the production of heat and to quicken the pulse; but this last symptom may be more than neutralized at a later stage by the increased strength which follows improved sanguification. In purpura and other conditions it certainly seems to quiet the pulse unless it irritates the stomach. It appears to give tone to the capillaries and to stimulate their contraction, to which may be partly ascribed its power over hemorrhages and discharges. As to its other effects on the circulation, the improvement in the tint of anæmic persons under a course of iron is sure to attract attention. It is even said to induce plethora, but perhaps this is going too far, or may only be a forcible way of saying that it is contra-indicated in individuals displaying a plethoric tendency.

The hæmatinic effects of iron may be measured by clinical observation, but of late years we have been able to adopt Malassez's method of counting the blood-corpuscles, and lately Dr. Amory has succeeded in projecting upon a screen a magnified image of the slide employed in this method for the purpose of demonstration. It has been shown again and again that the corpuscles, when diminished in number, rapidly increased, as determined by the hæmacytometer, under the influence of preparations of iron. It is true that, according to Hayem, the number was not always increased, and Claude Bernard found that the corpuscles were not invariably deficient in the blood of chlorotic patients. On the other hand, Hayem found the globules improved in quality, that is in size, shape, and color, under the influence of iron in cases in which there was no deficiency in their number. In anæmia and in chlorosis of moderate degree, when the number was not markedly deficient, he found alterations in size and shape as well as in coloring power, which last was sometimes diminished to the extent of half—that is to say, a given quantity only showed half the depth of tinge that it should have done. After a course of iron the corpuscles from these patients recovered their natural coloring power as well as their ordinary size and shape. Many observers have found large increase in number, and thus corroborated the views which had been entertained, and which were supported by chemical analyses, such as those of Simon, who found the globulin and hæmatin doubled and trebled in weight. It would seem as if the presence of iron in abundance excited a more active manufacture of corpuscles, just as other nutrient principles provoke more energetic construction of the tissues into which they principally enter; moreover, the blood-glands are perhaps stimulated by the iron, and then again, when the corpuscles have been deficient, their increase promotes other functions, which in their turn depend on the circulation of healthy blood. Thus, when given in appropriate cases, not only is the blood-forming process rendered more active, but digestion is improved, nervous tone imparted, and all the functions rendered more healthy.

In order to secure the hæmatinic effects of iron, only small doses are required, provided the medicine be given in a form which can be easily absorbed. The total amount present in the blood of a healthy adult is estimated at from thirty-seven to forty-seven grains, and the deficiency is believed scarcely ever to exceed fifteen or twenty grains, even in severe cases of anæmia. It has further been estimated that in health about a grain a day is usually taken up and eliminated, but in disease as much as four or five times this quantity has been assimilated during a short period. The blood will not take up more iron, nor form more hæmatin, than the normal amount, so that we cannot increase the richness of the blood beyond the standard of health. But it does not follow that unlimited quantities may be given. On the contrary, large doses are likely to derange digestion, and so defeat the object with which they are given. They also impede secretion, thereby hindering absorption and constipating the bowels. Something of course depends on the form in which the iron is presented, the preparation, the dose, and the degree of dilution; something, too, on the state of the stomach. The unabsorbed portion passes through the alimentary canal, becoming for the most part a sulphide, small quantities of which tend to constipate by checking secretion, though larger quantities may irritate, and so produce diarrhœa. Iron is also eliminated by the kidneys and liver, the rate of removal being in accordance with the rate of absorption, this taking place more easily when the dose is small and largely diluted and the preparation employed least astringent.

Diminution of iron in the system—anæmia—may be brought about by a deficiency of the metals in the food, by an impairment in the power of absorbing it, or by some hindrance to the process of sanguification. These causes resolve themselves into a diminution of supply. On the other hand, the same results may be brought about by increase in the waste; the loss by elimination may be in excess, or all the constituents of the blood, including the iron, may be removed, as in the case of hemorrhages. When the cause is not persistent, health returns under the influence of good food, but much more quickly when some iron salt is also given, and the rate of improvement is easily measured by counting the corpuscles. In the anæmia of convalescence, or after loss of blood, or in exhausting discharges, the effect is very manifest, and the more acute and the more direct the anæmia the more rapid and the more marked is the improvement. These effects have, however, been denied. Trasbot does not admit the hæmatinic power of iron, and so able an observer as Dujardin-Beaumetz speaks of its use as a therapeutical illusion. They stand, however, almost alone in this opposition to the general experience of the profession, supported as it is by both theory and experiment in all cases of direct anæmia, whether from loss of blood, excessive secretion, acute disease, interfering with food supply and bringing about as it were a degree of starvation, so that the blood is deficient in iron as well as probably albuminous constituents. There is no doubt that the restorative influence of good food may be greatly increased by the administration of iron, provided it is so given and in such quantities as not to interfere with digestion. In order to prevent relapse it is very desirable to continue the remedy for some time after recovery.

Idiopathic or pernicious anæmia, as it has been called, is not so amenable to iron, perhaps from there being some difficulty in assimilation, though some cases have been benefited by the hypodermic injection of dialyzed iron. But it may be said that there is a further profound interference with nutrition than the intense anæmia of these cases.

In chlorosis similar statements may be made to those referred to in anæmia. Here, also, diet, fresh air, exercise, and other hygienic influences are of the first importance and may sometimes effect a cure, but this may always be hastened, and sometimes can only be obtained, by adding a course of iron. Very often, too, aperients are particularly desirable; hence, perhaps, the popularity of formulæ combining these with iron, as in pilula aloes et ferri. Sometimes, however, astringent preparations are required, but in almost all cases attention to the *primæ viæ* is of the utmost importance, and soothing, stimulating, or other remedies may be resorted to according to circumstances, with a view of promoting digestion and the assimilation of the iron. In numerous cases fatty food is also of importance, and fresh air and exercise should never be overlooked. Da Costa has obtained good results from hypodermic injection.

Speaking of the difficulty of securing assimilation leads us to a remark on the effect of the metal on digestion. Too large a dose impedes the process and causes irritation, and these same effects are often produced by the stronger preparations, when the weaker ones are easily taken. Further, when no intolerance appears and the remedy even seems to promote digestion—which it perhaps does by acting as a stimulus to the gastric mucous membrane—after a time, perhaps, it may disagree, bringing on a sense of weight and irritation, or even more pronounced symptoms of indigestion. Perhaps this may arise from over-stimulation of the gastric glands, or, on the other hand, from a diminished secretion brought about by the astringent properties of the medicine. It sometimes occurs from the administration at improper times. Querance found in his experiments on dogs with a gastric fistula that iron improved the peptones when given with food, though it did not increase the proportion of pepsin nor diminish the time of digestion. The dogs gained flesh when thus given, but when administered without food, especially in metallic form, the iron did not cause sufficient secretion to bring about its solution, and therefore remained in the stomach undissolved and caused irritation of the stomach and bowels. To prevent disagreeable symptoms, then, the medicine should be given with the food, the lighter preparations so far as suitable to the case should be selected. Combination with other medicines may often be desirable, and in all cases absence of gastric irritation, biliousness or pyrexia is desirable. It may be added that in certain forms of atonic dyspepsia with general debility, and especially any tendency to anæmia, iron preparations are not only tolerated but are most beneficial.

As regards secretion the effect as already shown is astringent. In some instances, on the other hand, an opposite effect is spoken of, though very often that can only be indirect that is the consequence of the restorative action.

In hemorrhages the action is both local and general. As a styptic in bleeding from the tonsils after operation, as well as in epistaxis and other accidental hemorrhages, the perchloride or the subsulphate, as in Monsel's solution, may be safely employed. In hæmoptysis a spray is often very effectual and either perchloride or the ordinary sulphate may be used. It has been thought that this would lead, by coagulating the blood, to fresh congestion, but chemical experience has not found this to be so. Caution is necessary in administering iron internally in phthisis with a tendency to hæmoptysis, but there is little objection to its use as a spray.

In diphtheria the local and general use of perchloride has become as popular as in erysipelas. Here, too, the spray is generally better than the ruder swab, which is often too roughly applied. Internally the tincture or liquor has been employed even more extensively, and perhaps with better effect than locally. It was used in the earlier epidemics by Jodin and Aubrun, in France, and by Dr. Godfrey and Dr. Heslop in England, in 1857 and 1858. It has maintained its position as a valuable agent up to the present time, many observers being convinced that it is a direct restorative and exercises a remarkable influence in preventing blood-poisoning. Isnard believed that it might prevent exudation, just as it is admitted to restrain hemorrhage, and that by rendering the blood more plastic it diminished the liability to the blood-poisoning, and his idea seems to have become popular. To attain this object it should be begun early and given at frequent intervals. Aubrun gave a dose every five, ten, or fifteen minutes, day and night, for the first three days after the invasion, after which he gave it less frequently, as he believed that by that time the membrane would become detached or cease to be formed. Others urge frequent doses and some large doses, under the idea of quickly acting on the blood. Jenner considers it beneficial only in certain cases which seem suitable for it, and Dr. George Johnson regards it as the most successful of all medicine. The evidence in favor of its value is certainly great, but too much reliance must not be placed on this or any other medicine. The majority of cases seem very suitable for it, and it should be given with no sparing hand, but local applications or sprays of this or other medicines should not be neglected, and above all attention should be given to maintain nutrition, disinfection, or ventilation. The existence of albuminuria and even the appearance of blood- or tube-casts in the urine need not prevent the use of the remedy, though in such cases it is perhaps as well to combine it with nitric or hydrochloric acid or such other remedies as may be indicated. We may here recall an expression of Sir James Simpson's when recommending the tincture in surgical fever: he termed it "a renal purgative."

In scarlet fever similar favorable results have been obtained by the use of the perchloride internally as well as in the form of sprays; and the same may be said as to the angina of other exanthemata.

In bronchial affections iron is chiefly useful during convalescence, but it sometimes is also benefical in chronic bronchitis attended with profuse expectoration; it also restrains the amount of secretion in bronchorrhœa and somewhat modifies its character. Perhaps this is by an astringent effect on the mucous membrane, or it may be merely by improving the state of the blood or the general health, though it has sometimes been pronounced to be an alterative on the respiratory tract. Perhaps it is more of a tonic to this membrane, acting directly on the capillaries, which would account for the fact that it is often beneficial in emphysema. Either the phosphates or the more astringent preparations are most frequently indicated in affections of the air-passages.

In phthisis considerable differences of opinion have prevailed as to the use of chalybeates. Trousseau and others have expressed a fear lest iron should aggravate the disease, particularly in its early stages, and certainly when there is considerable pyrexia or any pulmonary congestion or a tendency to hæmoptysis it may be well to postpone this remedy; but when there is no marked febrile reaction, especially if there be any disposition to anæmia, a short course of a light preparation may be of great service, especially in conjunction with cod-liver oil or a considerable amount of fatty food. Iodide of iron has been largely used in such cases, but it must be remembered that this preparation should be considered as one of iodine, rather than of iron. In the later stages of this disease the good effects are much more manifest and the stronger astringent preparations may be employed to restrain expectoration, night-sweats, diarrhœa, and even the passive hæmoptysis which sometimes recurs. Under the influence of the tonic the appetite often improves, more food is taken, and apparently more assimilated. The taste for fatty food returns and the emaciation is arrested. Care should be taken to select a preparation which will not disagree with the stomach; the lighter ones are almost always well borne, but when night-sweats or other indications for an astringent are present, the perchloride is the most generally useful.

CHAPTER IX.

PHOSPHORUS AND ITS COMPOUNDS.

Some combination of phosphorus is found in all the tissues and in all the fluids of the body. It has therefore been said to be as important an element of organic structures as nitrogen or carbon, and as absolutely essential for the growth of a penicilium as a man. Experience shows its importance to vegetable life—bacteria and fungi require earthy phosphates and their value to corn-crops is well known. Phosphorus is an important constituent of lecithin ($C_{44}H_{90}NPO_9$, from λέκιθος, yolk of egg), in which it was first found, and it used to be called phosphorized fat, because one product of its decomposition is phospho-glyceric acid ($C_3H_9PO_6$). It is most abundant in the solids, on account of the large amount of tissue phosphates in the skeleton, but it is also a very important constituent of the fluids. It enters the body in the food and is removed in the excretions as phosphates.

As the salts have been long employed in medicine it will be convenient to consider them first.

Calcium phosphate (as well as the magnesian salt) is considered not only to impart solidity to the bones but to be an important agent in the consolidation of the other tissues. In bone it is associated with other salts but forms five or six times as much as all the other mineral ingredients combined, no less than fifty-seven or fifty-eight per cent. of bone being lime phosphate. It is not deposited as a mere granular powder, neither does it form such a chemical combination as to lose its identity. It unites more like a salt in a saline solution, or the pigment in colored glass. Of the fluids it is present in largest amount in the milk, obviously on account of its necessity for the growth of the infant's bones. In the blood-plasma it is held in solution apparently by albuminous ingredients, for Fokker found that earthy phosphates united with white of egg and became soluble in some proportion ; and this explains how blood and milk, both alkaline fluids, can hold in solution these salts which are naturally insoluble in alkaline or even neutral liquids. Lime phosphate is to some extent soluble in lactic or hydrochloric acid, and thus we see how small quantities taken into the stomach may diffuse into the blood, but large amounts would pass unchanged through the alimentary canal, so that if we employ it medicinally only small doses can be of service.

When excluded from the food the bones have been observed to soften, though Weiske's experiment already alluded to shows that this is not always the case. Milne Edwards found the bones of animals which had been fractured united more rapidly when this salt was given and clinical experience shows corroborative evidence of this. Chossat produced softening by withholding all lime salts. The lime phosphate is further important as a nutrient, being apparently essential to the growth of tissue. Even when excess of lime afterward preponderates, as in the shells of animals, this salt seems necessary to initiate the growth, for the carbonate cannot entirely replace it. Wherever growth is active there the lime phosphate is present in excess. Not only is this the case in normal growth but even in abnormally rapid cell-development the same excess is found. This phosphate is eliminated slowly. Neubauer and Beneke reckon 0.4 gramme a day by the renal secretion, a little more escapes from the bowels, but that may be only undigested residue of food.

Phosphate of lime has been largely used in medicine, first whenever there has seemed to be deficiency in the bony system, as in mollities osseum, delayed dentition, and in rachitis. It is true there is often no deficiency in the food, but still the presentation of an unusually abundant supply might act as a stimulant to its absorption. But this treatment has not been so successful as was hoped. Rickets, indeed, is not merely deficiency of phosphates, and all the symptoms of this disease are not produced in animals by excluding the salt from the diet, even when the bones lose their solidity. In rickets there is malnutrition of other tissues as well as the osseous, and although this salt may play some part in their nutrition, and therefore be rationally prescribed, cod-liver oil is as good or better. The phosphate generally suits better after the tenderness of the cartilages has subsided, that is, when the activity of the disease has passed by. The secretions are generally too acid and the earthy phosphate would moderate this condition, while the unusual degree of acidity would render it soluble. It would hardly seem necessary, therefore, to combine it with an acid, as some have proposed in this disease, in which excess of acidity would tend to dissolve the osseous salts and so promote their elimination rather than their deposit.

This phosphate is next employed in defective cell-growth and other manifestations of its insufficient action. It is true the salt may not be absent from the food or even decreased in quantity, but either it is not absorbed or the tissues for some reason are not using it. In such cases of malnutrition the phosphate is often excreted in excess. Beneke found this to be often the case in scrofulæ, and it has been observed in various conditions of ill-health. What we want is to find out why it is not assimilated and if possible remove the cause. To give it by the stomach when excretion is excessive could at best only provide for the waste, and this has been compared to the administration of sugar in diabetes. But the comparison is scarcely just, surely the practice rather resembles giving iron in anæmia, and we know that this is often useful when there is no deficiency of the metal in the diet. The presentation of abundance may, in the one case as in the other, stimulate the sluggish assimilation. The phosphate is itself sometimes useful in anæmia, especially in rapidly growing young persons, as well as in hyperlactation, excessive suppuration, and exhaustive discharges. Sometimes also in hectic and in chronic wasting diseases in which there is excessive elimination, a kind of phosphaturia which sometimes accompanies scrofula, large abscesses, leucorrhœa, caries and necrosis of bone—in a word, wherever growth or repair is unusually active, calling for the presence of the salt. This medicine has also been employed in chronic bronchitis and in phthisis, but in these diseases it has of late years been displaced by the hypophosphites.

As a good deal is taken in the food and as little is required it would almost seem that the best way of administering the remedy would be to take care that the diet contained a free supply. Then it may be added to the food, to farinaceous articles, or to milk, in the hope that the stomach will take up as much as may be required, and the surplus, which passes into the intestines, will be innocuous. It has been said that it is more easily dissolved when mixed with table-salt, and this would naturally be taken with the food. Some have thought that when obtained from organic structures it is more easily assimilated than when deposited as a mineral, hence the favor with which filings of bone and ivory have sometimes been regarded. Others have found the salt derived from the vegetable kingdom equally efficacious. Dusart and Blache recommend an acid preparation, lacto-phosphate, which is freely soluble and may often be useful, but as we have already hinted, there are cases in which the secretions are too acid, but in which the ordinary salt may seem indicated.

Magnesium phosphate accompanies the sister salt, but mostly in smaller quantities. Thus there is considerably less in bone, blood, and milk; but there is more of the magnesium salt in muscle, and in brain about twice the amount. Although one may not entirely replace the other, their action is probably very similar; this substance is eliminated by the kidneys at the rate of about 0.6 gramme a day, bringing up the amount of the two earthy phosphates together to about one gramme, while of the soluble alkaline phosphates about 4 or 4.5 times as much is excreted. The magnesium phosphate is sometimes employed in conjunction with the calcium and for similar purposes.

Sodium and Potassium Phosphate.—These are often distinguished as the alkaline phosphates, they are soluble to the extent of four per cent. in water, and so can be given in solution. They are important in the blood and intercellular fluids. There is this difference between them: the basic phosphate of soda gives to the blood its alkaline quality, and is, moreover, endowed like alkaline carbonates with the power of absorbing carbonic acid and so carrying off this gas; on the other hand, the acid phosphate of potassium is found in the muscular juice, where perhaps it promotes the transudation of nutrient matter (thus opposing the effect of the basic soda salt in the blood), and so the solution of worn-out tissue. It will thus be seen that the functions of these two phosphates are very different. They are eliminated to some extent in mucus and perspiration, but most of all by the kidneys. In the renal secretion a part of the alkaline sodic salt is replaced by an acid biphosphate, which gives to the fluid its acid reaction, although it contains no free acid. It is supposed that the urate of sodium is formed at the expense of the phosphatic base, uric acid taking the soda it requires and leaving the basic phosphate thus converted into a biphosphate, to be eliminated and render the secretion acid. According to Wood, the alkaline phosphates are increased during mental exertion, but the earthy are diminished, so that the total amount is not much changed. On the other hand, he found the earthy phosphates increased during abstinence from mental labor. The earthy salts are held in solution by the acid of the secretion, and when this reaction is absent or much diminished they are thrown down as a light precipitate. The alkaline phosphates being freely soluble never disturb the transparency of the fluid. It is in these salts that most of the phosphoric acid is excreted. Some phosphoric acid is probably formed in the body by oxidation, for phosphorus is taken in food, but is entirely oxidized in the system, and the phosphoric acid thus formed leaves chiefly in union with soda, with a little superabundance of acid. Daremberg found that phosphates were also eliminated in the expectoration, and in regard to this, that there were remarkable differences between cases of bronchitis and phthisis; he found in phthisis patients lost as much phosphates (and chlorides) by the sputa as by the kidneys, and in the same investigations he observed that a bronchitic loses about two per cent. of nitrogenous matter in his expectorations, while a consumptive loses about three times as much ("Doctor," 1877). Teissier and others have also drawn attention to the free elimination of phosphates by the kidneys in consumption, from which is derived one argument for their administration in this disease.

Sodium phosphate, administered in large doses, acts as a purgative, and the taste is not objectionable; it is, in fact, commonly known as tasteless, purging salt, and may be given in broth instead of table-salt, half an ounce to an ounce. In small doses it has often been employed instead of other phosphates. Dr. William Stephenson considers it to possess a powerful influence on the biliary secretion, especially in young children, and his observation is confirmed by Professor Bartholow and others. Three or four grains or more may be given for this purpose, according to age, with milk or other food. It may be used for the same purpose in adults, in twenty to thirty grain doses, with the food.

Potassium phosphate has not been so largely used. It is a white deliquescent salt, which crystallizes with difficulty. It may be employed for the same purposes as the sodium salt, and has recently been recommended in phthisis and scrofula, usually in conjunction with other phosphates. Dose ten to thirty grains, three times a day, after meals.

Free Phosphorus.

The history of the therapeutical use of uncombined phosphorus presents a curious illustration of the rapid and complete changes which take place in medical opinion. Extolled at one time as a most potent remedy for certain special and intractable diseases, it is soon after found utterly neglected, to be brought forward again after a lapse of time as almost a panacea, and soon after again to fall in disfavor to await another wave of opinion. Some years ago it was again studied, with the result of achieving considerable popularity, which, perhaps, attained its acme six or seven years since, though it is still largely prescribed, its place in therapeutics having become more generally recognized. This element enters the system in combination, perhaps with albuminates, but most of all with fats. The larger the amount of fat present the more rapid will be the absorption. It is supposed that some may enter the blood unaltered, and that some may be more or less oxidized, for in cases of poisoning it has been found in the form of hypophosphorous acid, of phosphorous acid, and of phosphoric acid, moreover, phosphuretted hydrogen has been detected. It is eliminated by the liver and other glands; by the skin, lungs, and kidneys, in an oxidized form, except in cases of poisoning, in which it may be excreted uncombined, as many observers have reported that the excretions have been rendered phosphorescent. It is easy to see that unabsorbed portions may pass through the bowels, but we should expect that in the renal secretion it would only appear as phosphates, unless under very exceptional circumstances.

The vapor of phosphorus is a local irritant to mucous membranes and will produce inflammation. On the periosteum it acts so powerfully as to produce destruction of bone, which used to be very common in the match-maker's disease. Internally, quite small doses, continued for a time, may so far disturb the digestion as to render a temporary interruption of the course necessary. On the nervous system it exercises a slight excitant influence, and some still consider it to possess a distinct stimulant action on special parts. Gubler found that zinc-phosphide set free hydrogen, even in artificial digestion, and this combined with the phosphorus and delayed the process. In large doses more distinct symptoms are produced, and these may gradually merge into the poisonous effects. The most remarkable of these may be summed up as profound alteration of the viscera, tissues, and blood, combined with the effects of corrosive poisoning. The gastro-enteric inflammation is usually attributed to the action of compounds, but this explanation seems doubtful, inasmuch as it has occurred after the introduction of the element directly into the blood. The duodenum suffers a good deal, whence the irritation is propagated along the biliary duct; to this may be due the jaundice which usually occurs, but others ascribe this to the action on the liver, in which the element has been found ten hours after death; others consider the jaundice as hæmic. The liver certainly suffers much in acute phosphorus poisoning, and it is difficult to distinguish the effects on this organ from acute yellow atrophy; in protracted cases, also, the organ atrophies and resembles the effects of disease. The other viscera are affected, especially the kidneys, with a kind of fatty degeneration, which process probably extends to all the tissues, since Wegner has shown it to involve the minute arterioles. The blood is rapidly deteriorated, great destruction of the red corpuscles is produced, the hæmoglobin is so altered as no longer to give its proper spectrum, and the reactions of fibrinogen and the fibrinoplastic substance are hindered. Ecchymoses appear over the body, and hemorrhages from all the mucous membranes take place; in fact, a kind of hemorrhagic diathesis is set up, combined with various fatty degenerations, which in protracted cases extend to the involuntary muscular system. The jaundice appears in from thirty-six hours to four or five days, according to Tüngel, Lebert, Wyss, and others. Albumen is found in the urine, and sugar has been detected. The bile acids also appear, and the biliary coloring matter as soon as jaundice sets in; also leucin and tyrosin, and (more important still) sarcolactic acid. On these points Munk and Leyden, as well as Lewin, Kohts, Virchow, Ossikovsky, and others previously cited agree, notwithstanding some doubts that have been thrown upon them. Albuminuria, however, is not often found in animals, even though the kidneys may be greatly damaged; and cases do occur in which the symptoms differ considerably. One is mentioned by Leidler, in which suppression of urine took place, and death in a few hours. Mayer says that where very large doses have been taken, both the urine and the blood may be phosphorescent. Casper says that generally delirium, paralysis, coma, and convulsions occur, but the nervous system often gives no especial signs.

The best emetic in cases of poisoning is sulphate of copper, as it combines with the element to form a less active phosphide, as shown by Eulenberg, Guttman, and Bamberger. Turpentine appears to be the best antidote, as first pointed out by the late Dr. Letheby, who was led to this discovery by noticing that the vapor of turpentine prevented the action of the fumes on the artisans exposed to them. After his suggestion MM. Andant and Personne soon published cases showing the value of the turpentine. It unites with the phosphorus in a spermaceti-like mass, which is soluble in ether, alcohol, and alkaline solutions, and can be eliminated by the kidneys without injuring them. It is the common commercial turpentine which is alone effective, probably because it is richest in ozone, having been exposed to the air. This turpentine also seems to prevent fatty degeneration of the tissues. To repair the damage to the blood transfusion is recommended by Jürgensen; of course all fatty matters must be withheld.

The question now naturally arises whether a substance which produces such effects as have been described can in any sense be called a nutrient. Certainly it can only be in a special sense that such a poison can be called a food. Yet, inasmuch as it enters into the tissues and promotes their growth it is to them a nutrient in the same sense as a phosphate is. Minute doses promote the growth of bone. Wegner has shown that under their influence the epiphyseal cartilages ossified more quickly and more completely, new deposit took place on the inside of the shafts of the long bones, and this to such an extent that in some instances the medullary cavity was filled up. The effect was more marked in growing than in adult animals. The new tissue is at first gelatinous. If the animal were deprived of lime the new tissue was formed notwithstanding, but remained soft; no excess of phosphates appeared in the bone, and no such action could be obtained from phosphoric acid unless given in proportional doses, nearly a thousand times larger. Hence Wegner concludes that it acts in its elemental form, not after oxidation.

Phosphorus is, moreover, a very important constituent of nervous tissue. Is it therefore a nutrient to that? The excitement caused by medicinal doses points in that direction. Gubler thought the excitement more marked than that produced by coffee or opium, and Dr. Ashburton Thompson compares it to the stimulus of alcohol without its subsequent depression. These statements seem to me highly colored. Dr. Gowers made a very important observation if it be confirmed in other cases. He found in lymphoma the number of the red corpuscles increased under the influence of phosphorus. This is very unlike what might be anticipated from the toxic effects by which the corpuscles are destroyed and the hæmaglobin so altered as no longer to give its characteristic spectrum. We see rather arrested nutrition combined with fatty degeneration than any nutrient effect, but this it may be said is because we have to do with a poison which may nevertheless in minute quantities supply a want or stimulate nutrition.

How then does it act? The irritation in the alimentary canal may be due to compounds. After absorption its affinity for oxygen has been conjectured by Lecorché and Eulenberg to bring about a kind of asphyxia, but this seems an untenable hypothesis, for turpentine, its antidote, oxidizes, and Crocq employed oxygen with success to prevent its poisonous effects. Gubler suggests that it acts by its ozonizing power, as though only a little oxygen could be lost to respiration, the remainder mixed with the ozonized, may be believed to be so much more active as to be able to stimulate the system and bring about increased activity, and so rapid waste, especially of the blood-globules. Phosphuretted hydrogen not only impairs digestion, Dybkowsky found it destroyed hæmoglobin and, like phosphoric and other acids, produced fatty degeneration of the tissues. This gas might readily be formed, the hydrogen being easily obtained from water, so other phosphides, or the more unstable hypophosphites might be produced in the system.

The use of phosphorus in diseases of the osseous system may be deduced from what has preceded. It has been rather extensively tried instead of the phosphates, but it is in diseases of the nervous system that it has recently obtained considerable repute, acting as a nerve tonic and stimulant and, as some maintain, a kind of food to nervous tissue. Rabuteau, in face of all the evidence that has been adduced, says he does not hesitate to assert that it has never cured anything, but has always been useless, and he never means to prescribe it. On the other hand, Dujardin-Beaumetz reports wonderful results, and his observations are confirmed by a number of London physicians, who believe that it is of essential service in neuralgia, neurasthenia, paralysis, spinal irritation, ataxia, etc. As a stimulant in the exhaustion of fevers it was in use more than a hundred years ago. Bayle considered it indicated wherever death was imminent from failure of vital force without much structural change, and particularly in putrid fevers and the exanthemata. He also recommended it in pneumonia where there was great depression, and in the recent revival of its use it has been employed freely in this disease, particularly when it attacks old persons or robust adults, in whom it is said there is more frequently nerve-depression than in the young or the feeble. It is during extreme prostration that the most marked effect is recorded, but it is also used at later periods, e.g., when red hepatization is complete and the fever and exhaustion seem to increase, but suppuration has not begun. After pus has formed it should be withheld. It may also be given in retarded convalescence, when there is no longer any febrile excitement. Dr. Ashburton Thompson advises large doses in this disease, and attributes the good results of the older authorities to their full doses, as the toxic effects were less known to them. Certainly, to act as a powerful stimulus minute doses at long intervals could not be relied upon, but it would surely be safer to repeat small doses than to run the risk of producing toxic effects. If phosphorus be used at all for this purpose there can be no reason for resorting to the heroic doses sometimes recommended. When administered in convalescence it is almost always advisable to give iron at the same time. In chronic pneumonia and bronchitis it is much less useful, though occasionally prescribed in the early or late stage, or when there is very little expectoration, and it should not be given when there is any tendency to hemorrhage. It has been recommended in pleurisy attacking persons injured by alcohol, but there is little if any evidence of its utility.

In phthisis, much as it has been vaunted, it should be used only with considerable caution. It is likely to bring on hæmoptysis wherever a tendency to that exists. It is not suitable in febrile conditions, and great care should be taken lest it interfere with digestion, on account of the extreme importance of promoting nutrition. It is reputed to relieve the cough and expectoration, as well as night-sweats and colliquative diarrhœa. Perhaps, when the cough is rather dry it is more appropriate than when the expectoration is free. When it agrees it will improve the appetite and perhaps promote digestion, and thereby relieve irritative attacks of diarrhœa; but when this symptom is due to ulceration it is useless, and at the stage when this occurs the remedy seems inappropriate. All the good effects of the phosphatic salts have been alleged to be dependent on the phosphorus, but this view cannot be maintained, and it is important to discriminate between them.

Phosphorus has been recommended on very unsatisfactory grounds in fatty degeneration, leukæmia, lymphadenoma, and other diseases in which it seems quite unsuitable, but into discussion concerning which we cannot here enter.

The form in which it is given is important, the first thing is to avoid offending the stomach or disgusting the patient with his cod-liver oil. Solutions in oil oxidize by keeping; an ethereal solution seems very unsuitable, as the ether may evaporate, and leave the solid phosphorus in the mouth or elsewhere. The pill of the British Pharmacopœia is very insoluble, and has frequently been known to pass through the system unchanged. Suet or other fat may be used as an excipient. The phosphide of zinc is preferred by some prescribers; whatever preparation be used, it should be given with the food.

Hypophosphites.

These salts have been introduced to supersede phosphorus as well as the phosphates, and it has been held that they are valuable, both with regard to the base and the acid, it being claimed, for instance, that calcium hypophosphite possessed the virtues of phosphorus on the one hand and the lime salts on the other. The hypophosphites were brought prominently forward by Dr. Churchill, of Paris, as curative in phthisis, but his statements have not been fully verified, and in this country are generally regarded as exaggerated. Some consider that they are oxidized and so only useful as phosphates; we must admit, however, that a distinction has been established, and that they in some degree derive their virtue from the loosely combined phosphorus. Toxicological records show that hypophosphoric acid may be found in the blood in cases of phosphorus poisoning, and we know of no reason why the same acid could not be absorbed in small quantities, except the admitted instability of the acid and its salts.

Clinical observation seems rather in favor of their use, and Dr. Thorowgood, who has had large experience of them, recently informed me that his conviction of their utility remained unchanged. Many years ago he selected for me some of the cases which he deemed suitable then under my care at the North London Consumption Hospital. These were placed on the hypophosphites without other treatment, with a view to comparing their progress with that of the other in-patients who were treated by the usual methods. I cannot say that much difference was perceived; but there is this difficulty in all such experiments—patients coming into a well-ordered hospital and provided with good food, medical necessaries, and every comfort, are likely to improve at once, and that in proportion to the want and exposure to which they have been previously subjected. The cases most likely to be benefited are those of early phthisis without pyrexia or any tendency to hæmoptysis, or else cases of chronic or fibroid phthisis. Perhaps, too, they may be useful in cachectic conditions, in which, however, free phosphorus would probably be more certain. Little benefit can be anticipated from them in bronchitis, emphysema, or pneumonic phthisis. They may be regarded as nutrients, or at any rate as substances which perhaps stimulate nutrition; when they thus act they promote digestion and increase the appetite, and in such cases of course are beneficial; whenever they impair the appetite they will to that extent only be injurious.

Dr. Thorowgood thinks that under the use of the hypophosphites (and phosphorus) he has seen pleuritic thickenings melt away, and consolidations of the lung of two or three months' standing disperse, while cases which looked to him and others very like acute tubercle in the lungs, have recovered under hypophosphites. In pulmonary diseases of inflammatory exudative origin, and apt to run into phthisis, he knows no remedy to compare to the hypophosphites. In a note to the *British Medical Journal*, June 21, 1884, he confesses, however, that, as with other drugs, some cases resist it, and he thinks it is the bacillus which sets the remedy at defiance, for it is in those cases in which the hypophosphites are most helpless that his clinical clerk usually finds "lots of bacilli."

CHAPTER X.

AIDS TO DIGESTION.

LEAVING now the ingredients of food, we pass to certain substances already mentioned, which as they are materially concerned in digestion may be considered as direct promoters of nutrition. The digestive ferments artificially extracted, and their allies or substitutes, have become largely employed as therapeutical agents. It is interesting to observe that although the process of digestion has only been understood in recent times there were attempts to utilize the gastric fluids at a very early period. Pliny tells us that liquids taken from the stomachs of sucking animals were used for medical purposes, and Asclepiades recommended such a fluid to assist the removal of large coagula of milk. Galen, too, tells us he had personally experienced the value of such a remedy, and observed that boiling destroyed it, as he had found that fowls' stomachs were rendered inert by cooking. The milk and fluid taken from the stomachs of sucking animals continued to be used for centuries, and was in the London Pharmacopœia of 1677; the issue of 1721 contained among its remedies the lining membrane of a fowl's stomach—which shows us how old is an idea which has of late years been again brought forward. In 1783 Spallanzani published his remarkable experiments on digestion, after which gastric juice became still further studied, and preparations from it were said to have been used by Laennec. The discovery of pepsin, of course, displaced the disgusting preparations, and gave an immense impetus to the use of such aids to digestion. In some parts of England, rennet has for generations been a popular domestic remedy.

Extract of malt may be first named, as it is commonly considered a substitute for ptyalin, and this it is so far as the diastase goes, but we must not forget that it is also a direct nutrient on account of the considerable proportion of dextrose and glucose. So that, when cod-liver oil cannot be taken, extract of malt is often advantageously given as a nutrient, in which case it should be given with the food. Of course we are here speaking of genuine extract, not of the porter or stout which is sometimes sold under that name. The extract mixes well with milk. For its diastatic use it should be given with farinaceous food. It is an invaluable emulsifying agent, especially for cod-liver oil.

Pepsin and Pancreatin.—These two ferments are sometimes prescribed together, but the practice is not to be commended. We have seen how different is the action of the two, which can never replace each other; in fact it would seem that pepsin is capable of digesting pancreatin, although it is doubtful if a converse process can be partially effected. This we know, the gastric juice is acid and in it pepsin is active; it seems therefore natural that the use of this substance should prove beneficial in retarded or difficult digestion. It is easy enough to introduce pepsin, and if necessary acid also, into the stomach at the time food is taken, so as to reinforce the gastric juice. It seems to me, also, that it is not necessary, as some have supposed, to introduce enough to digest all the food taken; for it may well be that the presence of some pepsin may stimulate further secretion. If we suppose the gastric glands to be incapable of any action, of course a small quantity of pepsin could only digest the due proportion of proteids. Probably the dose of pepsin should usually be larger than that generally prescribed; it should also bear some proportion to the proteids taken. Dr. King Chambers seems to have arrived at a similar conclusion from clinical experience, and therefore he says give it to those consumptives who cannot digest half a chop without it, and then it can be given until a whole chop can be digested, when he thinks it will have done all it can. I cannot quite endorse this, for if so much can be done in a few days as in some of his cases, it seems more might be effected still in others. One difficulty with which we have to deal is that in natural digestion the pepsin is not all poured on the food at once. By the movements of the stomach its contents are successively exposed to the action of the gastric juice as they come in contact with the walls. To imitate this it seems to me that we may sometimes take the pepsin in successive portions, but most patients think it hard enough to swallow a single dose of medicine with a meal and could scarcely be persuaded to sip it, as they mostly object to the taste of the liquids. At my suggestion Messrs. Savory & Moore have prepared a powder which may be taken in the place of table-salt; it is, in fact, a mixture of pepsin with salt, but has to be prepared in a special way, as an ordinary mixture of salt and pepsin is apt to decompose. The sphere of pepsin is to assist the solution of proteids in the stomach, and as soon as it passes the pylorus its power is arrested by the bile and other fluids and the acidity essential to its action is replaced by an alkaline reaction.

Pancreatin, on the other hand, comes into action at this stage, and the doubt naturally suggests itself whether it can reach the duodenum unaltered when administered by the mouth. Defresne and others have alleged that it may pass through the stomach without being destroyed, its action being merely arrested for a time and restored when it finds itself in an alkaline medium; but this is opposed to the teachings of our laboratory experiments, which rather indicate a probability of its destruction. To avoid this it has been proposed to administer it in an alkaline solution an hour or two after the meal. But this interval would scarcely suffice; at that time there would be plenty of acid in the stomach to neutralize an ordinary dose of alkali. To secure its passage unaltered the stomach should be empty, and if delayed so long it might possibly be too late to be of much service. Some have said that given on an empty stomach it would necessarily provoke an immediate secretion of gastric juice; but there is no proof of this, and we have already seen that iron introduced through a gastric fistula may not provoke secretion enough to dissolve it. Then, again, many physiologists, with whom we must agree, admit that water may at once pass through the pylorus. It has therefore appeared to me possible to obtain what we desire by a nice calculation as to time of administration. Perhaps, as we know that pancreatic juice begins to be poured out at the commencement of a meal, it is not unphysiological to introduce it immediately before eating, with a draught of water to assist its progress. A further precaution might be taken by enclosing it in capsules or pills with a coating easily soluble in a weak alkaline.

Peptonized Foods.—There is no doubt about our ability to secure to some extent the action of either pepsin or pancreatin upon the food before it is eaten, and the recent extensive adoption of that method may be regarded as a distinct advance. I can fully endorse the statements of Dr. William Roberts in his "Lumleian Lectures." I have derived the greatest benefit from peptonized and pancreatized foods. It is true these substances are only partially digested, but the completion of the process is rendered easy in the stomach, and assimilation thereby promoted. A little attention will enable any one to keep up a supply of pancreatized milk, avoiding on the one hand insufficient action, and on the other the development of too pronounced a bitterness. How easily an invalid's meal of a farinaceous food may be thinned by a teaspoonful of succus pancreaticus almost immediately, enough to make it drinkable, is by this time pretty well known.

Bile is still retained as a medicinal agent, though it is much less frequently employed than formerly. It is supposed to be useful not only as an aperient, but as supplying the place of the secretion when deficient. But a study of the nature and uses of bile and the amount usually secreted does not give much support to such a notion. The addition of five or ten grains of the purified fel-bovis to the large quantity of bile secreted in twenty-four hours could scarcely have much effect. Probably the bile-salts are of importance from their property of facilitating the passage of fat through membranes, and no doubt bile materially assists pancreatic digestion, but further information is needed on the subject, and the saline constituents of the secretion have not yet come into use as medicines.

Feeding by the Nose or Rectum.—Sometimes it is not merely necessary to assist the digestion of food, but to introduce it by unusual routes. Those who have much experience in throat diseases know how often deglutition becomes almost impossible. Stricture of the œsophagus, again, prevents the entrance of nutriment. Liquid food may be often introduced through a fine tube passed through the nostrils. Often it becomes our duty to institute feeding *per rectum*. In this case the use of peptonized foods is most important, and can scarcely be too highly valued. Lately dried blood has also come into use for rectal alimentation, and it has been proposed to revive an ancient practice of administering this fluid by the stomach. I remember many years ago that Dr. De Pascale tried to establish at the abattoir, at Nice, the plan of drinking fresh warm blood, and he bravely set the example morning after morning of quaffing a glass. He was followed by a few faithful patients, but the remedy did not become popular, and I suggested to him that it should perhaps be regarded as a ferruginous nutrient, inasmuch as it would be digested by the stomach, not absorbed unchanged. Dried blood presents less difficulty, perhaps, to the taste, and is very convenient for rectal alimentation. Dr. Andrew H. Smith brought the preparation before the New York Academy in 1879, and his favorable opinion was endorsed by a committee of the Therapeutical Society. Dr. H. F. Campbell holds that a reversal of the ordinary peristaltic movement takes place when aliment is introduced *per rectum*, though Dr. Austin Flint (*American Practitioner*, January, 1878), inclines to think that true digestion may take place from increased secretion of *succus entericus*. Dr. Craven has produced a reliable preparation which Dr. Sansom has used at the London Hospital (*Lancet*, 1880). Leube ("Ziemssen's Cyclop.," vii.) considers that all rectal foods should be pancreatized, and his emulsion has been extensively used with the greatest benefit. It is certainly desirable to employ the most easily assimilated preparations, and among these pancreatic emulsions combined with finely minced cooked meat, in the manner advised by Leube, stand pre-eminent. The *sanguis bovinus exsiccatus* does not contain the fibrin and some other principles, and cannot be said really to represent the fresh bullock's blood, though it may perhaps be more nutritious than a solution of the iron salts of the blood.

Forced feeding by the stomach has given some remarkable results to Débove and Dujardin-Beaumetz, who have demonstrated clinically the importance of nutrients in convalescents, in chronic diseases, and most important of all in phthisis. The systematic feeding in nervous diseases employed by Weir Mitchell is a further illustration of this, but the results in phthisis are perhaps more unexpected. All suitable nutrients may be thrown into the stomach in small quantities, at frequent intervals, even after that organ has become so unaccustomed to the presence of food, as to a large extent to have lost its digestive power, of which the best and most certain restoration is food. This is to be given in small quantities, which can be increased gradually.

Stimulants to Digestion.

Certain substances may be separated from previously considered digestives, inasmuch as they act rather by stimulating the digestive process. Among these may be first named our ordinary condiments (from *condio*, I season), such as pepper, salt, spices, pickles, and other agreeable or tasty additions to food. Some of these are sialagogues (σιαλον, saliva, αγω, I expel), and may therefore be useful by increasing the quantity of ptyalin. Others are the aromatics, so named from their fragrancy (αρωμα, an odor). The more distinctly medicinal ones are sometimes called carminatives (*carmen*, a verse or charm) and they certainly allay spasmodic pain and dispel flatus. Most of them contain some essential oil or volatile principle, which stimulates the gastric mucous membrane, and abates spasm of the muscular coat, so tending to restore the ordinary peristaltic movements. They are therefore used to relieve the pain set up by distention, and also to prevent that condition. Moreover, they are useful stimulants to the digestive process in atonic dyspepsia. Another class of remedies allied to the foregoing are the bitters, as we often call them, from their characteristic taste, and which apparently act as gentle stimulants to the stomach and thus tend to increase the appetite. For similar purposes other general tonics are also used, such as bark, quinine, nux vomica, strychnia, etc., each possessing its own characteristic action on the system. In the same way acids are employed, either alone or in combination with tonics. Allied to these, astringents may be also named as in turn useful. Then, again, we may employ alkalies as digestives, inasmuch as small doses administered shortly before a meal provoke an increased secretion of gastric juice, and so contribute in suitable cases to assist digestion and therefore nutrition, although their characteristic action is rather that of denutrients. Stomachic sedatives may also be named, such as bismuth and lime, which seem to act directly on the mucous membrane, and which therefore, in appropriate cases, are also promoters of digestion and so of nutrition. In connection with the last class we may mention general sedatives and anodynes, which may be called in to act locally on the stomach. From the foregoing illustrations it will be perceived that most diverse remedies, acting either upon the stomach directly or through the system, may, under various circumstances, be appropriately prescribed as likely to promote nutrition by beneficially influencing the digestive process, a fact which explains why some aperients, like rhubarb and aloes, are also reputed to be digestive tonics.

CHAPTER XI.

TRANSFUSION.

WHEN it is no longer possible to introduce nutriment into the alimentary canal, or when this way is insufficient, it is still possible to resort to transfusion, an operation more frequently performed to prevent death by hemorrhage. It was natural to use blood for this purpose, but other liquids have been successfully employed. In 1656 Sir Christopher Wren infused wine and opium into the circulation of a dog, and soon after Lower transfused blood from the vessels of one dog into those of another. In 1667 the first transfusion on a human subject was performed at Montpellier, in France, by Denis and Emmerez, who employed the blood of lambs and calves. In 1668 Fracassati, Manfredi, and Riva followed, employing the blood of healthy men. The early successes were so signal that the practice became abused, and sad catastrophes took place. Great opposition was excited, and the French Legislature in 1675 forbade the operation. In 1776 Kohler employed with success venous injection of emetics to remove a piece of gristle impacted in the gullet. His example was followed by Balck, and at intervals by others, until the discovery that hypodermic injection would be sufficient for this purpose. Heunmann injected an infusion of bark and ammonia into the veins of a patient who seemed dying from fever, and he recovered. The late Dr. J. Blundell revived the operation at Guy's Hospital about 1815, with brilliant success, for the details of which we may refer to his "Researches," 1824, and to the "Medico-Chirurgical Transactions," 1818. Dieffenbach in 1828 used defibrinated blood, and introduced the terms mediate and immediate transfusion; soon after this O'Shaughnessy proposed to inject saline fluids into the veins of cholera patients, and this was done in the epidemic of 1830 by Dr. Latta in Edinburgh, Majendie in France, Janichen in Moscow, and several others in London, among them Drs. Sweedie, Murphy, Girdwood, Craigie, Arthur; but most important of all by Dr. Little, whose remarkable recoveries in the London Hospital demonstrated the power of the remedy, and in a later epidemic his son, Dr. L. S. Little, also obtained excellent results. At the same hospital Drs. Woodman, Heckford, and others have also adopted this practice, and it has been shown that in many cases the use of saline fluids may advantageously replace blood, and this even in cases where the operation has been performed to prevent death from hemorrhage. The fluid employed by Dr. Little, and subsequently by his son, was composed of chloride of sodium, 60 grains; chloride of potassium, 6 grains; phosphate of soda, 3 grains; carbonate of soda, 20 grains; alcohol, 2 drachms; distilled water, 1 pint. Large quantities were introduced; in some of the most successful cases as much as four pints, at a temperature of 110° F. It was found that the blood-globules preserved their natural appearance.

When blood is employed for transfusion, small quantities are used, most frequently, perhaps, about six ounces, but it is better to inject cautiously three ounces, and if that is well borne, gradually increase it to double, or even to eight or ten ounces. The question of defibrinating has been freely discussed, and will be differently answered by different operators, a good deal depending, perhaps, on the instrument employed. It is certain that many of the corpuscles may be left after defibrination, but it seems almost equally certain that when thus supplied they soon undergo disintegration and excretion, so that they would appear not to be of the use that has been supposed. The fibrin, too, is regarded by some as useless and effete, and they therefore think it is better to avoid the possibility of introducing a fibrin ferment. Panum and others, therefore, employ simply blood-serum; at first sight it may seem that pure blood or its serum would be the most natural fluid, and moreover would prove a valuable nutrient, such as could not be supplied by saline solutions only; and this, if so, would of course render blood transfusions more effective in certain cases in which more than the saline ingredients of the blood are needed. Recent experiments, however, seem to show that the regeneration of red corpuscles is more active after the transfusion of salines than of serum or blood itself, and the operation is certainly simpler.

We have seen that the blood of animals has been successfully employed, but this is less appropriate, and it appears that the degeneration of the corpuscles is more certain, and perhaps more rapid in such a case. With a view of introducing a nutritious fluid, Gaillard Thomas, Hodder, and others have made use of milk, and many successful cases have been reported, but even this fluid is declared by Culcerq and others to be dangerous, inasmuch as agglomerated milk-corpuscles may give rise to fatty embolism. One reason assigned for the failure of certain milk injection is the difficulty of securing the purity and the freshness of the fluid. Dr. Meldon used goat's milk, as he was thus enabled, by bringing the animal into the patient's room, to use the milk without loss of time. Miglioranza, in the *Archives Italiennes de Biologie* (tom. iv., fasc. 2), condemns the injection of milk, and finds, from numerous experiments, that when thus employed the albuminoid and fatty matters are excreted by the kidneys, but that if any considerable quantity is present it produces violent vomiting and diarrhœa with prostration and often death. Even when previously filtered, milk produced these symptoms, with chyluria and fatty infiltration of the kidneys. Albertone proposed to use the serum of milk, and in fact did so, without ill effects. These observers consider that the cases pronounced successful by others must be due to the small quantity of milk injected, and Laborde has found that even such small quantities may give rise to serious mischief. It may be worth while to add here that Miglioranza found carbonate of ammonia injected into the veins produced excitement of the circulation, dyspnœa, hyperæsthesia, tetanic convulsions, and coma, in fact, urœmic fever or ammoniœmia, although, as we have seen, small quantities of ammonia have certainly been used without ill effect. Halford has indeed often injected solution of ammonia into a vein, as a remedy in cases of snake-poisoning, and recommends it to be done by means of a hypodermic syringe. It would seem, then, that milk requires to be digested before entering the circulation. Blood, however, does not, but there are difficulties about transfusion of this fluid, and it seems by no means certain that by introducing it we supply the nutritive elements which we anticipate ; but the saline constituents of this fluid may be employed with greater facility and are capable of producing rapid, nay, immediate and sometimes even startling effects. Schwartz maintains that the chief, if not the only use of transfusion is to restore vascular tension, and accordingly that the quantity of the liquid is more important than the quality. This, of course, refers to its use for hemorrhage, when the heart ceases to act because sufficient fluid is not poured into its cavities, but on injecting a saline fluid we re-establish vascular tension and so restore the circulation. In other cases, when the object is to supply something more nutritious, this reasoning will not apply. Milk is evidently no safer than blood, and perhaps for purely nutritive purposes, inasmuch as in such cases there is time to make preparations deliberately, immediate transfusion offers us the best resource, while in the appalling cases of hemorrhage, for which the operation has been most frequently performed, saline injections promise well. To render the operation desirable, as a means of renewing failing nutrition, what we need is a fluid which could represent blood, or perhaps rather its incipient condition—chyle.

Transfusion of defibrinated blood into arteries instead of veins was proposed by Albanese, and is recommended by Hüter and Asche, who have successfully practised it. The blood having to pass through the capillaries on its way to the right heart, sudden distention of the auricle is prevented, and the danger of admitting air or of thrombosis is avoided. This last is a danger more present to the mind in modern times, but is less likely to happen with defibrinated blood. Yet the statistics collected by Gesellius, show out of 146 cases without defibrination, $79 = 54.11$ per cent. were successful, while out of 115 cases when defibrinated blood was employed, 79 = 68.70 per cent. were unsuccessful. At the New York Practitioners' Society, December 5, 1883, Dr. Bull said that he had injected saline fluids into the central end of the radial artery, because after injecting the peripheral end serious consequences had occurred, such as embolism and gangrene of the hand, and, further, because it needed a great deal of force, while the fluid flowed into the central end without difficulty.

Injections into Serous Cavities.

It has been abundantly demonstrated of late years, that the peritoneum may be interfered with much more freely than might have been anticipated. It has also been shown that absorption takes place with great rapidity from this cavity, and a number of medicines have been introduced in this manner. It is not therefore surprising that Ponfick, Nussbaum, and others should have injected defibrinated blood; the practice, however, is not without danger. Ponfick first published three cases (*Berl. Klin. Woch.*, September, 1879) of pernicious anæmia so treated. Bizzozero and Golgi afterward showed that the operation enriched the blood in hæmoglobin. Burresi collected 38 cases, in 24 of which it was thought to have done good, in 7 the result was neutral, but in 7 it did harm. Foà and Pellacapi have traced the path of the corpuscles through the lymphatics of the abdominal glands and diaphragm into the bloodvessels. Much more recently experiments have been made by Dr. Bernardino Silva upon rabbits, in order to determine how far absorption of defibrinated blood would take place from the pleura. From ten experiments he concludes (*Revista Clinica*, October and November, 1883) that 1, absorption proceeds as well as through the peritoneum; 2, the effects of the injection of homogeneous blood are seen in an increase of hæmoglobin and in the number of red globules within four or five hours after the injection, and are prolonged for more than four days afterward; 3, the greatest increase in hæmoglobin takes place within the first twenty-four hours; 4, the absorption of hæmoglobin is greatest when the quantity of blood injected is small, if so much blood is injected as to produce atelectasis no increase of hæmoglobin is observed; 5, the transfusion of blood into the pleural cavity causes an increase in the excretion of urea—an increase, however, which is preceded by a diminution during the first twenty-four hours. This last conclusion, the author says, needs further confirmation. In a note at the end of his article Dr. Silva states that Professor Bozzolo has made an inter-pleural transfusion in the human subject, with beneficial results, in a case of anæmia complicated with cachexia, ascites, anasarca, and albuminuria.

Hypodermic Injections of Blood and Food.

Dr. Carlo Bareggi published in Milan in 1882 a prize essay on the hypodermic injection of blood, which he maintained might be advantageously used, either defibrinated or not, and which would certainly appear to be safer than the previously named operations. From experiments on animals he proceeded to clinical investigation. He found one gramme absolutely innocuous, but larger quantities produced in man slight febrile reaction, lasting from a few hours to at the most two days. He says the red globules are absorbed, in part at least, unaltered. Their course is from the lymphatic spaces in the connective tissue into the lymphatic vessels leading from the part, through the glands met with on the way (unless these are in an advanced stage of fibro-adipose degeneration, or in some other way profoundly altered), and thence into the receptaculum chyli and thoracic duct. They were found in the principal lymphatic trunks of the part twenty minutes after the injection was practised, and even after three days numbers of them in a good state of preservation were encountered in the thoracic duct. The greatest number was met with twelve hours after the injection, but even after fifteen days quantities of red globules, but little changed, were seen passing from the cellular tissue into the circulation. Absorption of the mass of injected blood proceeded rather slowly in individuals in whom the circulation was sluggish, especially in those in whom there was considerable subcutaneous adipose tissue, but more rapidly under normal conditions of the heart's action and in persons in good general condition. The red globules were unchanged after remaining many days in the cellular tissue, except in cases in which there was considerable febrile reaction. He considers hypodermic injections of blood capable not only of arresting the progressive deterioration in the quality of the circulating fluid, caused by insufficient nourishment or repeated losses, but also of increasing, in spite of such persistent influences, the corpuscular richness of the blood. But to obtain the best effects injections have to be repeated at intervals of five to fifteen days and a considerable quantity has to be employed each time, three and sometimes four ounces. And moreover we may observe that when iron agrees with the patient the increase in the red globules produced by that metal is almost, if not quite as great. Dr. Paladini has recorded (*Gaz. Med. Ital.-Lomb.*, 1883) a case in which, the patient living in a remote locality and being in a critical state, he injected with a common syringe about one hundred and thirty grammes (two syringefuls) of blood furnished by the husband. He threw it into the subcutaneous tissue, about four fingers' breadth to the left of the umbilicus, where the skin was lax enough to be raised in large folds. After the operation a lump the size of a hen's egg was felt, but disappeared in two hours. No pain or inconvenience was produced and the patient did well.

Hypodermic injection of milk and other alimentary substances has also been sometimes resorted to. Dr. Whittaker (*Clinic*, 1876) reported a case supported by injections of milk for two or three days, two drachms being given every two hours. Afterward, cod-liver oil was used, one day four ounces being introduced in eight injections. This caused no pain or inconvenience. Two small abscesses occurred after the milk.

CHAPTER XII.

WATER—DILUENTS—BEVERAGES.

We have seen the extreme importance of water and we take it up here because it stands, as it were, midway between constructives and destructives, or rather acts as one or the other in different circumstances. It is the one necessary beverage, the universal solvent and diluent. Without it the nutritive principles cannot be extracted from the food, still less can they reach the tissues. On the other hand, without it worn-out tissue cannot be dissolved, and therefore cannot be carried away from the place where it is useless. Water, then, may be regarded as the carrier into and out of the body; besides which, as the universal lubricant, it facilitates all molecular changes. As the carrier into the economy it presents the pabulum to the tissues, and is therefore a constructor, but as a solvent and carrier out of the body it aids disintegration and removal, though this last action, as we know, within due limits quickens construction. Rapid removal calls for rapid replacement, waste stimulates repair, and so while water directly promotes construction it also stimulates destructive changes and removes their products. It is, therefore, truly eliminant, dissolving all substances and increasing all secretions and excretions. This tends to loss of body weight, and unless fresh aliment be supplied, this will take place. If food be intercepted or if appetite fail the waste cannot be renewed. But moderate water-drinking usually increases appetite, so that more food is taken, nutrition is more rapid, and instead of emaciation the opposite results. In this case it would prove the best of tonics, but in default of food or appetite it becomes a very powerful disintegrator, and it has been believed that in this way it may be utilized to remove morbid deposits, though this power is more frequently attributed to medicinal water. Drinking in excess may itself impair digestion, still it is remarkable how large quantities have sometimes been taken without injury.

In health an adult will consume on an average nearly four pints of fluid per diem; some thirsty souls take much more, others do with less. A few make simple water their chief fluid; others consume much milk; a larger number take a considerable proportion of tea, coffee, cocoa, or other simple beverage; too many almost exclude the simpler fluids by the amount of alcoholic liquors in which they indulge. In this last case, of course, the effect of the alcohol masks that of the fluid, but we must not forget that the amount of liquid drunk and the time at which it is taken exercise some influence. Of course, the amount of liquid received has to be counterbalanced by that removed, so that the skin, lungs, and kidneys are called upon to get rid of the average four or five pounds of water a day, and of any excess that may be ingested. As the water is taken into the system, and is the vehicle for the removal of waste, it may be regarded as in some sense washing the tissues. It is commonly said that a very little water passes directly through the pylorus, but I agree with Küss in the opposite opinion, more especially when large quantities are drunk.

The water required to maintain the balance of the system is only a small part of that which is always at work in the economy. The quantity daily poured into the alimentary canal, and again reabsorbed by the lymphatics and termini of the vena portæ, is variously estimated at from twelve to twenty-four pounds—the latter quantity being, as we have seen, the estimate of Bidder and Schmidt. This water is, of course, the purveyor of various dissolved substances which are precipitated and left behind, and the solutions in question must contribute their share to the manufacture of the chyle. Water passes freely from the stomach into the vena portæ; thence it goes through the liver, where, becoming charged with the products of that organ, it brings them back into the intestines where they are precipitated. Hence we may understand the repute of free water-drinking as a cholagogue and as powerfully assisting the abdominal circulation. Apart, therefore, from its mechanical effect in the primæ viæ, which varies with the quantity imbibed at a time, water may be regarded not only as a solvent, but as the great diluent and eliminant. It dilutes the contents of the stomach and intestines, and thus acts as an aperient; it dilutes the bile, and thus, as we have seen, may act as a cholagogue; it dilutes other secretions, and thus acts as a diuretic and diaphoretic. Further, it dilutes the whole mass of the fluids; but probably the blood is only influenced thus for a brief period, as we find its specific gravity is maintained with tolerable uniformity. Yet the rapid introduction of liquid into the circulation undoubtedly tends to reduce the specific gravity of the blood, and that whether injected into the veins or absorbed from the alimentary canal. At the same time we must not forget that the excessive inflow excites rapid removal through the skin, kidneys, and lungs, in order to maintain the normal balance. Hence, water is diaphoretic, diuretic, and a stimulant of pulmonary exhalation. The importance of this balance being maintained is illustrated by the fact that a fall in the specific gravity of the plasma causes distention of the red corpuscles, as first noticed by Hewson [1] and confirmed by Owen Rees,[2] who also established the converse effect of raising the specific gravity. Further, medicines which hasten the removal of water from the system by quickening secretion indirectly produce the same effects. Water may further be regarded as a regulator of temperature, inasmuch as the surplus heat of the body is chiefly dissipated by its evaporation. Helmholtz calculates the loss of heat by the evaporation of the water of respiration at 14.7 per cent. of the total loss, and the amount dissipated in conduction, radiation, and evaporation by the skin at 77.5 per cent. A further loss of 5.2 per cent. is accounted for by the warming of the expired air, the balance of 2.6 being used in warming the excretions. Hence it is clear that the chief temperature regulator is the evaporation of water through the skin and lungs. With regard to the lungs, the greater the quantity of air that passes in and out the greater should be the loss of heat in a given time; but this effect may, perhaps, be neutralized by heat, producing changes in the lungs. Further the conclusion as to loss of heat in respiration is disputed, but as to the loss by the skin it is generally admitted.

The temperature at which water is drunk materially influences its effect. Absorption appears to take place most readily at or about the temperature of the blood, so that to secure rapid removal from the stomach by absorption it should be taken at about that point, and in small quantities at a time: any considerable bulk of merely tepid water will, indeed, as every one knows, be likely to cause vomiting, and, in fact, full draughts are constantly given as emetics. Cold water in small quantities stimulates the stomach and assists digestion, but a full draught is apt to impede that function; in weakly people "it lies heavy" for a long time, and the danger of drinking very large quantities, especially when the body is hot and exhausted, is well known. A glass of cold water early in the morning is an excellent aperient, stimulating the peristaltic action by its temperature, and promoting the secretions in the way already described. Cold water may be given even when the temperature is high, and it does not give rise to discomfort, but is usually very grateful. On the other hand, warm water is much more grateful in many painful affections, both of the respiratory and abdominal organs; it is better to increase the action of the skin, to promote other secretions, and possibly to stimulate the absorption of morbid deposits; as a cholagogue it is often efficient if taken regularly at bedtime, or night and morning. In laryngitis and bronchitis the use of hot drinks is well known as a popular remedy, and in a large proportion of respiratory diseases, both acute and chronic, warm beverages are more appropriate than cold, not only because they are more grateful to

---

[1] Works of W. Hewson, F.R.S., edited by Gulliver. London, 1846.
[2] Guy's Hospital Reports, 1841-2.

the patient, but because they promote perspiration and afford temporary relief to cough and dyspnœa.

The systematic use of warm or hot water dates from very early times, and has been largely carried out for ages at the various hot springs. It therefore excites a mild curiosity to find Dr. Ephraim Cutter claiming on behalf of another American physician, Dr. H. Salisbury, the merit of having introduced the proper use of this remedy in 1858. We may freely admit that gentleman's merit in carrying out a series of experiments in order to determine the best mode of employing this remedy, without forgetting previous labors in the same field. Dr. Salisbury recommends the hot water to be drunk regularly four times a day, from one to two hours before each meal, and an additional dose at bedtime. The quantity usually required is from half a pint to one pint, and sometimes a pint and a half for each dose, but the urinometer is to be regularly employed, and when the specific gravity of the urine rises above 1.020, more is required, and when it falls below 1.015, less may be taken. Of course the water should not be drunk so fast as to cause distention, but sipped during a quarter or half an hour. The effects are described as quite as remarkable as those of the most potent mineral springs. Thirst and dryness of mucous membrane disappear; the skin puts on a healthy appearance; the urine becomes free, pale, and resembles that of a sucking infant; indeed, all secretions become healthy, digestion rapidly improves, and a wonderful elasticity and buoyancy is perceived. Dr. Salisbury says: "If I were confined to one means of medication I would take hot water," and Dr. Cutter tells us that this is stated after drinking it for twenty-five years, and he himself from his personal experience and observation corroborates these statements.

An eminent medical friend of mine derived remarkable benefit from a plan which has some analogy with this. Every morning he had brought into his consulting-room a jug of hot water, which he consumed during the intervals of his consultations. At the same time he followed out an exact system of diet. I could record various instances of the value of warm water, and for that matter, of cold; but from what has preceded this seems unnecessary, and those interested in the subject may be referred to the large literature respecting mineral waters, as well as to that connected with hydropathy. As to the last, it is unfortunate that in this country it has been too much mixed up with quackery, but happily there are now many qualified medical men in charge of hydropathic establishments, and the follies formerly practised are, under their guidance, disappearing. On the Continent such establishments have always been regarded as appropriate spheres of professional work, and those engaged in it have therefore made valuable contributions to science. Water, both hot and cold, is an invaluable remedy, both externally and internally, and hydro-therapeutics, therefore, deserves earnest study, but we cannot countenance the exclusive claims which have sometimes been made on its behalf, and we deprecate all excesses—even excess in drinking so simple a liquid as water, whether hot or cold.

BEVERAGES.

As the vehicle of other substances, whether nutritious or not, water forms the bulk of all our beverages. Some of these contain so little else that they can only be regarded as simple diluents, but in others we find peculiar principles to which they owe their distinctive properties. It may be convenient to group them accordingly.

a. Simple Diluents.—Of course, pure water is the chief of these, and at one time some patients were restricted to it and nothing else, either solid or fluid, for days together. *Diète absolue* the French call this, which ought to be translated, not literally, but as want of food. Toast-water contains so little as to merit a place here, as may any drink in which there is but little nutriment or medicine, such as the slops, ptisans, teas, and other fluids familiar in the sick room.

b. Mucilaginous, such as oatmeal water, linseed tea, marsh-mallow tea, and the other demulcent drinks formed from gummy or farinaceous articles.

c. Saccharine may be grouped separately, as the presence of sugar may be an important consideration in determining their suitability.

d. Liquid foods, such as milk, beef-tea, soups, broths, constitute the most nutritious group, being capable, when properly combined, of supporting the system. They have already been sufficiently considered.

e. Aërated drinks, water containing carbonic acid enough to render it effervescing, forming an agreeable cold drink, the gas acting as a stimulant to the stomach.

f. Acidulous drinks, such as lemonade, etc., are often very grateful, and are believed to be refrigerant and somewhat astringent. They quench thirst, tend to check hemorrhage, restrain too copious perspiration, and to some degree abate hectic and other febrile states. Sub-acid fluids may to a considerable extent replace this group.

g. Salines.—In one sense the previous group resembles this, since the acids appear in the blood in the form of salts. But there are differences in the therapeutical effects, partly due, perhaps, to differences in quantity. Some of the saline domestic remedies are but convenient modes of administering medicines, and now and then it is necessary to beware lest under the name of beverage, active medicines should be too freely consumed.

h. Medicinal Beverages.—Some of the previous groups might be placed here, for both acids and salines are potent remedies, as, too, are alkalies and others which may conveniently be used as beverages. Soda water and mineral waters to wit, but only with due regard to the condition of the system.

i. Fermented Beverages.—Their effects may be referred to the alcohol they contain.

k. Tea and its Allies.—We come, at length, to those beverages which have obtained universal favor as ordinary articles of diet—tea, coffee, cocoa, mate or Paraguay tea, guarana, coca, or rather, as Christison suggested, cuca, African kola nut, etc. These are mostly used as warm infusions, and are always recognized as possessing a decided influence on the nervous system. These vegetables contain, first, a volatile oil, seldom amounting to one part in one hundred and fifty, but that is enough to impart the distinct aroma; second, an astringent styptic acid, allied to tannin, which amounts to thirteen to eighteen per cent. in tea, but only about five per cent. in coffee; third, a nitrogenous alkaloid, to which is attributed most of the effects; this is called thein, caffein, theobromine, according as it is obtained from tea, coffee, or cocoa; but all are believed to be identical. The amount of alkaloid varies in the different substances. In coffee it averages 0.75 per cent., but Payen found as much as 1.7 per cent., though no one else has detected as much; in tea it averages 1.8 to 2 per cent., but in some good specimens there is much more, Péligot once found 6.2 per cent., and Letheby found 4.94 per cent. in a sample of Himalayan tea; in cocoa 1.2 to 1.5 per cent., and in the leaves of this plant 1.2 to 1.26 per cent.; in mate 1.2, and in guarana the average is five per cent. The amount present, however, may not correspond with that extracted. Aubert found in a cup of coffee made with 16.66 grammes, from 1.5 to 1.9 grain of caffein, and in a cup of tea made from five to six grammes of the leaf, about the same quantity of alkaloid. Cocoa may be regarded as a nutrient combined with the alkaloid, but coffee and tea are rather nerve-stimulants. The alkaloid certainly seems to produce an agreeable excitation of the nervous system, without the subsequent depression which characterizes more powerful stimulants; moreover, the effect is more prolonged. These facts explain the high value so universally set on these beverages, as well as their occasional disagreement with a few individuals, some of whom, however, might very well partake of them if they would employ much weaker infusions.

Liebig noticed that the alkaloid was related to kreatinin, and so compared tea and coffee to soups; being related, further, to nervous tissue, it might be suggested that it would serve as a ready-prepared nerve-food. But it is not likely it enters into the formation of tissue. Perhaps, however, it may, on account of its near relation, serve as a kind of stimulus to change, as a true nerve-food would. Lehmann, in 1854, found coffee reduced the excretion of urea, and therefore he concluded that it retarded tissue change, and in this he seems to have been supported by the experiments of Böcker and Hoppe. On the other hand, Voit found no such adulteration produced, and Dr. Squarey's numerous experiments do not support the alleged decrease. The evidence as to tea points to a slight decrease, but is not very satisfactory. Dr. Edward Smith (" Phil. Transactions," 1859) found that tea and coffee increased the carbonic acid exhaled by the lungs, the quantity of air breathed being increased, the respirations being deeper and freer. This would indicate that they tend rather to increase than to retard tissue changes. He says that " tea promotes all vital actions and increases the action of the skin," but seems to think that coffee diminishes perspiration. Probably both infusions, taken warm, promote perspiration, but tea more decidedly so, and coffee certainly seems rather to stimulate the kidneys. In promoting perspiration they would indirectly diminish the heat of the body. The alkaloid is not the sole active agent, for the volatile oil has been shown to produce similar symptoms. It causes wakefulness, abates the feeling of languor, induces perspiration, and, according to Lehmann, lessens the excretion of urea.

We may, perhaps, on the whole, conclude that tea and its allies do really retard tissue changes, and so diminish waste, while at the same time they certainly excite nervous action, and thereby render work easier, both physical and mental, but more especially the latter. To the cerebral excitement is due the wakefulness, the persistent thinking, and the feeling of cheerfulness and exaltation which are characteristic effects of even moderate quantities; and poisonous doses expend their energy on the nervous system.

If these beverages really retard tissue changes, we can understand why they are reputed to abate heat. It makes some difference whether they are taken warm or cold; not that the temperature of the fluid itself would have much effect, but in the one case the sensation of coolness may be grateful, besides producing an effect through the nervous system, and in the other the promotion of evaporation by the skin and lungs increases the escape of heat. Are we then to regard such drinks as refrigerant, even when warm? Does the old nurse's tea really cool her as she says? Is the febrile patient's cry that it cools him correct; or is that idea only the expression of thirst, and the water, not the tea, the active agent? If so, may we not give the water and leave the additional ingredient to the patient's choice—tea or warm drinks, acid cool drinks, or fruits? Practically, we know that any refrigerant, we may say any pleasant tasting liquid, is grateful to a heated person, whether the heat be physiological or pathological. Thus we find in thirst with warmth of surface, produced by exercise, fluid is acceptable, and whether cold or warm is selected is often a matter of taste. In feverish conditions, no doubt, there is excessive production of heat, not merely deficient escape. Nevertheless, any increase in escape give proportionate relief, and drinking liquids promotes this. The cool taste of the cold liquids is grateful in itself, while the warm beverages, perhaps, act by their indirect effect on the skin, for the simplest hot drinks will sometimes at once unlock, as it were, the sweat-glands, and so afford relief. This is particularly noticeable in affections of the air-passages, in which, even when acute and attended with fever, warm fluids are preferred and are most beneficial. Perhaps this may partly arise from the increased elimination of water through the respiratory mucous membrane.

Cuca.—The leaves of the Erythroxylon cuca have from time to time had ascribed to them very remarkable qualities. In 1609 Garcilasso de la Vega, described the use of them by the natives of Peru, as a means of preserving strength during fatiguing exercise and privation of food. Dr. Mantegazza, of Milan, whose charming travels in South America are pretty well known, observed that it increased the frequency of the pulse, produced a strong tendency to muscular action and great mental vigor. Dr. Von Tschudi partook of the leaves after the manner of the Indians, and found they prevented difficulty of breathing and fatigue in ascending the Andes, as well as enabled him to pursue swift-footed game and to endure long abstinence from food. The Indians and Peruvians are said to run fifty miles a day without food or anything except their cuca. In 1876 Sir Robert Christison reported to the Royal Botanical Society of Edinburgh experiments on himself and students. The veteran professor having found that a fifteen mile walk without food or drink was sufficient to tire him, proceeded to walk sixteen miles in three stages. He chewed eighty grains during his second rest, and forty grains in the last stage; all sense of weariness vanished and he reached home without any fatigue or uneasiness. Although no food had been taken for nine hours, neither hunger nor thirst was felt, but he did justice to his dinner, had a good night's sleep, and next morning was free from fatigue. On another occasion, after ascending three thousand feet of Ben Voirlich, forty grains were chewed, and all fatigue vanished, the descent was made with ease, another twenty grains being taken when half-way down. Neither food nor drink was taken from half past eight in the morning to six o'clock in the evening, yet neither hunger nor thirst was felt, no drowsiness was experienced in the evening, and after a good night's sleep he awoke ready for another day's exercise. He concludes that cuca removes and prevents fatigue, and suspends hunger and thirst without eventually affecting appetite or digestion. It seems to have no effect on the mental faculties except liberating them from the dulness and drowsiness which follow great bodily fatigue. We should, perhaps, remember that Sir Robert's splendid mind was associated with a body always remarkable for its activity and power of endurance, and young, if weakly, persons could not hope to cope with the aged but athletic professor. On the publication of his report a friend of mine took a twenty mile walk with comfort and ease on sixty grains of cuca, and said he felt ready for another ; he took neither food nor drink from 8 A.M. to 4 P.M., and felt neither fatigue nor discomfort, in spite of inclement weather. Dr. Burness, author of the "Specific Action of Drugs," undertook similar experiments with the same result, and he informed me that cuca produced a rise in the temperature ; not only when walking, when it might be attributed to the exercise, but also when in repose. In the last case there was a rise in half an hour after a dose from 98.6° to 99.2° F., and within the next half hour to 99.4° F., after which it again gradually fell. Mr. Dowdeswell reported in the *Lancet*, 1876, results of an opposite nature, and some others have failed to find cuca banish fatigue in the manner described by Christison. The leaves contain a crystallizable alkaloid similar to, if not identical with caffein, besides a volatile oil, a peculiar waxy body, and an astringent substance allied to tannin. From sixty to ninety grains is sufficient for one trial ; it is uncertain whether it retards tissue change, but it probably does, as it seems to diminish the excretion of urea, and in other respects resembles in its properties the other caffein-containing plants, but it does not seem to diminish perspiration nor to produce cerebral excitement, although this has sometimes occurred, and very large quantities have been followed by hallucination.

Cucaine, the active alkaloid of cuca, has been given in one-eighth and one-seventh grain doses. In some respects it is said to resemble atropine. Dr. Aschenbrandt, during the manœuvres of the Bavarian troops, tried it on the soldiers, and reports that one-seventh of a grain sufficed to remove fatigue, hunger, and thirst, so that the men could go for hours without food or drink. He considers it a nerve-food. His observations corroborate those of Dr. Amess, who experimented in 1880 with chloride of cucaine.

CHAPTER XIII.

EXERCISE AND REST.

We have seen that exercise or work is a stimulus, just as aliment is, though acting in a different way. This is most familiar in the muscular system. We can increase a muscle by exercising it and supplying the body with aliment, and indeed, under the influence of exercise, nutriment may be drawn from other parts to the muscles engaged. Labor which exercises many muscles requires a proportionate diet. On the other hand, during rest in bed we need less food. A familiar instance of this is an accident, say a broken leg, which confines a healthy person to bed; if he takes the same food as in active life febrile symptoms will set in. It would seem, then, that exercise would be a nutrient, or rather would promote nutrition, provided proportionate food be taken, otherwise it would act in a contrary way. The effect of exercise is perceived in all organs, but the expression is mostly used in reference to the locomotor system. The muscles in contracting accelerate the flow of blood and so act on the circulation, which is also increased from the general effect. We have, therefore, increase of heat, more frequent pulse, fuller respiration, and increased secretion. Another more distant effect is the absorption of fat, whether from increased waste or from pressure exercised by the contracting muscle on the softer tissues. With the demand upon it the muscle grows, increases in volume, in firmness, in elasticity, acquires greater strength. As the muscular action is called forth by the nerves, they also are exercised. The oxidation of carbon and perhaps of nitrogen is also increased by exercise, so is the elimination of water, as we see in the perspiration produced, so that the body contains less water after a spell of work unless this effect has been neutralized by the person drinking. In every organ there is more rapid circulation, more rapid flow of plasma, as well as more active absorption, that is to say, there is more waste and quicker repair. Thus the products of action are at once removed, instead of accumulating in the organs. Further, there is a greater escape of heat by the evaporation of the increased perspiration. We see, then, that nitrogen and carbon are required. If not supplied the absorption of oxygen and the molecular changes would be retarded, as would also the elimination of carbon. What we have said applies to moderate exercise, immoderate work may produce exhaustion; moreover, the muscles require periods of rest in which to store up the material which is metabolized during contraction. It would seem, too, that oxygen may be stored up during rest. Both the heart and the lungs should be sound, in order to secure a free blood-supply to the muscles, and the rapid interchange between oxygen and carbon without too much fatigue to those organs. All this shows us that under favorable circumstances moderate exercise, especially such as calls forth the energies of large groups of muscles or many organs, is one of the best tonics, promoting nutrition and improving the health of the body; especially is this the case with outdoor exercise. Sometimes, however, we are only able to employ passive exercise, and in other cases special gymnastics.

Often instead of exercise rest is desirable, or even essential, and then, of course, we have the converse of some of the effects described, but not altogether, for rest again with proper food promotes nutrition by diminishing waste. During sleep, for instance, there is slackened destructive metamorphosis, but repair is still going on, and this accounts for the restoration of the wearied limbs. Children feel the want of sleep more than adults, partly because of their wonderful activity, still more because growth is so rapidly proceeding. But during rest not only is there this diminished molecular movement, but the respiration and circulation are slower, and the work of the organs less. As an example the heart beats less in the horizontal position, on assuming which in health a fall soon takes place of twelve beats per minute, or 17,280 in twenty-four hours; in disease the fall may be twenty, thirty, or forty beats per minute. The consequences of such a difference may be traced throughout the body.

Exercise and rest may alike be local or general, active or passive. A limb, a group of muscles, a single muscle, an organ, or other part may be subjected to the influence of either agent. When absolute repose cannot be had, relative rest may be procured. When the work of an organ is intermittent, there is a period of repose between its efforts. Thus the heart rests during the pause between its contractions, and we are able to lengthen that pause, while increasing the force of the beat, and in this way the rest of the organ is increased; as we have remarked, the recumbent posture greatly lessens the work of the heart, as it does also of the lungs. In respiratory diseases we cannot stop breathing, but we may regulate it by remedies. In chronic cases, fuller, deeper respiration takes place during exercise, which further secures other benefits, such as improved appetite and digestion, increased oxidation, and thereby augmented nutrition. But we more frequently need rest in diseases of these organs, and this is especially required in acute cases. Fever, of course, compels repose, as does often inflammation. In hæmoptysis the first remedy is rest, and often this is the only remedy required. In inflammation of the air-passages, besides avoiding irritation by securing that only warm and perhaps moist air passes along them, we avoid the irritation which is so injurious, and, as far as we can, help them to rest. Thus in laryngitis we impose silence as far as we can, by forbidding speaking, that is, we give all the rest possible to the vocal organ. In bronchitis the same means are of no little assistance. In capillary bronchitis we endeavor to attain the same result by an indirect method—such as relieving the oppressed heart, promoting the removal of secretion from the tubes, or arresting its outpour while maintaining the circulation. In pleurisy, besides rest in bed, we may bandage the chest or strap the affected side—a plan well reported of in acute general pleurisy, seen early, as well as when the disease is limited in area with little or no effusion. In pleuro-pneumonia caused by a fractured rib, the benefit of fixing the part is still more manifest, inasmuch as it prevents the repetition of the injury which would take place with every movement of the broken bones. In the intercurrent pleurisy of phthisis, and in the short cough and stitch so common at the commencement of that disease, temporary rest is also most beneficial, as it is in slight attacks of hæmoptysis during the progress of phthisis, but care should be taken not to deprive the patient too long of the general exercise of which he is in need, or the lungs of the free and full breathing which is so important. In advanced consumption cough and pain, of a most distressing kind, sometimes appear to be caused by adhesion of the pleural layers, and may be eased by such rest as may be secured by careful strapping.

CHAPTER XIV.

ALCOHOL.

WE come now to a thorny subject, for alcohol has been alternately abused and neglected. Wine has been used as a medicine from the earliest times, and its history as a remedy shows a series of oscillations almost as remarkable as those exhibited in the history of depletion. Asclepiades pandered to the Roman taste for wine, not only by giving it at once after the evacuants he ordered, but prescribing it in excess sometimes, even to the extent of intoxication, under the pretence of thereby procuring sleep, while in other cases he also ordered it for the opposite purpose. Other charlatans at different times have abused the use of alcohol, and upheld their practice by equally inconsistent dogmas. Even able and thoughtful physicians have sometimes been led away into excess or inconsistency in their use of this potent agent. It is perhaps curious to note that the most reckless prescribers of stimulants have often appeared to exhibit a touch of quackery in their conduct. Abuse of alcohol has naturally largely alternated with excess in the employment of depressants, especially venesection; but we must leave these changes in medical opinion and practice until we come to the similar epochs which have marked the use of bleeding.

Is alcohol a nutrient? Can it be regarded as a food? It has been held that this may be the case, because the excretion of carbonic acid and urea are lessened under its use. But the experiments of Parkes and Wollowicz show that moderate quantities do not interfere with the metamorphoses of the nitrogenous tissues, while it is, to say the least, doubtful whether the exhalation of carbonic acid is, after all, decreased. Another reason has, however, been advanced for considering it a nutrient, viz.: that inasmuch as it is very easily oxidized, it takes the place of alimentary material, which it thus spares, thereby permitting the accumulation of hydrocarbons in the system—that is to say it fattens. In the same way it would supply heat, its value for this purpose depending upon its ready combustibility; that is to say, it is preferred to other hydrocarbons, and therefore oxidized, while they are left unattacked in its presence, so that it acts by sparing real nutrient material. But does it thus supply heat? It is admitted on every hand that only small quantities can do this, for larger amounts have been distinctly shown to reduce the temperature of the body. Binz attributes this effect to a paralyzing influence on the vaso-motor nerves, which results in a dilatation of the superficial vessels, by which more blood is brought to the surface and there cooled by radiation; at the same time there is a retardation of molecular change by a direct depressing influence on the cells, and of course a diminution of the heat which would be liberated by cell oxidation. It is, of course, by no means impossible that such opposite effects may be produced by different doses, but it may be observed that we are here considering quantities which fall short of intoxication, and the line which divides the one effect from the other seems to be much narrower than is generally supposed.

In ordinary language a substance which, though it may be oxidized in minute quantities, yet in rather larger amounts depresses the temperature and retards the activity of cells is scarcely entitled to be called a food, especially when it does not enter into the composition of tissues, but if it be shown to promote indirectly their repair or to reduce their waste it may be perhaps grouped conveniently among indirect nutrients. It is admitted that, at any rate, some of it undergoes change in the body, for when minute quantities only are taken it can scarcely be traced passing out. What, then, becomes of it? Our information here is rather deficient. Out of the body it is easily converted into acetic acid, and some have supposed that such a change may be effected in the system; in that case the acid would unite with the soda in the blood and the salt thus formed would become a carbonate, which would be excreted by the kidneys in the same way as acetates are when taken into the stomach, but the effect would not be just the same, for in giving acetates by the mouth, we give the base as well as the acid, whereas, in the case supposed, acetic or carbonic acid formed from alcohol would have to obtain their alkalies from the system, and by uniting with them might thus increase instead of diminishing the natural acidity of the secretion. This would account for a slight increase of acidity which was observed by Parkes. Here, however, it must be observed that acetic acid has not been detected in the blood, nor, indeed, have other intermediate products of the combustion of alcohol. It has therefore been suggested that the oxidation is carried, so to say, at a single bound to its ultimate end, viz., the production of carbonic acid and water. In support of this view it is urged that before all the alcohol has escaped elimination ceases, or at any rate its presence cannot be detected.

It is impossible to determine from the pulmonary exhalation the proportion which may have been oxidized, for the carbonic acid is not increased, indeed nearly all experiments point to the conclusion that it is rather decreased. Unless, therefore, some other substance which would yield carbonic acid is prevented from doing so, the alcohol cannot be converted into that gas. The difficulties attending the investigation into the elimination by the skin are very great; we turn, therefore, to the kidneys. When only small quantities of alcohol are taken it can scarcely be detected in the secretions when larger quantities are taken elimination soon commences by the kidneys, as well as by the skin and lungs. So constantly is this the case that the late Dr. Anstie proposed to consider its appearance as a measure of the saturation of the system, or rather as a sign that as much had been absorbed as could possibly be oxidized. He found that about one and one-half fluid ounce always caused its appearance. Parkes and Wollowicz obtained this same result; they found in a strong, healthy man, accustomed to take alcohol in moderation, when between one ounce and two ounces were given him some was at once thrown out unchanged by the kidneys, and at the same time some ill-effects began to manifest themselves, though only in a slight degree. The man lost his appetite, there was slight but distinct narcosis, increased rapidity in the heart's action, quite a rapid rise of temperature and greater dilatation of the small vessels. The sphygmograph confirmed the analyses, and these symptoms are considered to mark the commencement of the poisonous effects. Here, then, we seem to be approaching a reply to the question, How much can be consumed in the body? In a strong man from one to one and one-half fluid ounce is the limit; it is possible some persons might oxidize more, but the probability would be that a larger number might be unable to dispose of as much. After that limit has been passed the kidney begins to throw it out, and symptoms of inconvenience or of distress may be produced by its presence. Parkes made his experiments on strong, healthy men, and the deductions he drew were most moderate. We can scarcely go wrong in following so judicious an observer, his experiments are only likely to be corroborated, and until some flaw in them can be pointed out it seems reasonable to conclude that about one ounce of absolute alcohol should be considered the limit which ordinary men should not pass if they wish to avoid danger. Now, that ounce of alcohol is contained in about two ounces of brandy, or five ounces of sherry or other strong wines, or eight to ten ounces of weaker wines, such as hock and claret. If we take the outside quantity it would only allow us to increase these amounts by one-half, and then we should have arrived at the maximum which can be disposed of by strong men. Of course, in weakly or unhealthy men the figure should be reduced. In women, also, on account of the greater susceptibility of the nervous system, as well as for the less doses they require of all medicines, the amount should be less. In children the injurious effects of alcohol are more rapid in appearing and more severe in character, so that while it is almost universally allowed that they ought never to partake of it as a beverage, it should only be given as a medicine with extreme precaution.

There is this to be said about the medicinal use of alcohol. We give other remedies which are rapidly excreted and we often have to give them in such quantities as may for a time distress the secreting organ. This may be particularly observed in acute cases, and it may be that to produce certain effects we might administer alcohol in doses which could not be oxidized. If we give it as an antipyretic, we must give enough to reduce the temperature, and that is more than the maximum which healthy men can dispose of in the system, and it is not alleged, so far as I remember, that larger quantities can be oxidized in disease. In chronic cases the use of alcohol presents a very different aspect. Here it approaches rather the dietetic employment, and consequently the quantity which the patient ought not to exceed should be rigidly kept within the limit of that which is likely to be oxidized in the system. For it should not be forgotten that the surplus is thrown out by the kidneys, not as so much useless, inoffensive matter, but these organs are irritated by the alcohol in its passage, and thus disease is produced. This has been denied, but although we may recognize certain pathological changes in the kidney as not necessarily induced by alcohol, yet statistics still show that drunkards are more liable than others to renal disease, some forms of which seem to be the consequences of alcoholic excess. But even this is not the most important. It is one of the less, for the alcohol in the system undoubtedly affects other tissues, in most of which fibroid or fatty degenerative changes are brought about by its agency. Further, these degenerations, though naturally looked for in drunkards, often appear in persons who consider themselves strictly moderate. And what else should be expected? They habitually take only a little more than can be disposed of in the system ; the effect of the surplus must usually be present, just as when an occasional excess is taken its effects continue until elimination is complete. So long as any surplus is in the system the spirit circulates in the blood, and the tissues are bathed in a fluid containing it. And here it is important to note that elimination is not so rapid as some have supposed. Subbotin found that it was not quite complete in twenty-four hours ; Perrin, Lallemand, and Duroy observed it to go on for thirty-two hours ; and Parkes and Wollowicz, after a large quantity of brandy, detected it on the fifth day, though exhalation by the lungs had ceased much earlier. This looks as if it might accumulate in the body ; at any rate, it proves that the alcohol may remain in the system a considerable time after the period at which it has generally been supposed that elimination was completed, and when, accordingly, it has not usually been looked for. This may possibly help to account for some of the effects produced on the tissues. Some of these effects we proceed to note.

On the mucous membrane it coagulates the albumen and corrugates the epithelium ; this it appears to do by absorbing water, and we may see the effect in the mouth.

In the stomach it produces a sensation of warmth or pain, according to the degree of dilution. It stimulates the mucous lining, and in small doses may thereby aid digestion, but in large quantities, or if not sufficiently diluted, it impairs that process, for it precipitates peptones and brings on congestion of the gastric mucous membrane. When this membrane is habitually subjected to such a congestive irritant the connective tissue between the glandules is increased and causes their disappearance, bringing about the chronic "catarrhal condition" of Dr. Wilson Fox. It diffuses easily into

The Blood.—Here it would seem the chemical changes are partially arrested. More fat accumulates, or at any rate becomes visible. Dr. B. W. Richardson says that the blood is deficient in oxygen when alcohol is present. This should diminish antiseptic resistance to disease. Dr. J. J. Ridge, who some time ago determined by experiment that alcohol was injurious to growing cells, has lately made a more surprising announcement. He says he has experimentally determined that alcohol, which in considerable quantity is admitted to be an antiseptic, is in very small proportions just the opposite, as he found minute quantities promote the decomposition of both mineral and vegetable matter, bacteria flourishing and being propagated to a greater extent when a little alcohol was added. As soon as the proportion of alcohol is less than one per cent. he found decomposition hastened instead of retarded by its presence.

The heart and vessels receive the full force of the stimulus. At first the pulse is fuller and more frequent ; it is so far a stimulant according to the general notion, and this is confirmed by the sphygmograph. It quickens the systole and shortens the diastole or the period of repose. Parkes found that brandy or wine alike augmented the rapidity of the pulse about thirteen per cent., at the same time increasing the force of the beat. The period of rest was shortened, so that the heart had more work and less repose. Now, such a stimulus as this would seem in one way to lessen the nutrition of the organ by an undue interference with its natural balance of exercise and rest ; for it is difficult to see how excessive wear is to be compensated during diminished repair. The effect on the heart may be compared to that on the voluntary muscles, which alcohol certainly weakens, while it certainly interferes with their co-ordinating power—doubtless through the nervous system. The arteries dilate more easily before the fiercer flow of the blood-stream as it is impelled by the excited heart, and the superficial vessels in their turn also dilate more fully, hence the familiar flushing of the skin. When this flushing becomes habitual the vessels become permanently dilated, and thus fix the mark of the evil habit upon the person in the turgescence of the capillaries and the altered condition of the skin, which gives to the individual the appearance of premature old age. Perhaps this yielding of the vascular system may serve as a kind of safety-valve; for otherwise the strong wave might rupture the vessels, or else, the force acting in the other direction, the heart might not be emptied by the contraction, and thus other inconveniences would arise. This would seem to indicate that when the arteries are rigid alcohol might be more dangerous, unless for some reason the heart did not react as usual under its influence.

Nervous System.—Perhaps the phenomena produced on the vascular system may be the indirect effects of the action on nervous tissue. Anstie and others have attributed them to the influence on the sympathetic, but others consider the vagus and the heart to be involved. On the cerebrum the first effect is exciting; alcohol certainly increases the rapidity of the flow of ideas, but by no means the power of controlling or directing them. This faculty is, in fact, diminished; close, concentrated attention is rendered difficult or impossible; very soon, however, a narcotic or anæsthetic effect comes on, and then the rapidity of ideas or impressions is naturally lessened. Even when, as sometimes, the imagination seems to be stimulated, it becomes uncontrollable. As to the special senses, they scarcely seem to pass through the stage of excitement, but under quite small quantities their acuteness or their accuracy is lessened. From this deadening of the senses, as well as the muscular power, it would seem to interfere with the movements of nervous currents, and it has been conjectured that it may do this by entering into a temporary combination with nervous tissue. But can it do this? and if so, would it be a nutrient or only a poison? As to the first question we are in doubt. We know that considerable quantities have been found saturating, so to say, the brain, or bathing nervous tissue, but this certainly does not necessitate a combination, and out of the body the tissue may be steeped in the spirit, which hardens and preserves it but does not appear to combine with it. With regard to the second question, it can only be regarded as a nutrient if it combine with the tissue as a necessary ingredient, or as something assisting its formation or disintegration by the usual steps. If, by any combination it may be supposed to form, it hindered the usual constructive and destructive changes, and at the same time impaired or prevented its function, it would have no title for such reasons to be called nutrient, though it might fairly be termed poisonous.

The Liver.—The degenerative changes produced on this organ are generally well known, but how much alcohol may be taken without setting them up we have no means of determining with any degree of certainty.

Respiratory System.—The muscles or nerves of respiration may be impaired or paralyzed, and death sometimes occurs in consequence. The carbonic acid and the watery vapor exhaled seems to be lessened, though different results have been observed in some of the experiments. Dr. E. Smith observed different effects with different kinds of spirits. It may be supposed that some variation might occur from the influence on digestion or circulation, or from other circumstances. There is room, perhaps, for further investigation, but it seems pretty certain that if not much lessened the carbonic acid is certainly not usually increased. The bronchial mucous membrane, and perhaps also the pulmonary tissue, must be injuriously affected by habitual excess of alcohol, and chronic bronchitis and lobar emphysema are among the penalties of heavy drinking.

Cutaneous System.—It is often said that alcohol is a sudorific, and hot grog is a popular dose, but the effect of the hot water should not be overlooked. Dr. Edward Smith thought alcohol lessened the amount of perspiration, but Weyrick found a large increase after all alcoholic beverages. No doubt the dilatation of the cutaneous vessels produced by alcohol tends to promote perspiration, but the effect is not such that it can be relied on as a diaphoretic; indeed, when taken cold, it is much more likely to prove diuretic. As to the sudorific influence of hot grog, tea or any other hot beverage is equally efficacious and less likely to be injurious, while a dose of sweet spirits of nitre in hot water is a domestic diaphoretic of much greater value.

Temperature.—The effect on temperature is less than might have been supposed from the vascular phenomena. Quite small doses do not affect the clinical thermometer. Experimenting on two healthy men, who were accustomed to take beer, and occasionally spirits, but who were strictly temperate, Parkes found neither a rise nor fall; Mainzer observed a slight fall in one instance and none in another; Obernier and Fokker also obtained no certain conclusions. But when larger doses are given there seems no doubt that the usual effect is to cause a fall. Binz, Ruge, Cuny-Bouvier, Richardson, Ringer, and Richards all seem to have met with this result. Lewis, Wood, and Reichert consider that the production and dissipation of heat are both increased. Parkes found a difference in respect to the food, no effect being perceived when the alcohol was taken with a meal, but taken fasting a fall occurred. Ringer found in a boy, aged ten years, who had never in his life before tasted alcohol, a constant and decided reduction of temperature. We may, perhaps, then conclude that in health a moderate fall usually follows the dose of alcohol; but no doubt this is much less, if at all perceptible in those accustomed to its use. In old topers the temperature is unaffected even by large quantities. Drs. Ringer and Richards once gave to an habitual drunkard twelve ounces of brandy in a single dose, which made him dead-drunk, but without the smallest reduction of the temperature.

We come now to the therapeutic uses of alcohol. Twenty-five years ago it was prescribed by almost every house-surgeon, so freely as to seem almost routine practice, just as their predecessors, twenty-five years before them, had employed bleeding and depressants. At present a considerable reaction has occurred, and alcohol is much more cautiously prescribed. It is said to be *stimulant;* this term is indeed applied generally to all intoxicating liquors, but we ought not to forget that it may just as properly be termed *narcotic*, for after the exhilarating influence it produces drowsiness and, in sufficient doses, coma. It has also been called *anodyne and anæsthetic*, and undoubtedly it lessens the consciousness of pain by deadening the perceptive power. Does it promote nutrition if the amount given be within the limit which can be oxidized within the system? or does it, by hindering waste, merely cause an accumulation of useless effete material? Opinions still differ on these points, but it seems clear that it can only be a nutrient indirectly, and to a very slight degree. Even so far as it is consumed, if it depress function, as it seems difficult to deny, would it not favor degeneration? If so, that would surely counterbalance the assumed advantage of its liberating force. To check oxidation is not necessarily to promote nutrition ; to economize blood, or even tissue, may not be so desirable as to quicken their renewal. It has sometimes been given rather freely as an *antipyretic*, and we must admit that in the febrile state the effect is greater in this direction than in health. Binz and Bouvier produced septic fever in animals, and then reduced their temperature by alcohol. On the other hand, one of the men who was taking considerable doses in Dr. Parkes' experiments took cold, and his temperature rose in spite of the alcohol. Others have found that it cannot be relied upon to abate ordinary febrile excitement, though there are many records in which it has seemed to moderate septic fever. It would seem to act more powerfully on animals than on men, but perhaps this may be due to the influence of habit. We must conclude that a high temperature is not necessarily a contra-indication, though the presence of arterial excitement and sthenic inflammation may be so regarded. Binz showed that it did not lower the temperature through the nervous system, so that the effect is probably due to its retarding metabolism. He gives it in order that it may be consumed, and so spare the tissues and supply animal heat. With such a view the amount administered surely ought not to exceed that which can thus be utilized, and yet it is too common for his followers to give much more. He says that small doses do not really stimulate, and compares them to small quantities of fuel laid upon a fire to prevent it going out. Exception might be taken to such comparisons, many of which have been introduced into discussions on this subject, but few, if any of which appear to me to have shed much light upon it. I have, therefore, in this chapter rigidly confined myself to a statement of some of the most important facts and the conclusions which seem to me to be fairly deduced from them. To obtain any considerable antipyretic effect it is necessary to employ large doses, so large as to be often dangerous, and therefore great caution is incumbent on all who use it for this purpose, and as the results are too slight in ordinary pyrexia, it ought not to be resorted to unless other indications than fever are present.

As an illustration of its stimulant action, its effect in rousing the heart and circulation in syncope is familiar to all. It would probably have been much less familiar but for the fact that some form of wine or spirit is present in almost all households, and therefore in a case of fainting it is the most ready remedy. As an anæsthetic or anodyne it is sometimes resorted to in neuralgia and other painful affections, but the practice is dangerous whenever the pain is recurrent, inasmuch as it is likely to lead to a too frequent repetition, and so may induce habits of intemperance. This danger ought never to be forgotten in chronic diseases, especially as in them it is often administered in the shape of a beverage, and it is easy to understand that when it seems to afford relief it may be resorted to more frequently than at first intended, especially by those who may not be aware of the ease with which it becomes a snare. The more caution is required, inasmuch as, after all, we have more potent remedies of this kind which are not dangerous in this direction; and if we had not, it would be better to endure the disease or to face death than to become the victim of intemperance.

Among acute diseases, pneumonia has been largely treated with alcohol. Indeed, this disease has been made the battle-ground between its advocates and opponents, just as it was in reference to bleeding. Now, however, it is generally admitted that simple pneumonia in healthy persons tends toward recovery, and therefore such a remedy should only be employed to meet certain conditions. As in all acute diseases with such a tendency, few would now resort to it during the early progressive stage, but when only the results remain it is often employed with a view to restraining waste, or to being itself, as some still suppose, a substitute for food. In such cases it is generally combined with milk, or alternated with that and other aliments, and possibly much of its influence may be due to its stimulating digestion, and facilitating the assimilation of the accompanying food. Wood, who is by no means averse to the use of alcohol, confines its use in acute pneumonia to cases "when so much consolidation has occurred as to render it doubtful whether the exuded matter can be removed," and then he thinks the demand for it may be "very great as a food and as an aider of digestion, and sometimes as a stimulant."

In phthisis, as a narcotic possessing some antipyretic action, it is sometimes given at night, but it should be combined with some form of food. Under its influence sometimes the patient sleeps better and the night-sweats diminish; for this purpose but small doses are required, and perhaps they are chiefly useful in promoting digestion; and the same may be said as to very small doses given early in the morning in conjunction with aliment, such as the popular rum and milk. Whatever spirit be employed for this purpose, for others are just as efficient, and to many more pleasant than rum, the dose should only be enough to assist digestion, and perhaps other aids to this process would be generally preferable. In chronic phthisis alcohol is sometimes prescribed at the same time as cod-liver oil, and by some under the impression that it will assist its digestion; but from what has preceded it would appear that ether would be a more efficient adjunct. In the sleeplessness of old age a small dose at night often procures relief, and the night-cap, as this is called, is therefore popular; but here again it should not be given without food, and very often nutriment alone will be equally, or even more effectual for the immediate purpose, and then, as every one will admit, should certainly be preferred. The small dose of alcohol probably acts by stimulating the stomach, as well as by exciting the feeble circulation; but in view of this latter action it is important that the arteries be sound. Where this is the case, stimulants of any kind may be given earlier in old age and feeble persons than in robust adults.

The importance of giving food with the alcohol has several times been mentioned, and it is to be feared that much injury arises from the neglect of this precaution. Even in extreme prostration small quantities of milk and alcohol can be taken at frequent intervals, but unless strict injunctions be given to the attendants they are apt to separate the one from the other, and the result may be that while the patient gets the stimulus the due quantity of aliment is not taken. The wine or spirit is trusted to; the heart responds for a time, but is not properly fed, and the consequences are necessarily disastrous. Patients who are strong enough to observe their regular meal-times should generally do so, as the stomach seems, from the force of habit, to work more easily in this way, and whatever stimulant be prescribed should be given with the food, unless urgent reasons exist for administering it at other times. Altogether, then, whether in acute or chronic disease, whenever stimulants are administered, it is an excellent rule to give them only in conjunction with some form of food.

CHAPTER XV.

DENUTRIENTS.

REMEDIES which hasten destructive changes next demand attention. We have already seen that water and exercise may do this. Nevertheless, they are indirectly promoters of construction unless taken in excess. It may further be said that some other agents more distinctly destructive may also be employed in such a manner as to become indirect nutrients— *e.g.*, fasting, which arrests the supply *pro tem.*, and low diet, which restricts it, give time for the removal of waste, and perhaps at the same time arrest, or at least retard metabolism; but when a new supply arrives its stimulus seems to provoke increased rapidity of construction, so that the loss is soon made up out of the fresh abundance. In the same way quickening excretion by evacuants, when carried far, causes waste; but a slight degree of the same process only stimulates to increased renewal. Further, the removal of the completely prepared nutrient fluid may act in a similar manner, *e.g.*, sudden hemorrhage takes away the nutrient fluid; but we see in accidents how rapidly loss of blood is made up for in healthy persons, and so it often is in disease. That a moderate loss of blood may really stimulate to increased nutrition is a familiar fact to many farmers, who when they find their cattle slow to fatten will bleed them. It may be, then, that denutrients stimulate construction, but they are usually spoken of, in consequence of their more direct effects, as weakening, depressing, lowering, etc. For the present, then, we have to consider them under this aspect, whether they act by removing nutrient material or by quickening destructive metamorphoses, or both.

One of the most important groups is

ANTIPHLOGISTICS,

So named because they are employed in inflammation (αντι, against, φλογοω, I inflame). This word, from the Latin *inflammatio*, itself derived from the verb *inflammo*, I set on fire, appears in almost the same form in all the languages of Latin origin. The German *entzündung* has a similar meaning, which is also found in other Teutonic tongues. The idea of kindling or burning is also found in the words by which inflammation is expressed in Hebrew, Sanscrit, Arabic, and perhaps other languages. Our word "fever," Latin *febris*, from *ferveo*, I am hot, is also as widely distributed, for not only is it found in all the Latin languages, but the German *fieber* shows that it has spread to the Teutonic. The synonym, pyrexia, from the Greek πυρ, fire, has a similar origin, and phlegmon, φλεγω, and phlogosis, once in common use, expressed the same notion (φλοξ, a flame). All these words have to do with the most striking symptom of the process, and show that the preternatural heat or sensation of burning has in all ages attracted attention. Charaka divided remedies into those which increased or diminished internal fire, while Susruta divided them into those which increase the strength by evacuating bad humors, and those which lower the exalted action of the humors. So extended, too, has been the idea of febrile action, that Hindu physicians say "that man is both born and dies in fevers (juvara);" while the Chinese do not seem to separate inflammation from fever, though they do distinguish several kinds of fever.

The antiphlogistic diet and regimen has always implied a restriction of food and the imposition of rest. Everything supposed to increase heat has been banished, and the patient placed on the mildest farinaceous fluid food, and very little of it; the beverages, also, have been simple diluents. At times such restrictions have been carried too far, and although judicious physicians have generally been found to inculcate the necessity of tempering the remedies to the patient, and even to the cause of the disease, there has very often been a danger of too strict adherence to the notion of combating inflammation by lowering measures. Of this a striking example was afforded when French physicians not only imposed their absolute diet, that is to say, denied their patients everything but pure water, but also practised bleeding and administered evacuants. But England was not much behind, for although there were always opponents to extremes of this kind, the general practice of the last generation was not inaptly described by the good and learned Dr. John Lettsom, in his rhyming pun on his own name.

> "When patients do to me apply,
>   I physics [= purges], bleeds, and sweats 'em;
> If after that they choose to die,
>   What's that to me? I lets 'em [I. Lettsom]."

Rest as an antiphlogistic agent ought to be both physical and mental, besides which local rest may be special. The object is to abate vascular action, to prevent waste of vital power, and avoid exhaustion. In severe cases, rest in bed is necessary, and it is always the best form of repose, for it usually tranquillizes the patient, prevents chills, and keeps him warm, *i.e.*, prevents excessive expenditure of animal heat, though sometimes we are glad to provide for the escape of its extra production. In severe cases of extensive inflammation, or of febrile excitement, loss of appetite settles the question of the amount of food, but sometimes solids can be taken, and need not always be forbidden; indeed, in some inflammatory affections of the respiratory organs, easily digested solids may be given all through.

BLEEDING.

The most potent antiphlogistic is depletion. In all ages some form of blood-letting, local or general, has been more or less employed, and though now so largely in abeyance, is still maintained by many to be a sure means of relieving pain, modifying disease, and even saving life. Still there have been remarkable changes in medical opinion. At one time excessive depletion has been practised, such as that of the last two centuries, so pungently ridiculed by Molière and Le Sage. At another time the natural reaction has led to an opposite extreme, as during the period of the Brunonian triumph and the still more recent attempt to banish venesection in favor of alcohol. Hippocrates, as well as his predecessors and disciples, bled, but he also fed, and much disapproved of indiscriminate starvation. He practised phlebotomy as well as cupping and scarification; indeed, he mentions one case in which he bled first in one arm, then in the other, till the patient, he says, had scarcely any blood left in his body, but he was perfectly cured. Erasistratus, according to Galen, interdicted venesection, but others writers say he did not absolutely forbid it, but used it less than his contemporaries, and distrusted it, relying rather on diet, regimen, and topical applications, in which he was directly opposed to his contemporary Herophilus. Asclepiades was not particularly averse to bleeding, although he rather trusted diet and regimen, and cold water both externally and internally. He is believed to have invented the shower-bath. He prescribed with great freedom wine, and hence was very popular with the Romans; and he said that pleurisy bore venesection on the Hellespont, but not at Rome. His pupil, Themison, was the founder of the Methodists, and, like his model, first weakened his patients by starvation, and then gave them wine as freely; for which practice he incurred the keen satire of Juvenal.

"Quot Themison ægros autumno occiderat uno" (Sat. 10, v. 221).

"How many sick in one short autumn fell,
Let Themison, their ruthless slayer, tell."

Celsus employed bleeding more generally, but in less quantities than Hippocrates, and expressing a fear lest copious venesection should exhaust his patient, recommended small bleedings repeated more frequently. He also used cupping, and still more scarifying; at the same time he prescribed abstinence at the commencement of the disease, but later on he cautiously returned to a generous diet. Galen raised a reaction in favor of the doctrines of Hippocrates. When he appeared, about the year 130, there was a great rivalry between the Dogmatists, Empyrics, and Methodists, the last of which were in the ascendency. It has been supposed that Galen began as an eclectic, as he professed to follow no one authority, but he very soon became the greatest exponent of Hippocrates, whose authority he placed higher than any. He revived the Hippocratic system almost in its original state, but he still dallied a little with the useless notions of the Dogmatists, who foolishly repudiated that patient observation on which the sage of Cos founded all his doctrines. The authority of Galen overshadowed, if it did not amalgamate, the various sects in medicine, so that the unity of the profession may almost be dated from his time. After him came the Dark Ages, during which little advance was made, but from the glimpses we get of the few men of genius who lived in those long centuries we find the same practice of bleeding prevalent. Galen seems to have been the first who performed arteriotomy in the manner which is now practised. This he did on himself, by opening the artery between his own finger and thumb when suffering severe pain in the region of the diaphragm, which was relieved after the loss of about a pint of blood. Before his time the operation was performed in a different manner. Oribasius greatly favored scarification by incisions, a much more formidable operation than that to which we now apply the word, a scalpel being used to make deep incisions after ligature of the limb and beating it with reeds to make it swell. He practised venesection as well. He was a man of considerable genius, and greatly distinguished for his therapeutical acquirements. Aetius, the great surgeon and advocate of the actual cautery, followed the same practice; and Alexander, who flourished about the middle of the sixth century, and who possessed more originality than either of the last named, bled in fevers, in coughs, in pleurisy, in mumps, and says that it is the chief remedy in quinsy, but should not be carried to syncope. He is the earliest writer who mentions a stone being coughed up from the lungs after causing the symptoms of consumption. He also recommended venesection in the paroxysms of stone. Paulus of Ægina bled from the jugular vein in diseases of the eyes, a practice not quite new with him, as Alexander had previously opened this vein for quinsy. Paulus describes the various modes of performing arteriotomy in his day, just as he does other more formidable surgical operations, his writings treating very fully of all the operations known at the time, among which, as connected with our subject, we may mention tracheotomy, which he recommends under various circumstances, and gives directions for its performance.

Rhazes, one of the most illustrious of Arabian writers, at the close of the ninth and beginning of the tenth century, practised bleeding quite as freely. In one case we find him opening the basilic and saphena veins at once, taking from each half a pound, and in three hours' time loosening the bandage and removing a similar quantity; and as soon as the patient had taken some nourishment, opening the saphena the third time, and after another interval of three hours removing a fifth half-pound. Avicenna, born 980, died 1036, followed Galen's practice. Albucasis describes various modes of bleeding and the numerous instruments employed; he carried venesection to its utmost extent at the commencement of small-pox, and to him belongs the credit of having first employed a cold regimen in that disease. We must pass over the other Arabians, whose works are mostly transcripts of the Greek writers, although they contain many interesting additions, especially in reference to pharmacy and chemistry, to which they devoted so much attention. From them may be said to date the origin of pharmacopœias, and they seem to have discovered some of our most useful purgatives.

The school of Salernum rose from the darkness which surrounded the Christian world at a time when clerical imposture and ignorance was only matched by the prevalent superstitions. This school, which originated in a Benedictine monastery, studied and translated both Greek and Arabian writings, and chiefly followed the teachings of Galen.

Passing on to the sixteenth century we see how prevalent bleeding must have been from the remarkable controversy which ensued when Brissot [1] proposed to restore the method of bleeding on the side of the disease in pleurisy, as it had been practised by Hippocrates and Galen, but which seems to have been reversed by the Arabians. Denys, then physician to the King of Portugal, opposed the innovator, and the dispute among physicians became so keen that it was brought before the University of Salamanca, and Brissot forbidden to practise until the question was decided. After a long hearing, the University pronounced that Brissot was right, and in strict accordance with Hippocrates and Galen; but the partisans of Denys appealed to the Emperor Charles V., denouncing Brissot's method as impious, heretical, and even Lutheran! Botallus (Leonard) flourished only half a century later, but during that time venesection had been becoming less employed, and indeed seems to have fallen nearly into disuse, but he produced quite a revival in its favor by his work on the subject.[2] He relates one case in which he bled the patient sixteen times, and the revival he established went on growing until the practice was carried to that extreme so ably ridiculed in the Dr. Sangrado of Le Sage. Sebastian Badus, or Baldus, of Genoa, who has the merit of being one of the earliest advocates of the use of

---

[1] Peter Brissot, born 1478, graduated 1514 at Paris, and died 1522. His book was only printed three years after his death; it was reprinted in 1622.

[2] De curatione per sanguinis missionem, de incidentæ venæ, cutis scarifiandæ, et hirudinúm affingen darum modo" (8vo, Antwerp, 1583).

cinchona, was also equally in favor of bleeding, and published a book on the necessity of bleeding in the exanthemata, in 1663.

But time would fail to tell of the many worthies, as Fabricius, Silvius, Cæsalpinus, and the illustrious Harvey, and we pass on to Sydenham, the father of English medicine, who required all his perseverance and genius to re-establish the example of Hippocrates. He bled in most acute diseases—in pleurisy, bronchitis, quinsy, small-pox, erysipelas, etc.—advising a full bleeding once in preference to repeated small ones. Pitcairn, a scholar and poet as well as a physician, continued the practice. Rutherford opposed bleeding in bronchitis and some other diseases, but he bled in pleurisy freely, but warns against excess.

Hunter's doctrines seem naturally to lead to free bleeding, and accordingly we find him employing it. Cullen succeeded to the Chair of Medicine in Edinburgh in 1766, published his "First Lines" in 1784, and from that time to his death held the very highest position and exercised the greatest influence. He employed large bleeding as a portion of the antiphlogistic system, which he prescribed in almost all inflammations and fevers, with a view to take off the "phlogistic diathesis" and to diminish the activity of the "sanguiferous system." He bled largely in pleurisy, as much as the patient could bear; nor did he hold his hand in erysipelas. Even in his lifetime some opposition was raised, for Dr. John Brown, who had once been an assistant of his, set up a directly opposite system of stimulation, especially in all asthenic diseases. Unfortunately he prescribed stimulants for himself, and indulged in them so freely that after a short career he died of intemperance. But his doctrine of stimulation obtained no little repute, expressing perhaps a natural reaction, and affording us an instance of history repeating itself in the rise and fall of doctrines.

Pearson, B. Bell, and a host of others followed Cullen. Armstrong regrets that the abstraction of 160 ounces of blood in six hours did not arrest a laryngitis, and he was in favor generally of bleeding to faintness. Cooper and Abernethy were imbued with similar ideas—the latter, who professed to be chary of taking blood, said, "You must either let the inflammation kill, or run the risk of killing the patient yourself." His pupil and biographer, George Macilwain, opposed [1] the antiphlogistic system. In 1836, Robert Williams, in his "Elements of Medicine," rejected bleeding in erysipelas and gave wine freely. In the next year Majendie opposed depletion in rheumatism; still, in 1843, W. P. Alison pronounced no proposition more certain "than that which regards the power of large and repeated blood-letting to arrest the progress of inflammation in its early stage." Nevertheless he complains that some teachers seem doubtful of its utility.

---

[1] Medicine and Surgery One Inductive Science. London, 1838.

Marshall Hall, who had long before begun to write upon the subject, attached still greater importance to venesection, advised thirty to thirty-five ounces of blood to be taken in pleurisy or pneumonia, and fifteen ounces in bronchitis. He also endeavored to set up bleeding as a means of diagnosis. He lived to see his views on this subject to a great extent abandoned, and to establish his fame by other more valuable investigations. Gooch, Travers, Kellie, Abercrombie, Copland, and even Hall himself pointed out conditions in which loss of blood was injurious. In 1837 Skoda's views and practice were brought to Edinburgh by G. W. Balfour. Pneumonia he considered naturally tended to recovery, and therefore did not require the customary large bleedings. From this time the practice gradually abated, and Todd, who in 1843 had said we cannot cut short rheumatism "as we can arrest an attack of pneumonia or of pleurisy," eventually laid aside the lancet and became the exponent of stimulating treatment in London, while Hughes Bennett became the leader of supporting treatment in Edinburgh.

We have now arrived at the period when the practice of bleeding last waned; for the present generation has not revived it, and venesection is now rarely practised. When I entered the profession as a student, about 1852, the Brunonian doctrine, as revived by Todd, was quite in the ascendant, and in London I saw nothing of venesection; but it survived in the provinces and I often saw it practised there, and even practised it myself. Having been taught in London that the practice was in abeyance, it was surprising to see how some practitioners held to its necessity in the country; and watching the cases under their care with no little scepticism, I had to confess how often the remedy seemed beneficial. They mostly admitted that in cities it was not appropriate, but who can forget how largely it had been practised in the most densely populated districts? Todd's practice continued to be widely followed until his death, his pupils spreading it in all directions, and he prescribed alcohol in increasing quantities. I have known a teaspoonful of brandy given to a little child, by his advice, every five or ten minutes for days in succession, and the considerable quantity prescribed for adults, both by himself and his disciples, is matter of history. Not only in so-called asthenic diseases, but in fever and inflammation, and especially in pneumonia, was stimulation carried to as great an excess as depletion had previously been. In erysipelas and serious surgical cases I have seen a bottle of brandy, besides twice as much wine, given in a few hours. And yet this was not many years after blood-letting had been practised with as great excess. Marshall Hall was then living, and yet a man who had broken his ribs had been almost literally bled to death in his own house, and he himself mentions a case in St. Bartholomew's Hospital, in which a man with broken ribs was bled twice on the first day, eighteen and twenty ounces respectively; on the second day he was twice bled to eighteen ounces; on the third day, as the pulse was jerky, the dresser only took a few ounces; by direction of Lloyd and Lawrence another twenty ounces was taken, after which "the pulse became a mere flutter" and he died. A similar instance has been related to me by the dresser of the case at the London Hospital; this also was a fractured rib. On admission he was bled according to the usual practice at that time; the next day he was seen by the surgeon, Sir William Blizard, who ordered him to be bled again; venesection was repeated before the next visit, then Blizard's attention was directed to his failing pulse, upon which he ordered a further bleeding; this was carried out, and shortly after the patient died, to the horror of the young dresser, who is still living.

When depletion was carried to such an extent in simple accidents, and in all cases of inflammation, it is not to be wondered at that it was employed as a preventive. People got bled as thoughtlessly as they now take purgative pills. It was a common practice to be bled every spring and autumn. Ladies were bled to improve their complexion, and the lancet was the almost universal remedy and preventive. A vein was opened to arrest hemorrhage. Women were bled in childbed. In every country surgery young apprentices wielded the lancet for the most trivial ailments. And yet of the thousands who for a long series of years were submitted to this practice, we have no evidence that any of them suffered from the customary depletion—which may be taken to show that, in health, at least, a moderate loss of blood is easily recovered from; and it is a common observation also, that in a good constitution the effects of hemorrhage from an accident rapidly pass away.

It should be pointed out that throughout the period when the abuse of bleeding was practised, there were always men, though in a minority, who refused to follow the fashion. At the London Hospital, one of Blizard's colleagues, Mr. T. G. Andrews, though he often bled his patients, always fed them, and very often prescribed stimulants instead of depletion, so that it became a proverb among the pupils that when Blizard said "bleed him," Andrews would probably say "feed him up," and so many severe accidents did well under the treatment that the pupils came to consider it the more successful method. Another curious divergence of practice at the same hospital, and at the same period, may here be mentioned: the late Dr. Billing bled fever patients when complicated with pulmonary congestion, but always in moderation; the elder Davies salivated them; their colleague, Dr. Frampton, did neither, but employed purgatives. My informant, who was clerk to each of these physicians in turn, tells me that the different methods were watched with no little interest, and that the mercurial treatment was perhaps the most popular of all. Dr. Billing told me that as the pupil, and afterward the colleague of Dr. Frampton, he had never known him prescribe venesection. Dr. Billing himself would bleed patients at the beginning of acute inflammatory diseases, but at a later stage he employed stimulants, and sometimes he would use them from the beginning, and that long before the Brunonian revival in London—in fact, as early as 1831. This astute physician lived to lay aside the lancet altogether, and though he never sanctioned either extreme, was by no means averse to the use of stimulants. Dr. Little, Billing's pupil, colleague, and successor, followed his practice, feeding some of his patients from the beginning, but he often prescribed depletion in the early part of his career. Before he resigned his physiciancy to the hospital venesection had become so rare an operation that, more than once when he prescribed it, neither house-surgeons nor dressers had seen it performed, and he had to show them how to do it. There are other physicians living with a similar experience.

If the abuse of venesection was not quite universal, the change in practice is complete. How can this be accounted for ? Some have thought that the epidemics of influenza in 1833 and 1837 are to be credited with a considerable share in bringing about the change, and believe that from about that period dates a change of type in disease; but perhaps the change is quite as much due to oscillation in medical opinion, for practice always follows doctrine, though sometimes *longo intervallo*. The revival of humoral pathology and increased interest in the natural history of disease greatly tended to reverse those theories which led to depletion. The change-of-type hypothesis looks like an invention to account for the reversal of practice; but Christison, Stokes, Watson, and other great clinical authorities have declared that they observed this change of type. On the other hand, as we have seen, there were always advocates of an opposite treatment, and similar reversals of practice have again and again marked the history of medicine.

Bleeding is now so seldom practised, and with such timidity, that it is almost outside our list of remedies; and yet there are not wanting physicians among us who now and then resort to it, and who believe that it is too much neglected, and that we are near a revival of the practice of our predecessors. Assuredly in some conditions venesection is able to give speedy relief, and the experience of two thousand years shows that the remedy is not more dangerous than some others we continue to employ. The most rational indication for its employment is oppression of the right heart with distention of the vascular system. The abstraction of a few ounces of blood is sufficient to relieve this condition. In croup and laryngitis, though it has been largely used, depletion has always proved ineffectual. The organ affected is small and within reach of other remedies. In acute pleurisy the severe pain and dyspnœa is certainly relieved by bleeding. In pneumonia coming on suddenly in robust patients, with hard and wiry pulse and intense oppression of breathing, venesection gives great relief, the pain abates, the fever falls, the pulse becomes softer and fuller, and the respiration comparatively easy. It will be seen that it is the embarrassment of the circulation which is the indication relied upon, in this as well as other cases. Venesection at once lowers the undue blood-pressure, and so ought to cut short all the phenomena depending upon that—and it is therefore in croupous rather than catarrhal pneumonia, that it ought to be of use. In complicated cases caution has always been enjoined. In pleuro-pneumonia the more moderate of our predecessors preferred leeches, but when the respiration was much oppressed they would open a vein, and, as we have said before, wherever the right heart is suffering from the vascular tension the abstraction of a few ounces of blood will relieve it.

Diseases of mucous membranes were always reputed to be least amenable to antiphlogistics, and bronchitis may therefore be said to be little influenced by depletion. In some instances, however, when the patient has appeared to be dying—his face blue and swollen, with purple lips, suffused eyes, gasping breath, and flickering pulse—venesection affords prompt relief, and that because it removes the distention of the right heart, and thereby enables it to carry on its work. Such a condition was graphically described by Dr. Hare last year in his presidential address, since published, on "Good Remedies out of Fashion," in which he said: "The fact is, that here the danger lay in the right side of the heart being gorged with blood, so that it was impossible for its stretched and distended walls to contract and to propel forward the thick and blackened blood. Oh, as you value your patient's life, as you value the blessed consciousness of being a minister who has done everything possible for his welfare, let me beg of you not to be contented with the futile treatment of to-day; relieve that poor, oppressed, distended heart, and all may be well!"

Lastly, in neuroses our predecessors bled cautiously in spasmodic asthma when the oppression was excessive, which may be taken to mean when the engorgement of the venous system oppressed the right heart. Thus all through the diseases of the respiratory system it is the right heart which is relieved by bleeding. It is true that agonizing pain and dyspnoea are also relieved, sometimes as if by magic, especially in pleurisy and pneumonia, and it was held that these diseases are arrested, or at any rate that their violence is abated or their duration shortened by timely venesection. But at the present day the one condition which seems likely to lead to the operation is when the right heart appears overwhelmed. The instant relief to this state is most grateful to the patient, often leads him to beg for a repetition, and to some extent accounts for the excesses that have been practised. In such cases even so strenuous an advocate of feeding as Dr. King Chambers would not hesitate to open a vein, though he would probably prefer a number of leeches, at the same time ordering support. Such practice, though some may think it paradoxical, is in reality rational. We have seen it followed by Billing, Little, and other conscript fathers of the London Hospital, and may now and then find it resorted to even to-day. How, then, are we to account for the prevailing disuse of so powerful a remedy? It seems to me that this is to a considerable extent because we now have other remedies by which we can often attain the same end; but few, if any, are so speedy.

Local blood-letting by leeches, cupping-glasses, or scarification is much less alarming, but the reaction against depletion has well-nigh banished these useful methods. Again, dry cupping is a most powerful remedy, but now much neglected. Here there is no fear of loss of the vital fluid, but it is only rarely resorted to, showing how much our ideas have changed. Again, instead of removing blood, surgeons sometimes endeavor to cut off its supply by tying an artery or by compression.

COUNTER-IRRITATION.

In dry cupping blood is not abstracted from the body, and it may therefore be classed with derivatives. Here we attempt to draw blood from the suffering part to the exterior, or wherever it may be less injurious. The foot-bath is a simple instance of a mild remedy of this kind. Much more powerful is a general warm bath or a vapor bath. These last, inasmuch as they attract blood to the whole surface of the body, should be the most powerful derivatives; but we must bear in mind that this only represents a part of their action, and therefore their other effects must be taken into consideration before employing them, effects which extend throughout the whole system. Often we prefer to apply our derivatives, revulsants, or counter-irritants to a small surface; for this purpose we may employ rubefacients (*rubere*, to be red, *facere*, to make), or those which cause redness of the part with heat and slight pain. Sinapisms, turpentine stupes, ammonia, and other familiar means may be employed for this purpose, but hot water applied by either sponge or flannel is more rapid in its action, much less irritating, can be renewed at intervals extending over considerable periods, and is often most effectual; the only point is to regulate carefully the heat, so as only to produce redness and not to scald or blister. Dr. Graves treated croup in this way, and was convinced of the great value of the remedy. A little less degree of heat is more frequently applied in poultices and fomentations; these undoubtedly relieve pain, even when deep-seated, and are at the same time soothing, probably through the nervous system. These milder applications should be applied over a larger area than the disease is supposed to occupy, and their influence is certainly not merely derivative or rubefacient. The chest may be enclosed in a linseed-meal jacket in pneumonia and other respiratory diseases; the effect of such a jacket is certainly soothing, and it deserves, therefore, to be regarded rather as a sedative than an irritant; just as fomentations are assuredly anodyne, though so far as the vascular system is concerned, both draw blood to the surface.

When more active counter-irritation is required, epispastics (ἐπί, upon, and σπάω, I draw), or vesicants (vesica, a blister) may be resorted to. These produce so much irritation that the part becomes inflamed, and the epidermis is raised so as to form a blister by the serum which escapes from the vessels. More pain is caused by these applications, and instead of the slight excitement produced by the rubefacients, there is depression, and if the area operated on be large, this is evinced by slower circulation and respiration, with a fall of temperature. A still greater degree of counter-irritation may be obtained from pustulants, issues, and the moxa. Here suppuration is set up, and the old humoralists taught that noxious materials were thus removed from the blood. In these last cases the effect is usually prolonged; thus it was once a common practice to keep an issue open for many weeks in certain intractable chronic diseases. For a more rapid and extensive effect in acute cases a blister was frequently dressed with an irritant ointment, so as to bring about free suppuration. The tendency now is to restrict counter-irritation to the milder methods.

The effects of counter-irritation have been referred, perhaps too exclusively, to the circulation. The more active a part the more blood flows to it—*ubi stimulus illuc affluxus*—and so the effort has been made to derive the blood to a less vital part. But more than this must be involved, the inter-arterial sympathies must be under the control of the vaso-motor system, through which, no doubt, various vascular areas related to each other may be affected one by another, and the vascular connection between these may be direct or indirect. Then, again, areas may be connected with each other through the nervous system, and such connection also may be direct or indirect. How very indirect it may sometimes be we may perhaps understand if we suppose the local irritation to pass to the nervous centres controlling the irritated skin, with so much intensity as to overflow, so to say, to adjoining or neighboring centres; these discharging laterally would then influence the periphery of other vaso-motor or trophic centres. Be that as it may, the practice is ancient, and if at times carried to extremes, no one can doubt that the less severe methods are of great utility.

The more severe the local irritation, and the greater the pain produced, the more cautious should we be in instituting the treatment. Even blisters, which unless too large may be considered of medium severity are sometimes injurious. In thoracic diseases they have been almost universally employed, and yet so able an observer as Skoda declared he had never seen them do any good, though they often did harm. Many will be ready to think that this must have been because in Germany they were applied to large surfaces and dressed with irritants; for moderate-sized blisters allowed to heal at once seldom do mischief, but large ones made to suppurate and other severer methods will increase fever and cause much nervous depression. Moreover, may not too severe counter-irritation aggravate the condition it is intended to relieve? for if a mild counter-irritant can affect a part in the manner we have indicated, it would seem possible that a violent remedy of the same kind might influence the organ so much as to exaggerate the inflammation. In support of this view it may be stated that blisters are not always intended to act as derivatives; *e.g.*, chronic synovitis of the knee-joint is sometimes treated by blistering, under which it is not uncommon to observe an increase of the effusion, evidently caused by the stimulus propagated from the surface, and it is only after the subsidence of the fresh inflammation that the swelling begins to diminish. At the same time, Dr. Herbert Davies' method of treating acute rheumatism is to freely blister round and near, but not immediately over the affected joints, so that this treatment may be really derivant, and certainly it does not produce any preliminary increase of the effusion.

Rapid rubefacients and vesicants are of course the more appropriate in acute diseases; more slowly acting counter-irritants are naturally preferred in chronic cases. The time required for the production of the effect thus becomes the chief indication in the selection of the application, excepting always its severity, in regard to which it is perhaps a good rule to employ the mildest which seems likely to achieve the desired result.

EVACUANTS.

Evacuants afford another mode of depleting, by hastening the removal from the system of some constituents no longer required, or even of others for which uses might yet be found. The most common and perhaps the most direct and powerful group of evacuants is

PURGATIVES.—Hippocrates had a very high opinion of purging, believing that it attracted " peccant " humors and discharged them from the system. He used drastics of the harshest kind, which also acted as emetics. Purgation has maintained its place until our own time, not without its ups and downs, it is true, but these have been less marked than those of phlebotomy. No doubt the use of milder aperients has helped to maintain the favor in which these evacuants are held. It may be said that slight cordial aperients are scarcely to be regarded as depletive, but even moderate increase of intestinal action is to some extent denutrient, since the removal of water even, or the mere hastening of the transit of the contents of the alimentary canal may suffice to prevent the absorption of some of the nutriment, and so deprive the blood of that amount. So much quickening, however, might only act as a stimulus, and fresh supplies coming in might therefore be taken up more readily; thus indirectly nutrition would be improved. This would be more marked should any undue accumulation be thereby unloaded. In the same way the bile is swept along, and its production may thereby be stimulated, while other excrementitious substances are got rid of. No doubt the portal system is at the same time relieved, its volume of blood being temporarily reduced, and later still the heart and general vascular system are similarly affected by the increased amount of fluids removed. We notice, therefore, that the blood-pressure falls, the respiration is easier, and the cerebral circulation is no doubt depressed, so that while the head may feel relieved by moderate purgation and the person enjoy a feeling of lightness, faintness may ensue in case of excessive action, and, of course, this is more likely to occur in the aged or feeble.

Quite mild cordial aperients (*aperio*, I open) may bring about most of the results we have mentioned, and even the regulation of the diet may be made to accomplish much, but for more decided antiphlogistic effects stronger cathartics (καθαίρω, I cleanse) are required; Abernethy's favorite blue pill and black draught is one of the most effectual. The saline laxatives (*laxo*, I loosen) especially promote osmosis into the intestines, while others, such as senna, act chiefly by increasing peristalsis, hence the value of a combination of these two groups, such as is well represented in black draught. When more sudden and violent action is required, drastics (δράω, I act) are resorted to, such as colocynth, gamboge, elaterium, or croton oil; these set up catarrh of the mucous membrane, more or less intense according to the dose and other circumstances. They thus remove a good deal of solid constituents of the blood as well as water, and are so irritant as to demand much more circumspection in their use. Those which cause profuse evacuation of water are distinguished as hydragogues (ὕδωρ, water, ἄγω, to drive), but some of these are quite mild, and apparently only increase the secretion of water into the canal, whence it is likely to be reabsorbed, unless combined with another purgative which acts by stimulating the peristaltic movements. These movements may indeed be increased by medicines which cannot be called aperient, but which on account of this property are valuable adjuvants to such remedies. Nux vomica is an example. We have seen that liquid poured into the bowel may be rapidly reabsorbed, and it is probable that some of our purgatives really stimulate the intestinal glands; this seems to be the case with mercury, and perhaps also with the salines. It is not improbable, that the drastics do so, but their other action overpowers this. Jalap and scammony seem to require the presence of bile to dissolve them.

It will be observed that the antiphlogistic action of purgatives is a complex one. They reduce vascular tension by withdrawing fluid from the circulation, and that in proportion to the amount removed. Then they not only take away water, but albuminous matter, and in this way impoverish the blood, thus they really deplete; further, they eliminate various effete matters and quicken tissue metamorphosis; moreover, they counter-irritate. It is not always easy to assign to these modes of action their relative share in the result, but the use of purgatives is, as we have seen, very ancient and almost universal. It is usual to commence the treatment of inflammatory diseases, even those affecting the respiratory organs, by this form of depletion, selecting the dose according to the strength of the patient and the severity of the case. Even in weakly persons it is well to unload the bowels, and this generally produces a good effect, the diminution of vascular tension tending to relieve the oppressed breathing and acting favorably on the liver and the heart.

Mercury.

Mercury has been named among the purgatives, and is still believed by many to be cholagogue (χολή, bile, ἄγω, to drive), if only indirectly; it is also regarded as antiphlogistic, and perhaps its sialagogue (σίαλον, saliva, ἄγω, to lead) properties further entitle it to a place among evacuants—it is therefore conveniently considered here.

Mercury has been used for a long period in many diseases. It was employed externally by Rhazes and Avicenna. Paracelsus is usually credited with being the first to give it internally, but probably he was not, for although he would have been reckless enough to employ anything which suggested itself to his mind, he was much more likely to appropriate what he found in use than to originate anything new. He would, however, be likely enough to use powerful medicines in full doses, and very frequently. Before his time mercury was employed by inunction in order to produce salivation, as we may learn from the writings of Almenar,[1] John de Vigo,[2] physician to Pope Julius II., Berengarius Carpensis, and Aloysius Lobera;[3] also from his contemporary, Fracastorius, the poet-physician, who in his remarkable poem, published in 1530, lauded fumigations with cinnabar; a little later, Rondelatius, 1583, gives directions and formulæ for such fumigations—these writers, apparently long anticipating the modern use of calomel vapor-baths. Almenar, like the Arabians, did not approve of salivating his patients, and tried to restrain this effect by means of purgatives and other treatment. Here again we observe an anticipation of the modern conclusion that salivation is to be avoided; he, however, recommended fumigations as most admirable in inveterate disease. We find Fallopius, in 1565, using mercury internally, in the form of his *pilula ex præcipitato*, and before then,

---

[1] Libellus de morbo gallico. 1512.   [2] Luisinus a aphrodisiacus, etc. 1566.
[3] Vide Luisinus; also Turner's Summary of the same.

1563, Bayrus tells us of a pill called *pilula contra morbum gallicum*, which was in great repute in his time, and of which the principal ingredient was a mercurial.

Passing over more than a century, we find the use of mercury gradually extending itself, until in 1700, Camerarius[1] and Caspar[2] are writing of it as a panacea, and Hoffmann[3] is recommending it in many obstinate diseases. Forty years afterward, Stenzel[4] and Havighorst[5] directed attention to the value of calomel as a resolvent. In 1769, Cavallieri[6] was advocating its use in rheumatic paralysis. By this time, and indeed earlier, the administration of mercury seems to have been common in various fevers, and soon after Lysons[7] published an essay on the use of calomel and camphor in continued fevers, which he followed up with further observations six years after his first publication.

This use in fever continued to grow, and we find Dr. Davis[8] chiefly relied on mercury and purgatives in the terrible Walcheren fever, and the value of mercury continued to be accepted until quite our own times. Thus we find Dr. Copland[9] remarking that death after salivation has been established is very rare; and Pereira[10] tells us his experience was the same, but Graves[11] pronounced the use of mercury to be injudicious in fever unless inflammation of some organ were set up. In 1778 Fowler[12] recommended mercury in small-pox, and two years later Langguth[13] in dropsy. In 1784 Dr. Houlston[14] urged the value of inunctions in dysentery, and Nevison[15] about this time in intestinal obstructions.

During this time many extravagances were committed in the em-

---

[1] Camerarius, R. J.: De panacea mercuriali. 1700.

[2] Caspar, J.: Dissertatio de panacea mercuriali. 1700.

[3] Hoffmann, F.: De mercurio et medicamentis mercurialibus solutis ad expugnandos sine salivatione morbos corporis humani rebelles. 1700.

[4] Stenzel, C. G.: De mercurio dulci præstantissimo pituitæ resolvendæ et evacuandæ remedio. 1742.

[5] Havighorst, J.: De singulari mercurii dulcis usu in desperatis quibusdam morbis. 1745.

[6] Cavallieri, G.: Storia d'una rheumatica paralisia curata con unzione mercuriale. 1769.

[7] Lysons, D.: Essay on the Effects of Camphire and Calomel in Continued Fevers. 1771. Further Observations, etc. 1777.

[8] Davis, J. B.: Account of the Fever of Walcheren. 1810.

[9] Copland, J.: Dictionary of Practical Medicine. 1858.

[10] Pereira, J.: Materia Medica, vol. i. 1849.

[11] Graves, R. J.: Clinical Lectures on the Practice of Medicine. 1848.

[12] Fowler, T.: De methodo medendi Variolam præcipue auxilio mercurii. 1778.

[13] Langguth: De mercurio dulci potentissimo Hydropis domitore. 1780.

[14] Houlston, T.: Observations on Poisons and on the Use of Mercury in the Cure of Obstinate Dysenteries. 1784.

[15] Nevison, A. S.: On the Use of Crude Mercury in Obstructions of the Bowels. 1786.

ployment of mercury, but many protests were also entered against its abuse, which, however, was destined to reach larger dimensions. The train had now been duly laid, and a small but very important contribution appeared in 1785, by Dr. Robert Hamilton,[1] which was destined to exercise very wide influence. He introduced the use of calomel and opium, which continued to be a favorite combination for nearly three quarters of a century; and, indeed, I have seen it employed by many practitioners even later than that. It is true that calomel had been employed in inflammations much earlier than this, and was in use even in America prior to that date, as shown by Dr. Beck;[2] but the combination with opium was certainly an important advance, and this is generally credited to Dr. Hamilton, who certainly succeeded in impressing its value on British physicians to such an extent that it became quite routine practice to give it in every inflammatory condition, and mercury was perhaps as much abused as blood-letting. Not that Dr. Hamilton alone is to be held responsible for this, for he was soon afterward supported by numerous writers, among whom we may mention Goy,[3] Rambach,[4] Maclean and Yates,[5] and, at the close of the century, Abernethy,[6] whose influence it would be difficult to exaggerate. It is not necessary to continue the history through the first half of the present century, as that will be more familiar to our readers; nor need we say that from time to time most powerful opposition to mercurials was offered; indeed, that much may be said respecting the whole history, for in 1562 it was denounced by D. Leoni (Luisinus) as a poison which produced tremors, convulsions, paralysis, and a host of other nervous symptoms; and his denunciations were reiterated in various forms and with many additions by a number of writers down to our own times, when the anti-mercurial crusade has been led by Dr. Hughes Bennett and his followers.

Mercury is speedily absorbed, produces important effects on the blood, and passes rapidly into the tissues, in which it lingers. It must exercise, therefore, a most important effect on nutrition. In the blood it destroys the red corpuscles, but in small doses this effect is not necessarily produced. Syphilitic patients improve under its influence, and if anæmic, the red globules increase instead of diminishing in number; but if the remedy should be continued, its usual effect will be manifest. We infer from this that the benefit must arise from the antagonistic influence upon the disease. It is only when given in quite small quan-

---

[1] Hamilton, Robert: Letter to Dr. Duncan, giving an Account of a Successful Method of Treating Inflammatory Diseases by Mercury and Opium. 1785.
[2] Beck, J. B.: Essays on Infant Therapeutics. 1849.
[3] Goy, J.: De virtute mercurii inflammationes resolvente. 1794.
[4] Rambach, J. J.: Usus mercurii in morbis inflammatoriis. 1794.
[5] Maclean, C., and Yates, W.: The Science of Life, etc. 1797.
[6] Abernethy, John: Surgical and Physiological Essays. 1797.

tities that the blood is improved, while larger doses impoverish it. This has been shown not only by analysis, but by the investigations of Wilbouchewitz (*Archives de Physiologie*, 1874), following the method of Malassez for counting the corpuscles. Dr. Keyes, however, has stated that in all persons, whether syphilitic or not, minute doses will temporarily increase the red globules (*Amer. Journal of the Med. Sciences*, 1876). But before then the impoverishing influence on the blood had been constantly noticed. Thus, as early as 1757 Huxham [1] had said that "a long and large use of mercury will turn the whole mass of blood into a mere watery colluvies." And modern writers have employed similar expressions: *e.g.*, Dr. Farre [2] speaks of mercury as "positively antiphlegmonous," and he also relates the following incident: "A full, plethoric woman, of a purple-red complexion, consulted me for hemorrhage from the stomach, depending on engorgement, without organic disease. I gave her mercury, and in six weeks blanched her as white as a lily." Pereira [3] classed mercury among his spanæmics as a resolvent or liquefacient. Headland, Gubler, Trousseau, and many others confirm the disintegrating and destructive effect of mercury on the corpuscles. On the general nutrition, although many observations show that it seems to exercise in minute quantities a favorable influence, it is only so in a state of disease, and may therefore be well called an alterative, while even in syphilis large doses not only impoverish the blood but impair nutrition, increase tissue waste, and bring about emaciation.

Mercurials certainly promote secretion and excretion, as they stimulate the glandular system and quicken absorption. They are generally held to affect, first of all, newly formed or ill-organized tissues. The precise *modus operandi* is not known, but we may, perhaps, conjecture that in some way they check the growth of young cells. On the whole they are certainly denutrients, though when they become curative of course their indirect influence is to restore nutrition.

Mercury is eliminated to some extent by all the excretions, but chiefly by the kidneys; the rate at which it is excreted has been carefully studied by Mayençon and Bergeret (Robin's *Journal de l'Anatomie*, 1873, and subsequently *Lyon Médical*), who found that after one centigramme of perchloride given to a dog subcutaneously, the urine contained mercury for twenty-four hours. When the dose was given daily for ten or twelve days, the metal was detected for four or five days after the dose was discontinued. In another series of experiments on rabbits, which were killed at different periods, they found the metal in half an hour after the dose in all the tissues, but most in the liver and

---

[1] Huxham: Essay on Fevers.
[2] Farre: Fergnson's Essays on the Most Important Diseases of Women. 1839.
[3] Pereira, J.: Materia Medica, vol. i., p. 175. 1849.

kidneys. After four days, sometimes less, a single dose could not be detected. Thus it would appear that a single dose may not remain in the system, but the same observers have proved that when repeated doses are given elimination will cease before it is all excreted. On one occasion forty-eight hours after a course of mercury had been discontinued it could no longer be detected in the urine; iodide of potassium was then given, upon which a large quantity of mercury was excreted, and the elimination continued for seventy-two hours, diminishing gradually.

As an antiphlogistic the use of mercury was for more than half a century almost universal. It is now only seldom used for that purpose, and when it is employed, with a view of counteracting the effect of the syphilitic poison, its impoverishing effect on the blood is so fully recognized that in these days nutritious diet is generally insisted on as at the same time necessary. The evidence of its antiphlogistic value is almost necessarily only clinical, and naturally it is open to serious doubt. Nevertheless, this evidence is so extensive that it is difficult to deny that the conclusion so long universally accepted and in accordance with so many daily recurring observations has some foundation in facts. It was long held that mercurials were most serviceable in inflammations of serous membranes; it was believed to check the tendency of fibrous exudations, and even to promote their reabsorption. Hence it was but natural to rely upon it in pleurisy and other allied inflammations, and yet this method of treatment has now become almost obsolete. Iritis is somewhat analogous to serous inflammations.

Pneumonia is a good example of the parenchymatous inflammations in which mercury was long held to be absolutely necessary, and this disease was for some time made the battle-ground between those who advocated and those who condemned this treatment, just as we have seen it was in reference to blood-letting, and, we may add, the antiphlogistic system altogether. In inflammation of mucous membranes mercury was held to be less desirable, and perhaps in the respiratory tract it never obtained the same credit as elsewhere; at any rate, in this locality it went out of fashion earlier, though in acute cases it was often resorted to. As efficient purgatives some of the preparations, especially in combination with other cathartics, are, as we have seen, frequently most useful.

With regard to special respiratory diseases, some physicians continue to employ mercurial preparations to some extent, e.g., Dr. Phillips says he has found small doses of gray powder cure coryza more quickly than any other remedy, "especially when there is much sneezing," and he says that "catarrh affecting the Eustachian tube is also well treated in the same manner" ("Materia Medica," "Inorganic," p. 662). It appears to me that in the latter case there is nowadays no necessity to resort to such treatment, and in the former few people will be inclined to adopt it.

In tonsillitis and in parotitis mild mercurials are still sometimes recommended; but by no means inconsiderable experience in these affections leads me to reject them as unnecessary, and the same observation may be ventured respecting various other diseases of the throat. Many practitioners still regard rapid mercurialization as highly desirable in acute laryngitis—they are under the impression that it will prevent the development of œdema of the glottis; but considering the great advances that have been made in our ability to control laryngeal diseases and to cope with their dangers, it is but natural to discard a remedy the value of which is at the best doubtful, and confidence in which is too likely to lead to the omission of more reliable measures.

In acute bronchitis where there is much congestion and but little expectoration, with severe cough at night, pyrexia, and dyspnœa, many authorities still give mild mercurials; but such treatment should certainly be reserved for robust patients in the prime of life, and even then should occupy only a subordinate position. Dr. Thorowgood has used blue pill with squill in such cases (*Practitioner*, 1878). Minute doses of antimony would seem more appropriate to such a condition, and if supplemented by warm, soothing inhalations are generally successful; the squill is better reserved for a later period, and then possibly mercury will be considered unnecessary.

In broncho-pneumonia ointment or the oleate is sometimes applied to the chest, or when a blister has been considered necessary it may be dressed with mercurial ointment.

In diphtheria there are not wanting able observers who still resort to mercury. Bretonneau used it freely, but with very little success, Trousseau employed calomel locally by insufflation, West believed it counteracted the tendency to form false membrane, and Stillé considers it urgent to bring the system under its influence as quickly as possible. On the other hand, so little benefit has followed the use of mercury that the majority seem for some years to have rejected it. With other methods of more decidedly and even more rapidly influencing the mucous membrane, and with the doubt that hangs about the power of mercury to do this, together with the probability that any excess would rather encourage than restrain the spread of the exudation, I have generally discountenanced its use and relied upon measures which seem to me more direct, and therefore more likely to be of service. Still there is such a mass of experience recorded in favor of mercury by English, American, French, and German authors that many may hesitate to omit its use. In such case let them be careful to proceed cautiously and to watch the effect, and to employ it rather with a view to its action upon the mucous membrane than for any hope of an undefined influence over the general disease. Dr. Jacobi confesses that he is less sceptical as to the action of mercurials than he was a quarter of a century ago ("Treatise on Diphtheria," 1880), and he recommends ten or twelve minims of the oleate to be rubbed along the inside of the arms or elsewhere when the surface becomes irritated, every hour or two, or else hypodermic injections of perchloride, which act very promptly. More recently, at the New York Academy of Medicine (*Medical Record*, May 24, 1884), Dr. Jacobi advised the use of the sublimate in pseudo-membranous affections of the respiratory organs, to be given in frequent doses, so as to bring the system under its influence speedily. He said that infants of tender age could bear half a grain a day for many days in succession, and he thought that the exudation might by this method be prevented from extending to the larynx. Salivation and gastro-intestinal disturbances are not frequent, and if the sublimate should not be well tolerated he would use inunction. As a rule, the administration of mercury is less objectionable the younger the patient. Doses of one-sixtieth to one-twenty-fifth of a grain may be given every hour, and continued from one to five or six days.

Others, however, who have tried this remedy are content with giving it in much less doses, *e.g.*, $\frac{1}{150}$ grain dissolved in water, every hour, continuing it both night and day, unless the patient sleeps. Others, again, use a spray every half hour or hour, some of which, of course, passes into the system. A quarter or one-eighth of a grain to the ounce of water (with a drachm of glycerine) will be strong enough for this purpose, and give a dose of $\frac{1}{100}$ grain in half a drachm or drachm respectively, of which a part will be lost.

Baerensprung's albuminate or Bamberger's peptonate may, perhaps, be substituted for the perchloride with advantage, for they are even less liable to induce stomatitis; they may be used hypodermically, or if taken by the stomach scarcely ever disturb it. At the same time the sublimate is an exceedingly powerful antiseptic. Bacteria are killed by it in dilutions of one in twenty thousand, according to Buchholtz, and some say one in three hundred thousand; at any rate, it is much more powerful than the other antiseptics which have been recommended, and may often be given more freely than is generally supposed. My experience corroborates those who have stated that the perchloride may be given in larger doses, or rather that the doses may be more frequently repeated without danger in various diseases. It is probable that this salt, or at any rate the albuminate or peptonate, passes into the blood unchanged and is not deleterious to the corpuscles. An infant under one year can take one-fiftieth of a grain well diluted every hour without obviously injuring the corpuscles, but no doubt smaller doses will usually be sufficient, $\frac{1}{100}$ or even $\frac{1}{150}$ grain, and it is perhaps better to proceed cautiously except in very urgent cases.

In Pneumonia.—The old treatment by mercury and antimony in the early state is now obsolete, but it is sometimes resorted to at a later period, when there is secondary fever and purulent degeneration is apprehended, or when removal of consolidation is delayed. In pleuro-pneumonia or in chronic interstitial pneumonia mild preparations are still sometimes resorted to, and strumous or other deposits are said to disappear under its influence. The question arises whether some of these deposits may have been syphilitic. The perchloride, albuminate, or peptonade, or else the iodide, are the best preparations in such cases, but I certainly prefer even to these small quantities of iodide of potassium or sodium. A blister is often effectual in these cases, and may be dressed with mercurial ointment.

In Pleurisy.—Mercury is still employed and has the sanction of Dr. Walshe. When there is considerable effusion modern practice seems rather to trust to the aspirator, but when it is very slight the stimulating effect of mercury on the absorbents is sometimes relied upon; even here, however, the simplest measures are perhaps equally effective, and certainly in the early stages there is no call for mercury; the fever can be restrained by aconite or salines, and pain, if severe, allayed by morphia, while rest, mild counter-irritants, and appropriate diet suffice to cope with slight cases, and severe ones cannot be arrested by the gentle mercurial treatment, which only is recommended.

EMETICS.—These very ancient remedies have not yet gone to the limbo of oblivion, as it is sometimes necessary to empty the stomach of poison or food, but otherwise they are not very frequently used; they may be divided into *the direct*, which act upon the stomach, and *the indirect*, which act through the system. Vomiting may occur in consequence of a mental impression, or from a disordered condition of the blood impressing the nervous system, producing central emesis. On the other hand, peripheral irritation in the stomach or some other organ may give rise to excentric or reflex vomiting. Some emetics act in both ways, *e.g.*, antimony, ipecacuanha, and apomorphia, though some doubt has been cast on the views generally held. Thus the purging which follows the use of antimony is certainly connected with the elimination of the poison through the intestinal membrane; and it has been asserted that the same thing occurs in the gastric membrane when the antimony is injected into the blood. So, too, Dr. D'Ornellas,[1] having injected emetin into the veins of animals, found the alkaloid eliminated by the gastric mucous membrane. Antimony depresses the circulation intensely, ipecacuanha not much, while ammonia is a stimulant. We do not employ emetics in respiratory diseases for the sake of depleting, but rather for the effects of the vomiting, and the acts associated with it. These complex, co-ordinated acts are said to be governed by a centre in the

---

[1] D'Ornellas: Du Vomissement; contribution à l'étude de l'action des vomitifs. 1873.

medulla, so that emesis may be produced by any powerful stimulant of this centre, but the respiratory centre must also be engaged, to which may be attributed the expiratory movements, such as sneezing and coughing, which so frequently announce the approaching vomiting; then there is a further, more powerful expiratory effort during the emesis, when the viscera are forcibly compressed and the glottis closed. Again at the close there is a further expiratory effort, so that the moment the glottis opens there is a blast of air outward which would prevent the entrance of particles into the larynx. So in emesis, the nasal passage is usually closed as in deglutition, though sometimes the ascending stream either forces its way or takes the muscles by surprise. There is also generally increased secretion of bronchial mucus, which is carried upward by the air-current. These movements explain the value of the act of vomiting in emptying the air-passages. The abdomen is also compressed, by which the gall-bladder is emptied, and sometimes the bile regurgitates into the stomach and is vomited; there is usually a free flow of saliva. The cardiac and vascular centres are depressed, hence we observe a feeling of faintness, and sometimes syncope occurs; the motor centres are lowered, and so there is a feeling of prostration. Profuse sweating is not uncommon, and this, too, is referred by many to a sweat-centre in the medulla.

The simple direct irritant emetics act without being absorbed, but the others require to find their way into the system. The irritants are therefore mostly more prompt, e.g., mustard is both efficient and quick, and has also the advantage of being always at hand; it is frequently successful even in torpor of the stomach, as in narcotic poisoning. Antimony and ipecacuanha act both as direct and indirect emetics and are of special value in respiratory diseases. Both produce nausea and depression, and the former great muscular relaxation. Apomorphia also acts both ways, but produces little, if any depression; it is exceedingly useful where time is of importance, as a hypodermic injection is rapidly effective; its only disadvantage is that the solution does not keep, but gelatine disks meet the difficulty.

Chouppe tells us[1] that apomorphia or tartar emetic injected into the veins of animals excited vomiting as freely after section of the pneumogastric as before, but this was not the case with emetin, from which he concluded that this last acted through the peripheral terminations of the nerve while the other two affected the centre directly. This view seems to be corroborated by Dr. Duckworth's experiments.

Passing over other indications for emetics, they are sometimes used for dislodging foreign bodies from the throat, for which purpose apomorphia hypodermically is most appropriate. In croup and diphtheria they

---

[1] Recherches Thérapeutiques et Physiologiques sur l'Ipéca. 1875.

are used to remove the exudation from the air-passages, and in this way they often greatly relieve the respiration, just as they also do in bronchitis when the passages are clogged with tenacious sputa. It has been supposed that they also restrain the exudative process, and to obtain this effect, nauseating doses are sometimes administered during the intervals between which emetics are given. It will be found, however, that such severe treatment is apt to be dangerous, and we should look only to their power to mechanically remove the exudation as an indication for their use in such cases.

In suffocative bronchitis an emetic will often produce relief when all other measures seem hopeless. Ipecacuanha is the most suitable for this purpose. A large quantity of mucus is evacuated, the intense dyspnœa gives way to quiet respiration, and the distress and anxiety subsides into quiet, refreshing sleep, the cyanosed face becoming natural. It is true that the secretion will again accumulate, but much has been gained and the danger of asphyxia is at least postponed, and sometimes quite removed; nutrients and ammonia may then be administered. In phthisis there is more danger of exhaustion. Dr. Paris ("Pharmacologia," 1843) mentions a case in which an emetic produced fatal syncope in a patient to whom it was given in the hope of dislodging the pus which embarrassed the respiration. Hippocrates recognized this danger, and said consumptives ought not to be purged by emetics ("Aphorisms," iv., 8). Emetics used to be frequently employed to induce relaxation in spasmodic affections of the respiratory system. In such cases the condition of nausea is indicated, and therefore we may use successive small doses at frequent intervals until vomiting ensues, or even short of it. For this purpose ipecac is the mildest and much the safest for children and old people. The plan of giving successive nauseating doses until vomiting takes place has often been used to cut short a cold, but many patients consider the remedy worse than the disease. It is not the act of vomiting which is here useful, but rather the free bronchial secretion which is hastened, and the lessening of arterial action which is produced. It is not emesis but nausea which is needed.

In acute laryngitis, tracheitis, and even bronchitis, this was once the favorite method of treatment, but it has been largely displaced by milder, or perhaps more direct methods, especially since the laryngoscope has enabled us to watch the local conditions.

NAUSEANTS.—Still smaller doses of the *indirect emetics, antimony* and *ipecacuanha*, may be also employed in order to set up a moderate degree of nausea, not culminating in vomiting. This condition may be excited with benefit in many respiratory diseases, especially when fever runs high. The relaxation, the depression of the whole system, more particularly of the cardiac centre, which are produced, all help, and show how our predecessors studied clinically and knew that these remedies would reduce arterial action and moderate inflammation. No wonder, then, that they were freely resorted to in the absence of some other remedies which are now available. More might now be made of them if more attention were given to dosage.

Antimony.—Tartar emetic depresses, sometimes even to a dangerous degree, but that depression is the key to its use. To obtain its contrastimulant effect, as it was called by Rasori, we do not require the massive doses he employed. Small medicinal doses repeated at intervals lower the pulse and increase the perspiration—act as diaphoretics and febrifuges; at the same time the respiratory mucous membrane is provoked to increased secretion, as is also that of the alimentary canal—but this last effect is not usually noticeable in doses too small to nauseate. As soon, however, as nausea is set up there is liability to purging also, unless some remedies to prevent this have been combined with it. At this stage the pulse falls considerably, the heart is much depressed, and fainting may occur; the antimony seeming to act directly on the substance of the heart as well as on the muscular and nervous system. As the cardiac contractions grow less frequent and less forcible they may become irregular, and in cases of slow poisoning the heart is arrested in diastole, its irritability after death being nearly or quite lost. Experimenters tell us that antimony does not alter the temperature. Dr. Ringer gave half a grain every ten minutes to a strong, healthy man for nearly seven hours, inducing great nausea and vomiting with profuse perspiration, but during the whole time his temperature remained remarkably constant, varying not more than 0.4° Fahr.; as great a deviation may constantly be observed in health. Whatever may be the case in health, a febrile temperature is easily reduced by even small doses, and this effect may often be obtained without exciting nausea if the doses and the intervals at which they are repeated be carefully regulated. After too large or too frequent doses, not only nausea but even vomiting may ensue, and this sometimes occurs against our wishes, on account of the unusual susceptibility of the patient; but generally we can obtain sufficient depression of the circulation, fall of the pulse, free perspiration, and bronchial secretion without emesis.

Antimony acts throughout the gastro-pulmonary mucous membrane. It irritates the stomach and bowels, and though no special lesion may be found on this membrane after poisoning, there are always indications of severe inflammation. The vomiting and the purging are probably connected with elimination, which is effected through the mucous membrane as well as through the kidneys. It possesses a special action on the bronchial membrane, even in moderate doses, as all clinical experience proves; this is why it is so effective in the first stage of bronchitis, but as soon as free secretion takes place its local work is accomplished, though, of course, if continued it would still be depressant and diaphoretic as well as expectorant. It must, however, be but rarely advisable to continue it after the secretion is established, as there is no doubt that it tends to weaken the respiratory movements as well as the circulation whenever it is continued too long. It is therefore desirable to obtain its good effect in the early stage of bronchial inflammation, and then to replace it by other remedies. Minute doses can be added with advantage to many febrifuges—that is to say, in sthenic cases.

In *pneumonia*, as a powerful cardiac depressant, nauseant doses were for a long time in favor. Rasori and the Italian school used much larger quantities, but the system is not without danger. In this country we are content with quantities which suffice to abate the arterial excitement and produce perspiration, and perhaps excite slight nausea. Laennec, Trousseau, and others used one-grain doses, Stokes, Watson, and Billing much less; and it may be said that the minute doses we have indicated are sufficient, and even these are at present seldom resorted to in this disease. In *bronchitis* attacking young robust adults, the remedy seems more distinctly indicated on account of its valuable effect upon the membrane. While it abates pyrexia and lowers the blood-tension, it brings about expectoration and relieves the oppression and dyspnœa, and when the cough is severe morphia or belladonna can be added to it. In the *capillary bronchitis* of children, with a pulse of 130 to 140, temperature 101° to 103° F., intense oppression of the breathing, dusky face and clammy skin, restlessness, delirium or coma, minute doses have sometimes proved remarkably efficacious, but the remedy should be withdrawn as soon as the effect is produced. One-hundredth of a grain every two or three hours will suffice for such a purpose. In *croup* nauseant doses have sometimes been employed, but are not required; if an emetic is needed to act mechanically, let it be given, but the practice of giving nauseants only debilitates the little patients, and should not be resorted to even to abate fever, which can be controlled by other remedies. In spasmodic croup nauseants are still less admissible.

Antimony is contra-indicated in old age and in infancy. In children it is likely to lead to collapse, and in elderly persons it may do this, though it often expends its energies on the stomach. Both young and old are very susceptible to its influence and likely to be too much weakened. In hernia, aneurism, or other diseases of the circulatory system, in cerebral congestion, in gastric irritation, in pregnancy, and wherever the act of vomiting is likely to be injurious, nauseants are to be avoided.

Ipecacuanha acts both on the digestive and respiratory tract, but does not produce much effect on the circulation unless nausea be produced, neither does it reduce the temperature. In poisonous quantities it affects the lungs, pulmonary lesions having been observed after death. By it we can stimulate the bronchial mucous membrane and perhaps also the skin, even in small doses, certainly in nauseant ones. The effect on the alimentary mucous membrane may be due to its being eliminated there, as D'Ornellas[1] and Pécholier[2] found emetin escaped by the stomach and bowels when it was injected into the circulation or the cellular tissue. As a nauseant it is depressant, but otherwise does not seem to affect the circulation, though Dr. Duckworth[3] produced cardiac paralysis by injecting emetin direct into the circulation. From half a grain to a grain every two, three, or four hours is sufficient for nauseating.

Some persons are peculiarly susceptible to the influence of ipecacuanha, the smallest doses producing violent disturbances of the respiratory system. A sort of spurious influenza, or bronchitis, or asthma, or hay-fever may be set up, according to the idiosyncrasy of the patient; the mucous membrane of the conjunctivæ perhaps participates in the affection. Sneezing, itching, or irritation of the nose, succeeded by a watery discharge, frontal pain, cough, and oppression of the breathing, may be produced. It is not always necessary for the drug to be swallowed to induce these attacks. Many years ago I knew a lady, the sister of a medical man, who suffered from violent asthmatic attacks whenever ipecacuanha was powdered in his surgery, although she remained in another room.

As a nauseant ipecacuanha may be used instead of antimony when it is desirable to avoid the depressing effects of the latter. As soon as the nausea comes on gentle diaphoresis ensues, and prior to this there is often increased bronchial secretion. To obtain this last result it is not needful to produce nausea, but with regard to the skin there is some difference of opinion; if, however, a little aconite be added, to act on the circulation, the diaphoretic effect will be marked. To the action on the alimentary mucous membrane we may attribute the favor accorded to this remedy in diseases of that tissue, and perhaps the frequency with which small doses are found useful adjuvants to aperient pills. The effect on the bronchial mucous membrane is the most important of all, but will be considered further on among the expectorants.

DIAPHORETICS.

As we have seen, the nauseants lower arterial action. Antimony may possibly act locally on the skin, as doubtless do some other diaphoretics, most of which have other actions besides that of promoting perspiration. No doubt nausea leads to sweating, but antimony certainly, and other nauseants probably, may be made to produce perspiration without dis-

---

[1] D'Ornellas: Op. cit.
[2] Pécholier: Récherches expérimentales sur l'action physiologique de l'Ipécacuanha. 1862.
[3] St. Bartholomew's Hospital Reports, vol. vii.

turbing the stomach; but as this group of remedies is most useful in respiratory diseases, it will be further considered with expectorants. Under the use of the nauseants the skin becomes relaxed, much as it is in collapse, or in a much milder degree, as it is in sleep, when we know there is a greater disposition to perspire, to which, perhaps, may be attributed the fact that colliquative sweats mostly occur during sleep. They may also act by reducing the circulation, by which they serve as antiphlogistics and febrifuges, for in pyrexia the secretions are mostly diminished and the skin is very dry, its function failing from excess of vascular action. In such cases whatever restores the balance liberates the perspiration, and so not only nauseants, but all febrifuges, all refrigerants, even such salines as are usually diuretics take a place among diaphoretics. These two classes continually interchange, for the skin and kidneys supplement each other, and whether a remedy stimulate the action of one or the other often depends entirely on whether the surface be kept warm or not. When a patient is in bed or keeps his room, or when the weather is warm, a medicine may act as a diaphoretic which, when he is walking about, or in cold weather, will produce diuresis.

Again, vascular fulness promotes secretion, so that drinking an extra quantity of fluid produces diaphoresis while the surface is kept warm, but in an opposite state diuresis occurs. Thus all beverages, especially when taken warm, may be grouped as diluent diaphoretics. Perhaps hot drinks also act reflexly on the sweat-centre, for we may often observe that almost immediately after taking a cup of tea or any hot liquid perspiration will break out.

External heat is a more powerful, and perhaps more direct diaphoretic, the influence being conveyed through the afferent thermic nerves. But here the vascular system is obviously also concerned, the blood being attracted to the surface and thus vascular fulness induced. The heat may be moist or dry, as in hot baths, vapor baths, the lamp bath, the Turkish bath, etc. Friction, exercise, and other influences which bring the blood to the surface also produce perspiration. On the other hand, sudden cold may do this, but that is only by the reaction which is set up. In the same way a draught of cold water instead of hot may bring on a sweat. Alcohol, by dilating the cutaneous capillaries will act as a diaphoretic, especially when taken hot, but perhaps some of this effect may be from its acting on the centre like narcotics, or rather on both centre and periphery, as well as by sending at first a more copious supply of blood to the vessels of the skin.

Another group of diaphoretics is supposed to stimulate the centre for sweat. Opium is one of these, but its diaphoretic action is by no means the most important indication for its use. It will be considered at length among the narcotics. Jaborandi and its alkaloid, pilocarpine, are supposed to stimulate not only the centre but the end-organs of the nerve and the sweat-glands. Perhaps other members of this group do the same. After entering the circulation and acting upon the centre they may be partially eliminated through the skin, and, acting as eliminants always do, stimulate the organ through which they are removed.

Diaphoretics are employed to restore secretion when arrested and to increase it when diminished, to eliminate from the blood liquid and even solid noxious matters, to moderate the circulation, to determine to the surface, to favor absorption, to arrest incipient disease, as catarrh, rheumatism, etc. Their action may be promoted by previous venesection, which, however, is nowadays scarcely ever resorted to; by the administration of diluents, except after the nauseants, when warm fluids would be likely to excite vomiting. When the temperature of the patient is high, cold diluents may be employed; otherwise cold drinks should be avoided, as they are apt to determine to the kidneys, and we know that the action is retarded by diuretics, by purgatives, and by exposure to cold. From the last statement, as well as from what has preceded, a cardinal rule may be deduced: to keep the patient warm, both by flannel clothing and by the maintenance of a suitable temperature in the room. It is especially desirable to keep the feet warm.

The amount of perspiration produced is by no means the measure of the benefit obtained. A single full free perspiration will indeed sometimes arrest a catarrh, but to make the quantity of liquid removed the chief object is a decided error. Very often the good effect is due rather to the change in the circulation than to the increased secretion.

It is usual to administer diaphoretics at bedtime, and the practice is convenient, as we thereby secure warmth and rest for some hours after the dose. But when the patient is confined to his bed this does not apply, and morning is then an appropriate time, for after the sleep the skin is in the condition of relaxation, which predisposes it to diaphoresis. This is well understood by many balneologists, who make their patients take a course of warm mineral baths in the morning and prescribe rest in bed after the bath for a time sufficient to secure a free perspiration. Powerful stimulant diaphoretics should not be resorted to in inflammatory and febrile diseases. In affections of the respiratory tract the nauseants perhaps take the first place, then the salines and refrigerants. To arrest a catarrh a single copious sweat will often suffice, and this may be secured by a hot bath or, where attainable, by the Turkish bath, but this is rather contra-indicated where there is much febrile excitement, for should there be a disposition to pulmonary congestion or œdema it would tend rather to precipitate an attack—the hot, dry air stimulating the respiratory system. Should dyspnœa or any considerable disturbance of respiration occur in the Turkish bath in a person predisposed to pulmonary attacks, it would be safer at once to quit the hot room.

NAUSEANT DIAPHORETICS.—There is no doubt about the power of antimony to promote perspiration, an effect which can be obtained without inducing nausea. With regard to ipecacuanha, this is less generally admitted, and many believe that it is a sudorific only when it excites nausea. It seems to me, however, that it tends to keep the skin soft and moist, though it does not produce a full sensible perspiration in doses insufficient to nauseate. To obtain a single copious perspiration Dover's powder is mostly effectual, but here the opium may be regarded as the principal sudorific, though the ipecacuanha doubtless assists its action. The compound is therefore one of the best sedative sudorifics, but though it may often be employed for this purpose it is to be avoided whenever narcotics are contra-indicated.

SALINE DIAPHORETICS.—The most commonly used of this group is the *ammonium acetate*, a solution of which was introduced by Boerhaave in 1732. Soon afterward one Minderer or Mindererus employed it, and it has since been known after him as Mindererus' spirit. It is only a gentle diaphoretic, producing no other sensible effect, and to secure this the surface must be kept warm; otherwise, it may act as a diuretic. Though in almost universal use, many have little faith in it. Cullen tells us he had seen four ounces of it taken, and shortly afterward the same dose repeated without producing any sensible effect. Perhaps the solution employed was not of proper strength, for at different periods various methods of preparation have been resorted to, and even when freshly made in small quantities, unless due care be taken, the results will differ. Sometimes the solution has been left with an acid reaction, sometimes the reverse, and it is obvious that the effects will be altogether different. The solution of the new United States Pharmacopœia contains about 7.6 of the acetate, which is about 1.0 stronger than the earlier editions, but very often, from carelessness, the proper strength has not been employed. A well-prepared neutral solution may be rendered acid or alkaline at will, or the solution may be given in an effervescing mixture. Citrate of ammonium has been lately used as a substitute for the acetate. Citrate of potash is also used as a basis for fever mixtures and is also most agreeably given in an effervescing form. It is, however, like other potash salts, more likely to act as a diuretic than the ammonium salt.

Spiritus ætheris nitrosi is often used in conjunction with liquor ammoniæ acetatis, and has for hundreds of years been regarded as a valuable diaphoretic, as well as diuretic. Raymond Lully mentions it in the thirteenth century, and Basil Valentine in the fifteenth describes an improved method of preparing it. Sweet spirits of nitre is now in daily use by the public, as well as by the profession, and yet some have little faith in it. Perhaps this is partly due to the uncertainty of the preparation, which is also apt to deteriorate by keeping. The British Pharmacopœia spirit is only about half the strength of the United States Pharmacopœia. It owes its activity to nitrite of ethyl, of which the British Pharmacopœia ought to contain two per cent., but many specimens contain less. There is no doubt that when inhaled the vapor may destroy life. Christison mentions that a druggist's servant was found dead in her bed one morning, after sleeping in a room in which a three-gallon jar had been accidentally broken.

Owing its activity to ethyl nitrite, we might suppose that the effects of spirits of nitre would resemble those of other nitrites, and this has lately been shown to be the case, particularly by Dr. Hay and Professor D. J. Leech (*Practitioner*, 1883). Comparing it with nitrite of amyl, it is indeed a feeble remedy, but possesses similar qualities. It unquestionably lowers arterial tension, quickens the beat of the heart, and dilates the peripheral vessels, and this is how it acts as a diaphoretic and a diuretic. In the same way the uncertainty in its action, which has often been complained of, may be explained. When the arterial tension is already low it is not a suitable remedy. We may presume that it causes diuresis by relaxing the renal vessels, but if the tension should be already low and the venous side of the circulation want relief, as is often the case in cardiac dropsy, then it would naturally fail, and curiously enough it is in cases of this kind in which its uncertainty has been most complained of. In the same way dilatation of the cutaneous vessels leads to perspiration, but that is only one factor in the production of diaphoresis. The use of the remedy is, then, to reduce tension, and for this reason, perhaps, large doses might be employed instead of more powerful agents, such as nitrite of amyl, nitro-glycerine, nitrite of soda. Like these its action is quick, but not quite immediate. Dr. Leech found no effect on the pulse at the end of a minute and a half, but in three minutes a very marked effect, and in eight minutes the tension was very low, and it was not fully restored at the end of an hour after the dose, fifty minims. He found this amount left a perceptible influence for one hundred and fifty minutes, and twenty-five minims affected the pulse for eighty minutes. These experiments coincide with our clinical experience of the value of this remedy, for as the tension falls there is a disposition to perspiration, and at the lowest point the patient sometimes breaks out into profuse sweats. We can now understand how small doses frequently repeated act favorably in febrile conditions.

There is this difference between the sweet spirit of nitre and nitrite of amyl—both diminish the tension and quicken the pulse, but this latter effect is not nearly so marked by the spirit as by amyl nitrite, and in small doses the acceleration of the pulse is often scarcely perceptible, which accounts for its not being much observed or commented upon, though very able physicians have called it a stimulant, an antispasmodic, and a carminative, while others have said that it acts on the nervous system chiefly as a diffusive stimulus, and others again look upon it as a stimulant diaphoretic, and say that it is specially useful in children instead of alcohol, while others again have noted that it sometimes increases pyrexia. All these observations are no doubt correct, and it is pleasant to observe that the clinical skill of our predecessors often enabled them to discover things which our modern methods of research confirm.

Various processes of making the sweet spirits of nitre have been employed and the results have unquestionably differed, to which fact may be attributed much of the distrust with which the preparation has sometimes been regarded. Moreover, as a popular remedy it has often been adulterated. Liquor ammoniæ acetatis, with which it is often prescribed, has also been an uncertain preparation. In each case it is important to have a definite compound. These two febrifuges act admirably in combination, and form perhaps the most popular of all those in use. Sometimes it is desirable to add a little ammonia in order to obtain a stimulant diaphoretic. In such case, however, a combination of the spirits of nitre with spir. ammon. aromat. is an elegant and effectual form. The stimulus of the ammonia keeps up the cardiac action while the ethyl nitrite diminishes the arterial tension. On the other hand, when it is desirable to moderate the heart's action, small doses of aconite may be advantageously added to the spirits of nitre.

Spirits of nitre is very rapidly eliminated, both by the kidneys and lungs. A single large dose may therefore act chiefly as a stimulant and diuretic. In order to diminish arterial tension we should employ small doses frequently repeated. These doses may have no effect upon the pulse but they will lower the tension and so lead to perspiration. As the action of the drug resembles that of nitrite of amyl and nitro-glycerine, perhaps it might be used instead of these more powerful agents. It might be well to try whether a large dose of sweet spirits of nitre might not sometimes prove an efficient substitute for amyl nitrite.

Jaborandi and Pilocarpine.—Though only introduced to European practice about ten years ago by Dr. Cautinho, jaborandi rapidly established an important position, and has already found its way into the American, German, and French pharmacopœias, and will no doubt obtain a place in the next British revision. Soon after a dose of jaborandi, or of its active principle, there is flushing of the face, ears, and neck, which progresses downward, travelling over the whole body. In from five to ten minutes after the dose the saliva begins to flow freely and soon afterward perspiration supervenes, and this lasts from two to five hours and is often so profuse as to soak through the clothes. As soon as the perspiration is free the flushing passes away. Féréol[1] noted an interchange between the amount of the salivation and the sweating, and this may

---

[1] Féréol: Note sur le Jaborandi du Dr. Cautinho. 1875.

sometimes be observed, for with profuse salivation there may be less sweating, and *vice versa*, but the rule is by no means absolute. Occasionally no salivation takes place, but it is scarcely ever that there is no sweating, though now and then we meet with individuals who seem to be almost insusceptible to the action of the medicine. It is curious that it has much less effect upon children than upon adults.

The amount of sweat transuded is sometimes enormous. The salivation, too, may be excessive, from a pint to a pint and a half being spit out, besides what is swallowed. Occasionally pain and swelling of the salivary glands follows, other secretions are also increased—the nasal, the bronchial, the lachrymal—but to a far less degree than the cutaneous and salivary; so we have watering of the eyes, sneezing, irritation of the nose, succeeded by coryza, and a loose cough with free expectoration. It is even said that the cerumen of the ears has been increased. Gastric uneasiness is often complained of, partly perhaps due to the saliva swallowed, which, indeed, is sometimes vomited. When jaborandi is taken the bulk of the remedy may also cause uneasiness, but gastric disturbance is also observed after pilocarpine. Gubler observed that diarrhœa may be produced, but this is not frequently the case. He also thought the remedy acted as a diuretic, and in small doses it may do so, but scarcely when the sudorific action is marked. It has no effect on the bile and it is doubtful whether it has on the milk.

The effect on the temperature is not so marked as might have been expected, an average fall of 1° Fahr. being established. Robin[1] thought that a slight rise preceded the fall; Riegel found no rise. More blood is sent to the skin, but the effect of this may be balanced by evaporation and radiation. The pulse rises some twenty to fifty beats, but Langley found an opposite effect produced in animals. The rise lasts from two to four hours, but the pulse is much weaker and a good deal of depression is produced; the blood-pressure falls temporarily, then perhaps it rises a little, and finally a fall is established. The heart, though the beat may be quickened for a time, is decidedly weakened, so that when this organ is unequal to its work jaborandi is to that extent contra-indicated. Metabolism is increased and the body weight falls; the excess in the secretions is not merely water, but the solids are increased; thus abundance of salts and ptyalin are carried away by the salivation and an excess of urea by the perspiration. The sweat is said to be at first acid, afterward neutral, and at last often alkaline. Robin found it contained excess of chlorides with some carbonates and phosphates, but much more important is the increase of urea to more than five times its normal amount. Hardy and Ball estimated that an average of seventeen grains was eliminated by the skin in their experiments, and some even higher estimates have been made.

---

[1] Robin: Études physiologiques et therapentiques sur le Jaborandi. 1875.

Jaborandi sometimes affects the sight a little, but without altering the size of the pupil; its local application, however, brings on contraction and impairment of vision, lasting from an hour and a half to occasionally twenty-four hours. Mr. Tweedie concludes (*Lancet*, 1875) that locally applied it causes (1) contraction of the pupil; (2) tension of the accommodative apparatus of the eye, with approximation of the nearest and farthest points of distinct vision; (3) amblyopic impairment of vision from diminished sensibility of the retina.

Respiration is not directly affected, and so we can use this diaphoretic in bronchial and pulmonary diseases, but the depression it produces on the circulation and the exhaustion which sometimes follows its action renders it unadvisable to repeat it too frequently. The narcotism which has been said to follow it is probably only that disposition to sleep which may be observed after profuse perspiration, however induced.

The action of jaborandi is promptly antagonized by belladonna: a hypodermic injection of one-one-hundredth of a grain of atropia will almost invariably arrest the sweating and salivation, and in case of excessive action it may be desirable to resort to this antidote.

The increase in the salivary secretion is probably due to a direct action on the gland or the ends of its nerves, as well as to a stimulation of the centre. Carville has shown (*Journ. de Therap.*, 1875) that section of the chorda tympani high up or low down after it has joined the gustatory nerve does not prevent the sialogogue action of the drug. Free salivation was produced even after section of both the gustatory and pneumogastric nerves and destruction of the upper cervical ganglion of the sympathetic. The sweating is also due to direct action on the nervous periphery, as well as on the centre, and perhaps the sweat-glands are stimulated in the same manner as the salivary. The sphygmograph shows lowered vascular tension. The cardiac depression is partly due to the action on the vagus and partly, perhaps, to direct action on the ganglia.

To produce a single copious sweat in almost any disease jaborandi may be employed. It may therefore serve as a substitute for the Turkish bath or the lamp bath. In renal dropsy it seems to be indicated, as we have seen that it greatly increases the elimination of urea. In respiratory diseases a single full dose will often cut short a cold in the same way as any other sudorific. In asthma it has sometimes been found to give relief; so it has in pertussis. When the temperature is high and there is a good deal of sthenic excitement, it is well to combine it with an arterial sedative. In pleuritic effusion, when there is no cardiac weakness, jaborandi may be employed. In diphtheria somewhat contradictory statements have been made. Guttmann regards pilocarpine as a specific; he recommended it in every case, however severe, septic or otherwise. Soon after Dr. Jacobi stated his experience at the American

Medical Association (1881). In septic cases he believed the treatment accelerated death by hastening the cardiac failure, but in many cases he held that the membrane was softened and separated by the copious secretion produced by the pilocarpine. He therefore attributed the recovery, 1, to the macerating effect; 2, to the timely withdrawal of the alkaloid. It seems certain that the remedy greatly increases the secretion of the respiratory mucous membrane and renders it more fluid, thus tending to disintegrate and separate the false membranes. It is therefore a kind of expectorant, and will again be noticed under that head, but there is evidently danger of its depressing influence on the heart, calling for great circumspection in its use. If we could find a drug which would antagonize its action on the heart without interfering with its influence on the mucous membrane we might combine the two. Atropia unfortunately dries up the mucous secretion, and we must therefore look to ammonia, ether, etc., to sustain the heart. In a robust patient pilocarpine may be employed without fear, provided it be discontinued in time, but after its influence has been obtained the use of steam will keep up the effect, and where there is depression at first this old but most powerful remedy should perhaps be trusted altogether. In reference to croup and diphtheria, we must not forget the peculiar insusceptibility to jaborandi and its active principle manifested by children.

It has been stated that small doses—one-twentieth grain—of pilocarpine will check colliquative sweats in phthisis. A similar statement has been made as to other diaphoretics. Supposing this to be the case, why should we not resort to the far more reliable atropine or any direct antihydrotic?

CHAPTER XVI.

ANTIPYRETICS.

We have already referred to the effects of refrigerants and salines, as well as to acids and various beverages, in restraining febrile action. We have also considered diaphoretics and some other methods of affecting the cutaneous circulation. We have seen that the skin is the great regulator of the escape of heat; and we may act upon it directly, as we do, for instance, when we reduce the external temperature or change the medium in which we are living by placing the body in a bath. Cold baths and the local application of cold water or of ice may be considered external refrigerants; and the introduction of cold or iced water into the body by beverages or injections will affect to some extent the general temperature, but not so much as might be imagined. Cold, then, whether by the medium of water or air, is the first of the direct antipyretics, and appears to act principally by withdrawing heat from the system. Such heat may of course be due either to overproduction or diminished dissipation, causing accumulation in the body. It is quite conceivable that fever may be reduced by the mere abstraction of excessive heat, or by a diminution of its production. For whatever disturbs the balance between the production and the loss of heat seriously affects us. When the temperature of the atmosphere rises there is a greater dissipation of heat by perspiration, and by exhalation from the lungs, as well as by a cooling of the surface from the blood flowing rapidly through the cutaneous vessels. At the same time, the call upon the heat-producing function being diminished, it is probable that less is liberated. So, again, during exercise the skin flushes, perspiration sets in, and quickened circulation and respiration sends large quantities to be cooled at the surface; thus, as we have said, the skin is the chief regulator, but the lungs assist its action.

When, on the other hand, the external temperature falls, a converse compensating process takes place. We now have less loss of heat, perspiration is suspended, and the cutaneous vessels contract, the respiration and circulation being retarded. In this case, in response to the call upon production, more heat is liberated, and so the balance maintained. The effect of a change in the external medium from air to water will largely depend on the difference between the body temperature and the bath.

In health we know that the compensating powers of the system are such that short baths do not much alter the temperature ; and so we find that our most powerful antipyretic medicines have but little effect on robust, healthy men. This may be because in health the compensating arrangements easily maintain the normal condition, even although it may be admitted that some of these medicines may diminish the production of heat by retarding metabolism and decreasing excretion. Quinia and the salicyl compounds diminish the excretion of urea and probably resorcin, chinolin, kairin, and other such substances may act in a similar manner ; if so, it would seem that they should be regarded as restraining the production of heat. But it is quite possible that this is not the true interpretation, but that in some way they may increase the dissipation of heat. Dr. H. C. Wood, at the International Medical Congress, 1881, in London, stated that in experiments which he had made jointly with Dr. Reichert on healthy dogs, extending over two years, quinia nearly always caused a slight increase in production and always a great increase in dissipation. The augmented production seemed to him only an indirect result of the excessive dissipation.

But whatever may be the effect in health, an increased heat-production seems to be the great cause of pyrexia, and as quinia, the salicyl compounds, the phenol derivatives, and other antipyretics reduce the temperature, it is generally held that they do so by restraining the excessive production ; while the cold-bath treatment, as it manifestly abstracts heat, is credited with acting only in that manner, though there are facts which seem to show that this method also checks heat-production. This is shown in the diminution of the nitrogen eliminated, as determined by several investigations ; *e.g.*, Barth, in 1866, found that under the action of baths in four cases of typhus and two of typhoid, the elimination of urea, phosphates, and chlorides was lessened. This has been corroborated by Schroeder, while in 1879 Bauer and Kuenstle investigated the comparative effects of baths, quinine, and salicylate of soda, and found that the flow of urine was increased by them all ; but by far the most complete research of this kind was published by Dr. Sassetzky in *Virchow's Archiv.*, 1883. He analyzed both the urine and fæces in order to estimate the total elimination of nitrogen. Further, in order to determine how much assimilation is interfered with, and to what extent metabolism would account for the nitrogen eliminated, the quantities of nitrogenous ingesta were also investigated.

Sassetzky selected cases of pneumonia, typhus, and relapsing fever, excluding typhoid lest the intestinal lesions should disturb the result. In fifteen cases the nitrogenous excreta were estimated daily, both in the urine and fæces. Further, the phosphates and chlorides of the urine were determined and also the solids and nitrogenous substances ingested. Analyses were made in every case during three periods of from three to eight days each; viz., one period in which antipyretic treatment was employed, one without treatment, and one after pyrexia had quite subsided. In nine of these cases baths at 72.5° F. were administered four times a day for fifteen minutes. In four cases two ten-grain doses of quinine were given in the evening, and in two cases salicylate of soda was used.

The general effect was diminished elimination of nitrogen. This decrease was noticed under all methods, but to the greatest degree under the baths. So, too, the quantity of urine was increased by all methods, but most of all by the baths. Further, the assimilation of the solid and nitrogenous constituents of milk was improved, and this too most of all under the baths. This was shown by a very great diminution in the elimination of nitrogen by the fæces. Less water was ingested under the treatment, and the loss of water by the lungs and skin was decreased, except under the salicylate, under which there was an increase of loss through the skin.

Among the causes of pyrexia, we may assume, then, that increased production of heat in the tissues is the most important. In fevers the increased metabolism, the excessive tissue degeneration, and the rapid oxidation manifest themselves in the rapid emaciation and the great increase of urea and other excretions. This conclusion is confirmed by other phenomena; thus a chill—which is a sudden suspension of, or at least interference with the regulating function of the skin, causing the vessels to contract and so shivering, chills, or rigors to be felt, and bringing about arrest of perspiration—hinders the escape of heat, whereupon the temperature rises, even though the production be only maintained at the usual rate; but if that be quickened, of course the rise must be more considerable. It may be thought that increased local production, which we observed in inflammation, may also cause pyrexia. No doubt this is the tendency of such a condition, but the additional heat thus eliminated would probably only produce a slight effect on the general temperature. Another cause of pyrexia is external heat, as we see in cases of sunstroke or thermic fever. Again, the penetration into the system of minute organisms or other poisons capable of setting up processes allied to fermentation, tending to destroy the tissues and otherwise affecting the normal functions, rapidly raise the temperature, as we see in septic conditions, when what may be termed disinfecting antipyretics are naturally resorted to. It is supposed by some that quinine cures ague by exercising a toxic influence over such organisms, and it has been conjectured that the salicyl compounds exercise a similarly injurious quality on a hypothetical microzyme in rheumatism. But others, while freely admitting the clinical value of these remedies, do not countenance that theory of their action. Another cause of high temperature would seem to depend on some interference with the nervous system. Thus in injury or disease in the upper portion of the spine very high temperatures have been recorded; and this fact may perhaps be attributed to a direct effect on a centre for heat, if such exist.

Cold.

The ancients were not averse to the use of cold, and in fevers they administered cold drinks freely; but the practice was disused and eventually warm drinks only were permitted. Perhaps this strange reversal in modern times was largely due to the ascendancy of Boerhaave's doctrine of lentor in the blood as the cause of fever. Even Cullen hesitated to permit fever patients to indulge in cold drinks; and yet, a general view of the history of medicine would show that in all countries, and almost in all ages, fever patients have been treated successfully and pleasantly by the external and internal use of colds. Asclepiades, who is credited with the invention of the shower-bath, employed cold pretty freely; but less reverence is due to him than to his predecessors, in consequence of the touch of charlatanism which characterized his practice. Suetonius tells us that Musa successfully treated Cæsar Augustus by cold baths; and Horace relates in his epistles (book i., 15) that, under the same advice, he was to discontinue his warm baths and take them cold, even in frost. Travellers in the East tell us that this method continued long after Europeans had learned to consider it dangerous. The ancient Britons and all Northern peoples had no fear of cold water, and would plunge into it new-born children, a practice which continued general down to the fourteenth century or later.

In the fifteenth, sixteenth, and seventeenth centuries many works on the use of cold water were written. In 1721 the Faculty of Medicine of Paris awarded a prize to M. Noquez for an essay in which he advocated water almost as a universal specific. The next year there appeared in London a curious work by the Reverend Dr. Hancocke,[1] which gave rise to a number of other pamphlets on the same subject, one of which, issued in 1726, extended over nearly three hundred pages, and was entitled, "Febrifugum Magnum, Morbifugum Magnum; or, the Grand Febrifuge Improved, being an Essay to make it Probable that Common Water is good for many Distempers, that are not mentioned in Dr. Hancocke's 'Febrifugum Magnum.'" Meantime, Dr. John Smith's "Curiosities of Water"[2] had appeared, 1723, and a translation of Van der Heyden's "Arthritifugum Magnum," and Dr. Short's "Rational Discourse." In 1729, Professor Cyrillus, of Naples, is stated in the "Philosophical Transactions" to have

---

[1] Febrifugum Magnum; or, Common Water the Best Cure for Fevers and probably for the Plague. 1722.

[2] Smith, J.: Curiosities of Common Water; or, Advantages thereof in Distempers. 1723.

recommended water, and even powdered ice or snow, in fevers, both internally and externally. In 1730, Boudon[1] collected in two volumes thirteen essays on the subject by various writers. In 1734, De Hahn[2] gave an account of a terrible epidemic at Breslau, in which the mortality was not reduced until cold affusions were used, and in which he himself suffered and was restored by cold sponging. Hoffmann[3] (1747), Smollett[4] (1752), C. Lucas[5] (1756 to 1758) in a very able essay, Percival[6] (1769), Englehart[7] (1776), De Hersfeld[8] (1776), and many others followed. In 1779, Dr. Wright presented to the Medical Society of London an account of a number of cases, including his own, in which he had systematically employed affusions with the greatest benefit. On reading these, Dr. Currie,[9] physician to the Liverpool Infirmary, determimed to adopt the practice, which he did with great success, as testified in his "Reports," the first of which was published in 1797 and the second in 1804. From this time the modern practice has been frequently dated, and Currie has been regarded as the reviver of the method. But Dr. Jackson,[10] in 1798 and in 1808, must also be credited with no little influence, and although he differed in some respects from Currie, in many others confirmed his views. Indeed, Dr. Francis Adams, the accomplished translator of Hippocrates, expressed his regret that the profession did not follow Jackson rather than Currie. Dr. J. E Stock[11] in 1805 produced his "Medical Collections on Cold." Giannini[12] and Mallonay[13] (1805) recommended cold baths in all forms of fever. Fröhlich[14] (1820) regulated his practice by the use of the thermometer. In 1819 Dr. Bateman, in his treatise on cutaneous diseases, pronounced cold water

---

[1] Boudon: Les vertus médicinales de l'Eau commune ou Recueil des Meilleures Pièces, etc. 2 vols. 1730.

[2] De Hahn: Unterricht von der Kraft und Wirkung, des frischen Wassers in die Liebe der Menschen. 1734.

[3] Hoffmann, J. A.: De usu et virtute aquæ simplicis. 1747.

[4] Smollett: Essay on the External Use of Water. 1752.

[5] Lucas, C.: Essay on Waters. In three parts, treating of 1, Simple Waters; 2, Cold Medicated Waters; 3, Natural Baths. 3 vols. 1756–1758.

[6] Percival: Experiments and Observations on Water. 1769.

[7] Englehart, F. J.: Dissertatio sistens aquæ frigidæ interno. 1774.

[8] De Hersfeld, St.: De Aquæ communis differentiis usu et Viribus. 1776.

[9] Currie, J.: Medical Reports on the Effects of Water, Cold and Warm, as a Remedy in Fever and Febrile Diseases. 1797. Second Series. 1804.

[10] Jackson. Outline of the History and Cure of Fever. 1798. Exposition of the Practice of Affusion; Cold Water on the Surface of the Body as a Remedy of Fever; to which are added On the Effects of Cold Drinks, etc. 1808.

[11] Stock: Medical Collections on the Effects of Cold as a Remedy in Certain Diseases. 1805.

[12] Giannini: Della natura delle febbri. 1805.

[13] Mallonay: De usu aquæ frigidæ in febribus. 1805.

[14] Fröhlich, A.: Gründliche Darstellung des Heilverfahrens in entzündlichen Fiebern überhaupt und insbesondere in Scharlach, etc.

"the most effectual febrifuge," and commented on its "certainty, safety, and promptitude" in scarlet fever. In 1826, Dr. Macartney[1] in his lectures pronounced water to be "worth all other remedies put together," and his opinions exercised no little influence over his contemporaries.

Then came a period in which the scientific use of cold was less in favor, this potent remedy having been taken up by ignorant empyrics, and hydrotherapeutics thereby degraded into hydropathy. Priessnitz opened his establishment at Gräfenberg in 1828, and many enthusiasts and speculators embarked in the business; no wonder, therefore, that a remedy thus exploited, and the reckless and ignorant use of which often led to catastrophes, should have been looked upon with some suspicion. Nevertheless, from time to time articles and treatises of scientific value were produced, such as those of Dr. J. Arnott, who from about 1849, when he advocated congelation in inflammation, continued to show the use of cold under various circumstances, and still later, Esmarch, whose treatise on "Cold in Surgery" has been translated for the New Sydenham Society.

We come now to the last revival of the use of cold as an antipyretic, which dates from the appearance of Ernest Brand's[2] work (1861) on typhoid; which, although perhaps too enthusiastic, restored, so to say, the proper and systematic use of cold baths, extended the area of their application, and excited the attention of the profession. Jürgensen, of Kiel, followed his method, and in due time published a work on typhoid (1866) embodying three years' experience. His results were such that the Kiel system spread through most of the German hospitals, and as gradually extended over other countries. Liebermeister and Hagenbach[3] followed in a joint work in 1868, which added greatly to our reliable evidence on the subject, and the following year Küchenmeister published a very handy summary of trustworthy results, and Ziemssen and Immermann,[4] published an important work on typhoid, of which a much enlarged and more complete edition appeared in 1874. The Franco-German War contributed largely to the extension of the method in Europe; but Germany is still the home of the treatment, and the works named the most important contributions to its literature. M. Peter (*Union Médicale*, 1777), Glénard,[5] Béhier,[6] Raynaud,[7] Pécholier,[8] Dumontpallier,

---

[1] Macartney: Lectures in Trinity College, Dublin. 1826.

[2] Brand, E.: Die Hydrotherapie des Typhus. 1861.

[3] Liebermeister and Hagenbach: Aus der medizinische Klinik zu Basel. Beobach. u. Versuche über die Anwendung des Kalten Wassers bei fieberhaften Krankheiten. 1868.

[4] Ziemssen and Immermann: Die Kaltwassenbehandlung, des Typhus abdominalis. 1869.

[5] Glénard: Traitement de la Fièvre typhöide par les bains froids. 1875.

[6] Béhier: Du traitement de la Fièvre typhöide par les bains froids. 1874.

[7] Raynaud: Application de la méthode des bains froids au traitement du Rhumatisme Cérébral. 1874.

[8] Pécholier: Sur les indications du traitement de la Fièvre typhöide. 1874.

Homolle (*Revue Générale*, 1878), Féréol (*Union Méd.*), and others have adopted the method in France; Barduzzi[1] and others in Italy. In England we have adopted the method, but with considerable caution, and certainly without enthusiasm. Dr. Wilson Fox, in 1871, proved the value of the method by rescuing a patient from hyperpyrexia in the course of acute rheumatism, and a number of other patients have been saved by the same plan. Sir Spencer Wells reduced the fever previous to the removal of a suppurating ovarian cyst, after the temperature was brought down to a hundred, having previously varied between 102° and 104° F. (*Medical Times*, 1872). Baths have since been employed more or less assiduously in a number of the London hospitals, but the treatment is by no means universally adopted.

Baths gradually cooled down from 95° to 72° F., or lower, are sometimes employed, as recommended by Von Ziemssen, a method often called after his name, though he has since almost abandoned it; as this plan takes longer to produce the same effect. Liebermeister recommends a full length bath of 68° F., or lower, in typhoid for adults, for a period of ten minutes—if kept in longer it is unpleasant and may do mischief. If the patient should be feeble and continue cold, or be collapsed, the time is reduced to seven or five minutes, and he thinks this better than using a long tepid bath. Very feeble patients may begin at 75° F., but the effect is much less, or Ziemssen's plan may be tried. In common with other German observers, he insists that the baths to be effectual must be given often. The temperature should be taken every two hours, and whenever it rises to 102.2° F. in the axilla the bath is to be renewed. In children, inasmuch as the body is larger compared to the weight, we may wait till the temperature reaches 103° F. in the axilla, or 104° F. in the rectum, as we may also in persons who are known to resist heat well; but in those in whom the resisting power is less it is well to commence earlier, with perhaps a shorter bath, or else a little higher temperature. He regards it as a delusion to suppose that a few baths will do any good. When the disease is severe, the interior of the body is only slightly cooled, and that for a short time, by a single bath. It is therefore necessary to repeat the baths, sometimes every two hours, and this is to be done both night and day, so that twelve may be required in the twenty-four hours, and some of his typhoid patients have taken more than two hundred during the illness. Usually from four to eight per diem have sufficed, and from forty to sixty in the course of the illness, especially if antipyretic medicines were also given.

There is no doubt that the baths are disagreeable to the majority of patients, and it often requires persuasion, or even the exercise of authority on the part of the physician to secure his orders being properly car-

---

[1] Barduzzi: Dell' Idroterapia nelle Febbri Tifoidee. 1874.

ried out—the more, perhaps, as the method is troublesome and very exacting for the attendants. Later on, having experienced the benefits, the patients desire to continue the baths after the necessity for them has passed by. They feel hot, and desire the refreshing cold they have previously enjoyed. With a rectal temperature of 101.5° F. they may be indulged, but when it falls below that cold sponging will be found sufficient, or cold packs, or other local modes of abstracting heat.

The most important contra-indication admitted by Liebermeister is hemorrhage from the bowels. He does not suspend the baths on account of pneumonia or hypostatic congestion, which often disappear under the treatment. Great weakness of the heart's action also contra-indicates the baths, for when the circulation is so reduced that the surface is cold, even though the interior may be hot, there is no likelihood that further cooling of the surface will much reduce the internal heat, while it might cause still more obstruction of the peripheral circulation. In slighter cardiac weakness we may resort to the gradually cooled bath. Wunderlich does not admit that hemorrhage is any contra-indication (*Schmidt's Jahrb.*, clvi., p. 101). He treated sixteen cases in which there was intestinal hemorrhage with only two deaths, and neither of these were from the hemorrhage, but one was due to perforation and one to pneumonia. Bäuer and others follow Liebermeister in discontinuing the baths when hemorrhage sets in, and this seems reasonable, inasmuch as they would be likely to increase the hemorrhage, and the necessary moving of the body might be dangerous. Menstruation is not a contra-indication.

In *pneumonia* cold baths have been freely employed, particularly by Jürgensen, who is perhaps the most enthusiastic advocate of their use wherever there is fever. Not only in croupous pneumonia—which he regards as an infective fever, looking upon the inflammation of the lung as merely the chief symptom, and the pyrexia the essential thing to treat—but even in catarrhal pneumonia he resorts to this method. He employs usually baths from 77° to 86° F. for twenty or twenty-five minutes and then cold affusions. The younger the patient the greater the need, in his view, of active interference, and accordingly in young children he is more careful than in older patients to treat every fever attended with catarrh by the general abstraction of heat. Cold packs, he declares are more severe, if thoroughly carried out, and give rise to much greater discomfort. As he considers the pyrexia of croupous pneumonia to be the first point to attack, he puts aside the fear of catching cold as mere nonsense, and even says that if he had no water in a given case he would not hesitate to expose a patient to cold air until the necessary cooling had been obtained. In opposition to all prejudices, he insists that the proper treatment consists in the abstraction of heat, and in this he is largely supported by Liebermeister and others. The fear that an overworked heart may be paralyzed in consequence of the increased demand made on both the heart and respiratory muscles during the bath he admits to be reasonable, but thinks it may be prevented, that is to say, he does not deny that fatal collapse may occur, but the administration of a preliminary stimulus will prevent this. He gives two tablespoonfuls of claret immediately before and immediately after each bath, and if the heart be not working satisfactorily he employs port, madeira, or champagne. When the bath is quite cold he gives one to three spoonfuls of the stronger wines five minutes before, repeats the dose while in the bath, and again immediately after. Children may require relatively larger doses; in all cases the quantity is to be regulated by the pulse, and he is well assured that those who would avoid accidents must not spare stimulants. The abstraction of heat is intended to guard against cardiac exhaustion, and the danger of this arises from the fever, so that in strong persons in a moderately severe attack the treatment adopted by the German school is almost the same as for typhoid. Jürgensen, for example, with a rectal temperature of 104° gives baths of from seven to twenty-five minutes, though he employs at the same time quinine; but in feeble patients, in the aged, and in the corpulent the temperature rarely reaches that point, but most commonly varies from 101° to 103° F. These are the cases which require special care, and in them it is best to take advantage of the usual rise and fall during the twenty-four hours, giving a tepid bath from 77° to 78° F. from twenty to thirty minutes at a time early in the morning, say from four to seven o'clock, when the temperature is at the lowest, and at the same time a dose of quinine. Young children may be treated with the wet sheet, which in this country is usually considered milder, but Jürgensen believes that it is more troublesome and gives rise to more discomfort. The application of cold cloths to the chest only he considers to be quite useless in children over one year of age, as sufficient heat is not abstracted by this method, though other observers place more confidence in the local application of cold. In some cases of pneumonia the temperature rises so high that the most energetic and repeated efforts to lower it are called for. In these he has often given baths at 61° or 60° F. for ten minutes at a time, and then, the temperature returning quickly, he has given the bath at 43°, 42°, and even 41° F. His own child he treated when nineteen months old in this manner for several days without the slightest indication of collapse, and has repeatedly resorted to the plan during the last few years. But we must remember that he regards the simultaneous administration of wine as an essential part of the treatment—giving adults from half a bottle to a bottle a day, including of course the doses given before and after each bath. This is the ordinary red wine of Germany, and we may use claret, but in severe cases, as already stated, he does not hesitate to use stronger wines, and if the heart show signs of failure even hot grog. Further, the simultaneous use of quinine is most important, as this remedy above all others reduces the temperature without injuring the heart.

This brings us to a consideration of the danger of cardiac failure, which in some cases of pneumonia is most urgent, and demands active treatment. In consequence of the congestion in the pulmonary circulation the right ventricle suffers while the left is comparatively empty. In severe cases blood accumulates in the veins of the systemic circulation, in consequence of the obstructed right ventricle. Pulmonary œdema may occur, and generally does in fatal cases. The respiratory and cardiac muscles at the same time receive only an insufficient supply of blood. It is obviously desirable to prevent such a train of symptoms, and when this is impossible, the earlier the treatment is directed toward them the greater will be the probability of relieving them. It has been generally held that the indication here is complete for blood-letting, which at once should relieve the heart, and clinical evidence shows that such relief is prompt and almost certain. Nevertheless, great objections have been raised to this remedy, the chief of which is that the symptoms are likely to return, and a repetition of blood-letting is undesirable. Efforts have therefore been made to stimulate the heart to increased work until the obstacle in the pulmonary circulation is overcome. For this purpose stimulants have been largely used, and even Jürgensen employs them largely, though he advocates the bath treatment so earnestly; indeed, he says that the most dangerous enemy to the heart is the high temperature, and this may be safely and quickly lowered by the bathing. But then we must add stimulants, which he believes not only spur the heart to perform increased labor, but directly enable it to do so. As each vigorous beat forces more blood from the overfilled right ventricle into the left, it benefits the heart by supplying it with more oxygen, while it removes the accumulated débris of oxidation. The quantity of blood is left unchanged, increase of exertion has relieved the temporary embarrassment; but he says that a bold use of stimulants will maintain life three or four days or more after the heart has shown signs of exhaustion. Strong wine he gives with the first indication of cardiac failure, and if the symptoms continue alternates the wine every hour or half hour with full doses of camphor, and if sudden or severe collapse occur, musk, one or two grains at a dose, with a tablespoonful of champagne at intervals of ten minutes to half an hour; when the patient cannot swallow, hypodermic injections of camphorated oil, of which he speaks highly. He also gives hot grog to obtain a more rapid effect than from champagne, or brandy or other spirit diluted with strong tea or coffee.

This, it will be seen, is a stimulating treatment, of which much has been said. If stimulation can tide over the immediate danger it seems rational to employ it, but it should not be forgotten that stimulation is not nutrition, and does not deserve the name of "support" which is too often applied to it. There is nothing unreasonable in using the spur or the whip, but neither can supply the place of corn to the struggling horse, though Jürgensen, like other advocates of stimulants, would persuade us that they supply the place of both. Nor is it so unphilosophical to resort to stimulants, after a previous blood-letting has relieved the over-burdened heart ; and so, again, it is not unreasonable so to use stimulants as to enable the heart to bear the bath treatment, which doubtless increases its labor for the time being, but then diminishes it. If the heart can only be tided over the temporary extra work the baths may be employed, and this may generally be managed unless the collapse be extreme. Massive doses of quinine should be administered at the same time. This is by far the most potent antipyretic ; it does not injure the heart, and is not open to the objections which may be made to stimulants, though we readily admit that the free use of wine or spirits in such urgent circumstances bears no more relation to their use as beverages than does the exhibition of a dose of morphia to the pernicious habit of opium eating. Since, however, alcohol has been admitted to be antipyretic in considerable doses, there has been a tendency to employ this fact in support of the system of stimulation. (See previous chapter on Alcohol.)

In *scarlet fever* cold has always been a favorite remedy. Currie employed it, as did Jackson, Laycock, Trousseau, and various others before and since, and they all report well of it, the majority employing affusions. More recently G. Meyer[1] puts a child into a bath of 93° to 73° F., according to the intensity of the fever, for about ten minutes whenever the temperature rises to 102° F. This is sufficient to insure a reduction lasting for several hours. Others prefer cold packs or cold sponging, or the application of cold compresses to the trunk and ice to the throat, but these measures do not reduce the temperature so decidedly, and require to be continually renewed. When, therefore, the temperature runs high the bath is a much more potent remedy.

In *diphtheria* favorable results have been reported from the use of cold baths, and as much may perhaps be said for a great many other remedies. This disease is not, as a rule, characterized by the very high temperatures which urgently call for the treatment, and in this country the method has only been resorted to now and then to meet the special indication of an unusually high temperature. But the local use of cold to the throat by means of ice-bags or ice-poultices has been much more largely employed, though with less success than has been looked for.

With respect to the success of the antipyretic system, statistics certainly show a reduced mortality to that which prevailed prior to its adop-

---

[1] Jahrb. f. Kinderk., vii.

tion in the most important German hospitals. Roughly it is claimed that in typhoid the mortality has been reduced from sixteen to eight per cent. by systematic antipyretic treatment, of course including besides cold the use of quinine and other agents. If we exclude mild cases, there is a mortality of ten or eleven per cent., as against twenty-five to thirty under the expectant plan. And a still further sifting of the statistics would, it has been claimed, show a still more favorable result. Brandt has, indeed, claimed that if we exclude cases already moribund when admitted, none would die; and Glénard says that there would scarcely be a death in five or six thousand cases thus treated from the commencement. These enthusiastic statements have scarcely produced the effect that might have been expected had they been generally accepted; but all observers agree that the mortality has been greatly reduced by antipyretic treatment, and that the patients are left stronger than when they are abandoned to mere expectancy. Goldtdammer[1] contrasted 2,068 cases treated in Berlin between 1868 and 1876 with 2,228 cases treated between 1848 and 1867. The deaths were 13.2 per cent. in the later period, instead of 18.1 in the earlier, showing a difference of five per cent. in the mortality in favor of the period after the adoption of the antipyretic system. There was also a difference in the time spent in hospital, amounting on the average to 6.3 days in favor of cold. He admits, however, as others have been obliged to do, an increase in the proportion of relapses, and supposes that this may be due to the bathing hindering the natural destruction of the poison in the body. To effect this, Immermann, who also adopts this view, prescribes a daily dose of salicylate of soda, and since he adopted this remedy he has found a great reduction in the number of relapses.

As to respiratory diseases, Jürgensen's statistics are very remarkable. He claims a reduction in the mortality of at least one-half in four hundred cases treated in the same hospital, compared with the same number prior to the adoption of the method. He admits, indeed, as nobody can deny, that patients recover from pneumonia under all sorts of treatment, and that mild cases require little more than regulation of the diet, but even in them he prefers active interference to mere expectancy, and thinks that for the future it must be regarded as settled that all prostrating measures are injurious, and that all antipyretics which depress the heart, if given in doses sufficient to reduce temperature, are to be avoided. Most prudent physicians will be inclined to adopt the last conclusion, but it may be observed that treatment which reduces the pulse, and which would, if pushed too far, tend to paralyze the heart, may, when cautiously employed, produce decided benefit, and therefore be far better than impotent expectancy, while it is not attended by the dangers which

---

[1] Deutsch. Arch. f. klin. Med., xx., 1877.

are admittedly present when very active measures are employed. Thus we have treated pneumonia successfully by moderate doses of quinine and other antipyretics, and especially by aconite. But, as already remarked, similar statements may be made in support of other methods, and we have seen rapid and complete recovery follow venesection and counter-irritation, as we have also the directly opposite method of stimulation. Such facts to some extent account for pneumonia having been a kind of battle-field for contending theorists, and perhaps we may learn from them that while healthy human beings may be expected to shake off a serious disease, all treatment to be effective should be directed rather to the condition of the individual patient than to the particular disease by which he is attacked.

COLD AFFUSIONS.—Though these have been mentioned incidentally, we have thus far treated chiefly of baths. Affusions, however, have been employed from early times, and are often very useful when circumstances forbid a resort to the more powerful measure. Affusions are less disagreeable, and often they are pronounced pleasant, but they do not lower the temperature much, or for a long period, and they are often used as a sudden stimulus to the respiration, and to other functions rather than as antipyretics. To produce a decided effect a considerable quantity of water should be employed, from one to three gallons, and that very cold ; it should be poured from a moderate height, in a full stream, over the nape in such a way as to run down the back and chest—this plan may be adopted in the late stages of pneumonia.

COLD PACKS are well borne, even by very weak patients, especially when they do not include the feet or legs, and four consecutive packs, of ten to fifteen minutes' duration, may be regarded as about equal to a bath of ten minutes. But when using them in this way, a shorter time is perhaps desirable, otherwise the reaction which ensues leads to perspiration, and we have a different remedy in use. Some quite young children take the packs with pleasure, but others resist them, fearing them as much as the bath, which is certainly the least trouble in children, and very effectual.

COLD SPONGING.—Cold sponging will produce an antipyretic effect, provided it be repeated pretty constantly, but even then cannot be regarded as a substitute for bathing. A single sponging, even with very cold water, although it may be agreeable to the patient, can scarcely be expected to produce much effect on the thermometer. Surgeon-Major Welch thinks (*Brit. Med. J.*, 1884) the best course is to frequently sponge with " chilly " water, not wiping it off, and merely cover the body with a light sheet, or not even that ; in severe cases he makes the application continuous by a wet sheet packed close to the body, and suggests the use of a "punkah" to assist. The use of wet towels, continually renewed as fast as they become warm, is easier, perhaps, and more effectual; this plan has been largely employed and is especially applicable when it is desired to limit the cold to the trunk.

Dumontpallier uses a refrigerating envelope, and Leube has shown that a considerable reduction of temperature may be obtained by placing the patient on large pillows containing a freezing mixture of ice and salt. Dr. Mundie (*Brit. Med. J.*, 1884) placed a patient whose temperature was 106° F. on a Hooper's bed filled with cold water, with only a single blanket between the bed and patient. In six hours the temperature fell two degrees, and next day it was 103° F. Cold may also be applied much more locally by means of compresses, ice-bags, ice-poultices, etc. If long continued they seem to lower the temperature of the locality to a not inconsiderable depth, and are therefore believed to protect the subjacent organs, and to influence their circulation. For this purpose they are often employed on the head or chest. Leiter's coils afford a very convenient method of applying cold locally. These plans scarcely affect the general temperature. Riegel (*Deutsch. Arch. f. klin. Med.*, 1872) employed a couple of ice-bladders, one on the thorax, the other on the abdomen; but other German authorities failed to obtain much reduction of temperature in severe pyrexia, even by employing three bladders.

COLD BY SPRAYS.—Prof. Preyer, of Jena, lately brought before the Jena Medical Society (*Berlin. klin. Woch.*, May 5, 1884) some experiments on guinea-pigs, showing that the temperature could be easily reduced by this method, which he thinks applicable to man. He used a spray of simple water, and found that the rapidity and degree of the effect may be regulated by varying the temperature of the water and the frequency of its employment. Thus, when the temperature of the water used was between 40° and 45° F., the rectal temperature of the animals was reduced about two degrees within from five to ten minutes. If, after the discontinuance of the spray, the minute particles of water entangled in the hair be allowed to slowly evaporate, the temperature continues gradually to diminish for several hours. If water at the temperature of 70° F. be employed, the refrigeration is manifested within twenty minutes, but is not so pronounced.

COLD DRINKS reduce the heat by the amount that is necessary to raise them to the temperature of the body, and the same may be said of ice-liquids. Cold injections would act in the same direction; some effect may thus be obtained, though not very much. But there is this advantage about the introduction of cold fluid, that it does not seem to provoke the reaction that occurs when it is applied to the surface, and, therefore, the reduction obtained is not neutralized by a subsequent rise. Cold drinks, consequently, so far as they are desired by the patient, are as useful in febrile cases as they are always agreeable; and it is something to wonder at that the earnest desire of the patient should so often have been frustrated. We must admit, however, that cold drinks have sometimes been injurious when the body has been overheated, and there are clinical facts and even physiological experiments, which seem to show that active pulmonary congestion may be thus induced. Perhaps it was the observation of such cases, and of some other evils caused by drinking large quantities of cold water by overheated persons, which brought about an undue fear of cold beverages.

COLD ENEMATA have been shown by Foltz to produce a distinct fall, and Rutenberg's experiments support his statements. He administered a pint at a time, at 50° to 55° F., every two, three, or four hours, or at longer intervals during sleep, if the temperature had fallen, the number given during the progress of the case varying between thirty and three hundred.

It will have been observed that it is in Germany that *cold* has been most thoroughly and systematically tried. The method has, however, extended to other countries, though much more partially. In France, for instance, comparatively few have adopted it. Féréol, for instance, employs it in grave cases; Homolle declares it is not a specific, but thinks it will do more when we understand better the conditions in which to use it; Reynaud's experience was rather favorable than otherwise; Dumontpallier, however, completely adopts the method, and Gignoux, of Lyons, has lately published a paper[1] giving details of five cases of grave pneumonia in which the treatment was carried out with a resolution equal to that of German authorities. Four of the patients recovered, and M. Gignoux thinks very highly of the baths. In America, Dr. Austin Flint admits that Liebermeister was justified in saying that typhoid has lost a great part of its terrors. In England there has been no enthusiasm, and though cold has been tried in our hospitals it has hardly been so fully tested as might have been expected, and very often the manner in which it was tried would have been pronounced by German authorities insufficient. In hyperpyrexia, indeed, considerable confidence is felt in the plan, and as no other holds out much hope of success it is thoroughly carried out; but in less dangerous cases it is often less thoroughly applied. Dr. Collie, in the Homerton Hospital, never gave more than three baths in the twenty-four hours, none of which were longer than ten minutes and for children seven minutes; he discontinued bathing at the end of the second week of fever, and did not use it for old people or young children. In the London Fever Hospital the bath has been tried in many cases and satisfactory results reported. Dr. Cayley admits (*Brit. Med. J.*, March 1, 1884) the treatment reduces mortality and fulfils the physiological indications and produces marked alleviation of the symptoms, while complications are rendered neither more frequent nor more severe. Dr. Affleck in the same journal (May 17th) points out some of the objections, and

---

[1] Observations de Pneumonies traitées par les bains froids. 1884.

concludes that the applicability of the method is a narrow one, and he thinks it is not to be recommended as a general principle of treatment. In exceptional cases and in the rare event of hyperpyrexia he would certainly employ it. Dr. Ord, with anything but enthusiasm, says he has not known harm done and perhaps in some cases death has been averted. While Sir W. Jenner states that the published records have not carried conviction to his mind of the advantage of baths. In an address at the Midland Medical Society (1879) he said, respecting typhoid, "while admitting, without reserve, that heroic measures fearlessly but judiciously employed will save life when less potent means are useless, the physician whose experience reaches over many years will, on looking back, discover that year by year he has seen fewer cases requiring heroic remedies, and more cases in which the unaided powers of nature alone suffice for affecting a cure; that year by year he has learned to regard with greater diffidence his own powers, and to trust with greater confidence in those of nature." Professor Gairdner has even gone farther and condemned the system (*Glasgow Med. Jour.*, 1878) in a series of important cautions in respect of the so-called "antipyretic treatment in specific fevers."

QUININE.

We shall not discuss here the ordinary tonic effects of small doses, which are closely allied to those of other vegetable bitters, and which have been sufficiently alluded to. We may, however, remark that perhaps the simple bitter might more frequently be preferred with advantage, for it often happens that even small doses of quinine will disagree with the stomach when other tonics will not. Moreover, larger doses may give rise to nausea and vomiting, especially when the mucous membrane is irritable, and then if persisted in gastritis may be set up. Sometimes also it irritates the intestinal mucous membrane. No doubt it is easily dissolved in gastric juice; but perhaps when massive doses are given some of it may pass through the pylorus especially when taken in powder. Even in solution it may not be all absorbed from the stomach and then the alkaline juices would precipitate it. The biliary acids also form very insoluble salts with quinia, so that if it pass into the duodenum we might expect to find it in the fæces, and there it has been detected; it is not precipitated in the blood. Although this fluid is alkaline, probably the carbonic acid holds it in solution.

On the blood quinine exercises a great effect. In 1867 Professor Binz[1] announced that it had a direct action on the white corpuscles,

---

[1] Binz: Experimentelle Untersuchung über das Wesen der Chininwirkung, 1867. Also same author's Das Chinin, nach den neuern pharmacologischen, 1875, and numerous journal articles during the last few years.

checking their amœboid movements and arresting their tendency to pass through the capillaries in inflammation. There seems to be no doubt that it checks the amœboid movements, but the diapedesis being, like the local accumulation of the leucocytes, a consequence of the inflammation, it may certainly be said that it might be expected to cease as soon as anything arrested the inflammatory process. It requires a toxic dose to act on the white corpuscles, which under the influence of such an amount diminish in number sometimes to about a quarter. The red globules have been said to be enlarged under the influence of quinia. Manassein[1] observed in the lower animals that the red globules were lessened in size during pyrexia. In this state quinine restores them, but so with other antipyretics; even cold will do this, so that the change appears to be due to the lessened temperature rather than to any direct action of quinine on the corpuscles. Binz, however, has demonstrated [2] that quinine lessens the ozonizing power of the blood, and to this property some would attribute its antipyretic value; but such interference hardly accounts for all the reduction of temperature and the antiseptic influence of the alkaloid must be of considerable importance. The effect on the ozonizing power is seen in the impediment offered to the action on guaiacum. When ozonized oil of turpentine is dropped into tincture of guaiacum no change of color is produced, but if a drop of blood be added it strikes a blue color immediately, because the blood acts as a carrier of ozone from the turpentine to the guaiacum. Now, quinine prevents this reaction, one part in twenty thousand being enough to cause a perceptible hindrance. A similar impediment to the reaction of indigo, turpentine, and blood also occurs, one part of quinine to a thousand of the mixture delaying the characteristic color-changes from green to the clear yellow of isatin for about an hour. Binz has shown that all the salts of quinine act in the same way, and that the action is on the red corpuscle—on the hæmoglobin; in fact, the same result is obtained when crystallized hæmoglobin is substituted for blood. Accordingly the theory has been propounded that the antipyretic effect is due to the impediment offered to the ozonizing power of the blood, but this seems scarcely to offer us a key to all the medicinal virtues of quinine. A further influence on the blood has been observed: it retards the change to the acid condition, which ordinarily takes place in shed blood, and which Binz regards as due to oxidation.

Summing up the effects of quinine on the blood, it evidently interferes with oxygenation, impairs the carrying power of the red corpuscles and hinders them from supplying oxygen to the tissues, while it also impedes

---

[1] Ueber die Dimensionen der rothen Blutkörperchen unter verschiedenen Verhältnissen. Berlin, 1872.

[2] Archiv f. exper. Path. und Pharmakologie, 1873.

the function of the leucocytes. Pursuing it further, we find that it reduces the nitrogenous excreta—both urea and uric acid being considerably lessened, and this in health as well as in pyrexia. Ranke,[1] who first noticed this, G. Kerner, Zuntz, Rabuteau, and others all agree in this, though they differ in their estimates of the extent to which it occurs; but they appear to have experimented with different quantities, which may, perhaps, account for the variations. In different experiments, with varying doses, the decrease of urea was one-eighth, one-fourth, and four-tenths; of uric acid, one-half and four-fifths. It might have been anticipated that carbonic acid would also be decreased, but this has not been satisfactorily shown to be the case. This reduction of nitrogenous elimination being accompanied by a fall in the temperature points to diminished metabolism. Under the influence of quinine the tissues receive less oxygen, or else are rendered incapable of incorporating it, or, as we might say, of overcoming its affinity for the corpuscles. This idea is confirmed by other facts such as these, that quinine arrests many fermentative changes, hinders fungi from absorbing oxygen, quenches the phosphorescence of certain infusoria, and destroys the ozonizing power of vegetable juices. Moreover, in an atmosphere of ozone, quinia will prevent the peptonizing of albumen. But above all these is perhaps what may be termed the

Antiseptic Action.—Dr. J. Pringle, in 1750, brought before the Royal Society some "experiments on septic and antiseptic substances, with remarks on their use in medicine," and these were afterwards added to his observations on "Diseases of the Army," 1764. He showed that cinchona hindered putrefaction, but he thought that its febrifuge action must differ from its antiseptic property, on account of the rapidity with which it was produced, though he added the suggestive remark that all the medicines which he had found useful in intermittents were "powerful correctors of putrefaction."

The discovery of quinia in 1820 gave a great impetus to the neurotic hypothesis, as the few alkaloids then known were all considered to be nervines; but Binz has returned to the antiseptic view with great success. In 1867 he showed that quinine is directly poisonous to the fungi which inhabit various fermenting and putrefying liquids. Colpoda and paramecia are killed by a solution of one in eight hundred immediately, they die in a few minutes in a solution of one in a thousand, while one in twenty thousand proves fatal after a few hours. Penicillium, vibrios, and bacteria also perish in the presence of quinine, but for the last two a stronger solution is required to kill them. One per cent., according to Bochefontaine,[2] being necessary, and some resisted this for a long time.

---

[1] Ranke: Ueber die Ausscheidung der Harnsäure beim Menschen im physiol. Zustande und in einigen Krankheiten, sowie unter dem Einflusse des schwefelsauren Chinins. 1858. [2] Archives de Physiologie, 1873.

In 1880 Krukenberg found organisms which died after a few hours' exposure to a solution of one in one hundred thousand—a solution, said Binz, at the International Medical Congress, in London, in 1881, "far more dilute than that furnished by our blood after ordinary doses of quinine."

If the pyrexia-producing process is closely related to fermentation—whether set up by living organisms or by complex chemical compounds—we might anticipate the antiseptic influence to be important. Accordingly Binz and his followers regard this explanation of the antipyretic power as sufficient. They do not expect quinine to act as a fire in destroying microzymes, but if it enfeebles them they anticipate that the task will be completed by nature, and there are indications that time is required to produce the effect. Now, in this respect one of the most valuable properties of quinine is that it remains in the body for several hours, and further, as it is not poisonous to human beings even in considerable amounts, we can keep up its influence as long as may be necessary. Further, it is present in the body as quinia, and perhaps some powerful disinfectants do not act as antipyretics simply because they do not circulate in the state in which they are administered, but are decomposed as soon as they are swallowed, while others cannot be used because they are caustics or poisonous. Again, the effect of alcohol is evanescent, but arsenic remains for a long time in the system, while the carbol derivatives, though they distinctly reduce the temperature, are apt to give rise to disagreeable and even dangerous symptoms. They are, however, certainly antiseptics.

It is difficult to deny that there is a close relation between the power of destroying micro-organisms out of the body and the remedial value of antiseptic medicines. But the action may not be precisely the same, and some observers are convinced that it is not. Buchner[1] estimates that for quinine to act as an antiseptic in the tissues it would require a dose of three ounces (!), but surely the estimate of Binz is more reasonable; for certainly in most cases, if not all, we should only have to make the blood an unfavorable fluid for the growth of the organisms, and we know that minute quantities of antiseptics will effect this. With so powerful a poison as corrosive sublimate we may well fear that it would be dangerous to attempt to introduce enough to prove fatal to organisms in the blood, but even those who will not concede that it acts antiseptically have been constrained to admit that beneficial results have been obtained by its use. Arsenic, also, a very powerful poison, though with slight claims to be regarded as an antiseptic generally, is acknowledged to be a most powerful anti-malarial and anti-periodic, and there are not a few who believe that these qualities are really due to its deleterious influence on bacteria. We must admit, however, that to push these powerful

---

[1] Centralbl. f. klin. Med., 1883.

poisons to the extreme in deference to a theory as to their anti-parasitic properties would be a practice fraught with danger, but the same can scarcely be said of quinine and some other antiseptics which fulfil the indication suggested above, viz. : that they ought to be capable of circulating in the blood for a considerable period without inflicting serious injury on the patient.

Notwithstanding all that can be said in favor of the antiseptic theory of the action of quinine, it does not satisfy all observers. The late Professor Gubler maintained to the last his opinion that the effect of quinine should be attributed to its giving tone to the sympathetic system. Others are content to refer its action to the nervous system indefinitely. Gubler regards it as a direct stimulant of the auditory nerves, as well as of the great sympathetic ; Binz says it produces partial anæmia of the brain, a view directly opposite to that of Briquet, Hammond, and many others. Dr. Hammond's experiments at the time seemed to me to show that it produces, rather, congestion of the brain, and this conclusion is certainly most consonant with clinical experience when it is given in moderate quantities. The noises in the ears, the deafness, the sense of fulness in the head, the flushing of the face, and other symptoms of cinchonism, all suggest cerebral congestion, and are relieved by remedies appropriate to such a condition and sometimes disappear after an attack of epistaxis apparently provoked by the quinia. The effect on the spinal cord has not yet been satisfactorily cleared up.

Briquet found that large doses of quinia greatly lowered arterial pressure ; when he injected it into the jugular vein in full quantity it instantly arrested the beat of the heart which was afterward found to have lost all contractility, the left side being full of scarlet blood. So when it was thrown into an exposed heart in such a way as to go into the coronary arteries, the same effect was produced even more quickly. When a frog's heart is immersed in a solution the beats at once become slower, irregular, and soon stop altogether. Large doses are constantly proved by clinical observation to lower the frequency and force of the beat. It may, indeed, become rapid and feeble after toxic quantities, but this is only an evidence of cardiac feebleness. It is, then, a cardiac sedative in large doses, acting by depressing the heart-muscle and probably its ganglia, and not through a cardiac centre ; collapse would, therefore, seem to be a probable effect of poisonous doses. Small tonic doses apparently exercise no perceptible depressing influence over the circulation, but, by virtue of their restorative power, rather the reverse.

Large doses also depress the respiration, though small ones may perhaps accelerate it. In fatal cases death has generally seemed to be due to respiratory and cardiac failure.

Elimination is effected chiefly by the kidneys, but quinia has also been detected in the saliva, the sweat, the tears, the milk, and even in the serum of dropsical effusions. Its presence in the sweat may perhaps account for the cutaneous rash which sometimes appears; while its presence in the milk is a fact of some clinical import. It appears in the secretion in about half an hour after it has been taken, and the elimination is rather slow, lasting for two or three days. In six experiments made by Binz, only two-thirds of the dose was excreted in forty-eight hours; but Dr. L. Thau,[1] in three experiments, recovered nearly the whole of the alkaloid in the first forty-eight hours, so that most of the remainder might be allowed for the elimination by other channels. Dr. Thau further determined the rate of elimination, finding that from one-third to a little less than half escaped in the first six hours, and about three-fourths was eliminated in the first twelve hours. As the alkaloid thus tarries in the system for a considerable period, it is obvious that the dose should not be repeated too often; small doses are constantly given more frequently than is necessary, and massive doses should certainly only be administered at considerable intervals.

Though quinine is a most powerful antipyretic it does not lower very much the temperature in a healthy person, but it will prevent the ordinary rise of temperature caused by exercise. Its power over fever-heat is now generally admitted, a committee of the Clinical Society in 1870 reported ("Trans.," vol. iii.) that it had a very decided effect, commencing in from one to two hours after the dose, and lasting from a few to many hours. The large doses employed in Germany cause the fall to commence early, to descend far, and to continue for a long time. In consequence of the lowering of temperature, quinine has been employed in many febrile diseases, and may be tried whenever the rise becomes serious, except perhaps in inflammation of the brain and its membrane. In typhus, typhoid, scarlatina, acute rheumatism, erysipelas, and septic diseases it has been tried again and again in varying doses, sometimes being largely used, at others unduly neglected. The massive doses in vogue in Germany seem to have been first resorted to by W. Vogt,[2] who was soon followed by Wachsmuth,[3] Liebermeister, and others. The last-named author considers large quantities most important; *e.g.*, he gives in typhoid 22 to 45 grains at once, that is, he may divide this quantity into three or four doses to be swallowed at intervals of ten minutes, and he insists generally that the whole of it must be taken within half an hour, or at the utmost within an hour, and he considers it useless to look for the full effect if it is extended over a longer time. When the dose is distributed over only half a day, the reduction of temperature is relatively slight; on the other hand he does not repeat such a dose for twenty-

---

[1] Thau: Die Ausscheidung des Chinins beim Gesunden und Fiebernden. 1868.
[2] Vogt.: Schewizrische Monatsschrift f. prakt. Med., 1859.
[3] Wachsmuth: Archiv f. Heilkunde, 1863.

four and usually for forty-eight hours. The decline of temperature goes on for from six to twelve hours, then a rise begins, but usually it does not reach the previous point during the second day. Those who fear this dose, and instead of 30 grains, for instance, give two doses of 15 grains, do not meet with such striking results. Liebermeister stated a few years ago that he had then given some ten thousand doses of 20 to 45 grains without once observing any injurious effect. Some German physicians have gone much further, but his maximum at that time had been 45 grains. The object in view, he considered, is to reduce the temperature to the normal point or near it, and if the first dose does not accomplish this he increases it, but if it brings about a fall below the normal—a circumstance not uncommon—he diminishes the amount of the second dose. Such a distinct remission of the fever is best secured by giving the quinine at night, as the full effect then reinforces the ordinary morning fall. When the fever is abating the indication for quinine is much less than in the continuous stage; its value consists in its power to produce a temporary intermission, and when that already exists of course the remedy is less distinctly indicated. Most German physicians give it in conjunction with cold baths; it obviates the necessity of administering the baths so frequently, and the reliance placed upon it may be estimated by Liebermeister's statement that if he were compelled to the disagreeable alternative of adopting either cold or quinine to the exclusion of the other, he would mostly take the quinine. Should the stomach be unable to tolerate these huge doses of quinia they are said to be equally effectual when given per rectum.

All antipyretics are said to act more energetically on children than adults, but German physicians give quinine very freely even to infants. Hagenbach gives to children under two years of age from 10 to 15 grains; between three and five years, 15 grains; between six and ten years from 15 to 23 grains; between eleven and fifteen years from 23 to 30 grains. Jürgensen reckons a grain and a half for every year of the child's age up to five, after that he gives from 7 to 15 grains. With very high fever he advises as much as 15 grains to a child in its first year and 75 grains to an adult, and says that he has repeatedly resorted to both these quantities.

Quinine will abate the pyrexia of acute rheumatism, and prior to the recent use of salicin in this disease was much more frequently employed. In what perhaps Germans would call moderate doses, namely, from 20 to 30 grains during the twenty-four hours, it has in my experience kept down both the temperature and the pulse, while the effect on the joints seem to be beneficial. The late Dr. Billing, who regarded rheumatism as a neurosis, relied upon quinine, which I have given in seven to ten grain doses three times a day for weeks together. It has often been administered in pyæmia, but without much effect, though many seem to trust it in all septic diseases. Its marvellous power over malarial disease is regarded by many as due to a toxic effect on a bacillus. As soon as we admit that any disease is caused by the intrusion of a micro-organism it is only natural to anticipate that it may be cured by a medicine which is fatal to that organism ; but, as already insisted on, to obtain such a result, it is necessary for the remedy to remain long enough in the system, and of course in the doses required it must not be injurious to the patient. It is, further, easy to understand that a substance which is fatal to some microzymes may have no effect upon others, although possibly some substances may be found to be poisonous to a large number of organisms. Quinine will not cure relapsing fever, and we find that the spirilla of that disease is not injured by the alkaloid.

In respiratory diseases quinine has been employed both as an antipyretic and antiseptic. Whooping-cough has often been thought to depend on a microzyme, and it has been held that quinine will cure it. Letzerich announced, in 1871, that he had found a fungus in the lung in whooping-cough. Acting upon this, Henke employed quinine sprays in hope of reaching the organism. Helmholtz proposed to treat hay-fever by a snuff containing quinine, which certainly sometimes succeeds, as also does a solution employed with a syringe or douche. The local effect on the mucous membrane is by some credited with the result, rather than any influence over organisms. But other antiseptics seem also effectual in this disease.

In *croupous pneumonia* the use of quinine is combined with baths by most German physicians. Jürgensen always employs the two, arguing that quinine reduces the temperature without injuring the heart, from which organ the chief danger is to be apprehended, since death usually occurs through cardiac insufficiency. He therefore rejects both tartar emetic and veratria, and usually also digitalis, on account of the danger of collapse from their effects on the heart. And yet, as we have seen, Briquet's[1] experiments, which were made on dogs, show that quinine is a cardiac poison, and he has been corroborated by Schlockow,[2] and by A. Eulenberg.[3] Their testimony can scarcely be put aside, though Jürgensen does not allude to it, but appears to take it for granted that quinine is a tonic to the heart. But assuredly his doses are not tonic ones. He insists positively on amounts which would seem to incur the risk of being toxic. Thus, in moderately severe pneumonia, he prescribes 30 grains in solution, and in severe cases 75 grains, and even this he does not regard as the limit, and these quantities he orders always in a single dose. This is often followed by vomiting ; if retained for half an hour or three-quar-

---

[1] Briquet: Traité thérapeutique du Quinquina et de ses Préparations. Paris, 1853.
[2] De Chini sulfurici Vi physiologica nonnulla Experimenta. 1860.
[3] Reichert's Archiv f. Anatomie, 1865.

ters, he considers this of no consequence, as most of the alkaloid will have been absorbed. But if delay seems dangerous he repeats the dose, on the plea that it is better to give too much than too little, and whenever vomiting occurs in a quarter of an hour he gives a second dose, and has never seen any injury result. Ice, etc., may be used to allay or prevent vomiting. In order to obtain the full effect of the remedy these huge doses need only be given every second evening; but nutrients and wine as well as cold baths are required to complete the treatment. Very satisfactory statistics have been published as to this heroic treatment, but we cannot forget that pneumonia is a disease in which the tendency to recovery should not be overlooked, and in which masses of figures may be cited to demonstrate the success of the most opposite methods of treatment. I have treated pneumonia by moderate doses of quinine, keeping the patient slightly under its influence, and when further antipyretics seemed desirable, have used aconite, being careful all along to supply nutrients, but not necessarily wine. At other periods I have tried the stimulant plan, once so much in vogue, and still earlier venesection. I have not yet been driven to resort to the huge doses of quinine commended by German physicians.

In *catarrhal pneumonia* the danger arises from respiratory rather than from cardiac failure, so that the treatment by moderate doses of quinine seems to be rational, and aconite may be employed, when it is also indicated, to restrain the heart, but the use of nutrients, of respiratory stimulants, and other adjuncts may be essential. But what, then, shall we say of cold baths, and huge quantities of quinine, which are employed by the advocates of energetic antipyretic treatment? If quinine depress the respiration so much that in poisonous doses death seems to take place from failure of this function, is there not a risk in pushing doses which appear to be dangerously near to toxic ones? This theoretical objection seems to be ignored by those who give such quantities, but they are constrained at the same time to resort to cardiac stimulants, and they insist on the necessity of administering wine freely. It appears to me that this method is less applicable in catarrhal pneumonia than in fevers, while there is also, to my mind, a strong objection to formulating so precisely the treatment of any disease; and this objection is, perhaps, more forcible in reference to respiratory diseases than to the specific fevers. It is the condition of the patient for which we are called upon to prescribe, rather than the nosological position under which his sufferings may be grouped.

In *phthisis* antipyretics are only temporarily useful to reduce high temperatures, and quinine is the only one to which we need resort. Small or moderate doses suffice for this purpose, and it is, perhaps, undesirable to give it continuously. Digitalis is sometimes employed, especially in Germany, but it is apt to irritate the gastric mucous membrane, and for this reason is unsuitable. In no disease do we more need the assistance of the stomach, which is too apt to fail us, and therefore whatever interferes with its function is likely to do more harm than good. Nutrients must form the foundation of all treatment, and rest or a change of diet will suffice very often to put on one side the suggestion of an antipyretic. If it be required, a mild febrifuge or a dose or two of aconite may now and then be employed, and these will not interfere with digestion. In acute miliary tuberculosis the pyrexia has a large share in the phenomena, and cardiac exhaustion is frequently the cause of death. Can we control the fever? Quinine seems the most likely antipyretic, and has been most frequently employed, except alcohol. But it seems useless to dwell upon the appropriate management of this fatal disease.

THE SALICYL COMPOUNDS.

SALICIN, SALICYLIC ACID, AND THE SALICYLATES.

We are certainly indebted to Dr. Maclagan for the modern revival of the use of salicin as an antipyretic, or rather as a specific for rheumatic fever. His earnest advocacy at once led to extensive trials, in which his statements have been generally confirmed, and of course the extensive use of salicin soon led to the employment of salicylic acid and its compounds, all of which are being found useful. As intimated in the word revival, salicin is not a novelty. The willow has, in fact, been employed medicinally from early ages. The Greeks regarded it as a useful astringent, but it was not much employed in modern times, except as a popular remedy. In 1763, Mr. Stone communicated to the Royal Society ("Phil. Trans.," vol. liii.) a paper on the use of salix alba as a remedy for agues, and during the next few years several writers published their views on its medicinal value, e.g., Meyer,[1] Guenz,[2] Koning,[3] Hartman and Luders,[4] and Akerberg.[5] A little later Dr. S. James[6] introduced into practice the broad-leaved willow bark (*salix caprea vel salix latifolia rotunda*) and pronounced it to be a most admirable substitute for cinchona. His statements were confirmed soon afterward by White,[7] who had employed it

---

[1] Meyer, I. J.: De salicis fragilis uso medico. 1770.
[2] Guenz, J. W.: Disp. II. de cortice salicis cortici Peruviano substituendo. 1772.
[3] Koning, P.: De cortice salicis albæ ejusque in medicina usa. 1778.
[4] Hartmann, P. I., et Luders: Diss. de virtute salicis anthelmintica. 1781.
[5] Akerberg, M.: De usu corticis salicis in Febribus Intermittentibus. 1782.
[6] James, S.: Observations on the Bark of a Particular Species of Willow, showing its Superiority to the Peruvian, and its Singular Efficacy in the Cure of Ague, Fluor Albus, Abscesses, Hemorrhages, etc., with case. 1792.
[7] White, W.: Observations and Experiments on the Broad-leaved Willow Bark. 1798.

largely in intermittents. Loeben [1] had in the meantime used the crack-willow (*salix fragilis*) in putrid fevers. About the commencement of this century G. Wilkinson [2] went so far as to claim for the willow bark a superiority to cinchona ; and his observations and cases are full of interest, although it must be confessed he failed to establish his conclusions. Salicin was obtained in an impure state in 1825 by Brugnatelle and Fontana, and Buchner, 1828. Leroux succeeded in 1829 in separating the alkaloid pure. It has been found in several species, at least fourteen of salix, besides eight of populus, occurring in the bark, the leaves, and the flowers. At intervals from this time for about forty years salicin was commended by various authors,[3] and obtained no little repute as a febrifuge, as a substitute for quinine, and as an antiperiodic. It was also found to be tonic and to be much less irritating to the stomach than quinine, so that it was considered as not unlikely to take the place of quinine should that alkaloid, as was feared, become very scarce. The willow bark, besides tonic and febrifuge properties, also possessed some astringency, owing to the presence of tannic acid : it was therefore frequently classed with astringent bitters, and besides being used as a substitute for cinchona, than which again it is less irritating, it was employed in derangement of the digestive organs and as a tonic in convalescence, and, in consequence of its astringency, also in chronic mucous discharges, and even in passive hemorrhage. It was, moreover, said to be anthelmintic, and occasionally even employed as a local astringent. We need not be surprised that with such a history the willow established itself in some districts as a popular remedy, and Dr. Maclagan need scarcely have gone to a colony, or even to a malarial district, for the hint which he turned to so good an account. It appears, however, to have fallen into disuse by the profession, though about thirty years ago it was partially revived, and I then saw salicin freely tried as a substitute for quinine. It was then cheaper, but it was found that so much larger doses were required to produce the same effect that its employment was even more costly.

---

[1] Loeben, J. : De usu corticis Salicis fragilis variis in morbis, præcipue in Febribus putridis. 1793.

[2] Wilkinson, G. : Observations on the Cortex Salicis Latifoliæ. 1803.

[3] Desmartis, T. P. : Propriétés médicinales des diverses espèce de Saules. 1852.

Blaincourt, J. B. : Essai sur la Salicine et sur son emploi dans les Fièvres Intermittentes. 1830.

Besser, O. L. Von : De Salicinio. 1831.

Richelot, G. : Mémoire sur les propriétés fébrifuges de la Salicine. 1833.

Elliot, V. L. : Considératious sur la Salicine et ses propriétés fébrifuges. 1834.

Kenzler, B. : Experimenta circa Salicinæ virtutem febrifugem. 1835.

Blom, P. J. : Beobacht und Beitr. über die Salicine. 1835.

Duhalde, Halmagrand et Gaucheron, MM. : De l'administration du Cyano-ferrure de Sodium et de la Salicine dans les Fièvres d'accès, comme succédané du Sulfate de Quinine. 1861.

This, no doubt, partly contributed to its again being neglected, until Dr. Maclagan thoroughly established its value.

Salicin is decidedly antipyretic; it is easily tolerated by the stomach; it remains in the system a considerable period; it is also antiseptic, and probably antiperiodic. It does not, at any rate in reasonable quantities, produce the extreme depression, and even collapse, which is apt to occur under the influence of salicylic acid and its salts. It is more easily tolerated than quinine. We can scarcely regard it as poisonous, since, according to Huseman ("Die Pflanzenstoffe"), Ranke took three ounces in the course of three days without suffering any inconvenience. Perhaps his specimen was not very pure, for certainly much less quantities have given rise to toxic symptoms, including rather severe delirium. Ordinary medicinal doses, say from ten to thirty grains, if frequently repeated, will soon produce noises in the ears, headache, and giddiness. But toleration is very soon established. I have taken half-drachm doses every hour for five or six hours in succession, and then continued the medicine at longer interval for some days; with the result of reducing a high temperature and rapid pulse, and with no inconvenience except singing in the ears, and that not so marked as when produced by quinine, though it continued for some days after leaving off the medicine.

Salicylic acid is much more powerful as an antiseptic. Wagner says it is more active than carbolic acid as a disinfectant for application to wounds. The presence of one part of acid in two thousand will arrest fermentation, while half that quantity, one in four thousand, suffices to check the amœboid movements and retard the outwandering of the white blood-corpuscles. A proportion of one in a thousand produces stasis in the vessels and destroys the leucocytes. A solution of one per cent. checks the action of ptyalin on starch, while in artificial digestion the action of pepsin is greatly interfered with by the presence of 0.2 per cent. The acid is a local irritant, and as such is very apt to distress the stomach; it is, too, rather insoluble, and for this reason, as considerable doses are required, often given in powder in wafer paper. This is the easiest mode of administration, but very often the stomach cannot tolerate it. The acid is the quickest of the salicyl antipyretics; it is much more rapid in its action than quinine, though the effect does not last so long, and it is far more depressing. After a very transitory stage of excitement the fall of temperature and depression of the heart's action with lowered blood-pressure and relaxed vessels are produced. The respiration is also much disturbed, and if the medicine is not suspended collapse may ensue. A couple of doses of fifteen to twenty grains will often bring down the temperature several degrees in the course of an hour or two.

The salicylates are much less irritating to the stomach, and are therefore frequently preferred to the acid; of course, the salts, as such, are neither antiseptic nor disinfectant, but they may be decomposed in the body, when the free acid would act in this manner. As antipyretics they are nearly as powerful as the acid, and much more easily tolerated. The salicylate of soda is most commonly used, but I often prefer the potassium or lithium salt—the last having special advantages. Salicylate of quinia and the other cinchona alkaloids have also given me satisfaction. The salicylates may also be administered alternately with quinine, or a single large dose of the latter may be employed during a course of frequent doses of the former, for in order to obtain their full effect the salicylates have to be given at short intervals.

Just as in the case of quinine, the salicyls have very little effect on the temperature in health, though they very speedily reduce the abnormal heat in fever. Here the effect is rapid; it is most remarkable, perhaps, in acute rheumatism, though in other febrile states it is also very marked. There is, however, a tendency to relapse after the salicyls, which renders it necessary to continue the remedy for some time after the relief has been obtained. Moreover, there is during the administration a danger of collapse being brought on by these compounds, except in the case of salicin, which therefore usually deserves the preference.

All the salicyls are readily taken up into the system. In the blood the acid would almost necessarily become salicylate of soda, though a portion of it seems to unite with glycocol to form salicyluric acid. It has also been conjectured that some of the acid is again set free by the carbonic acid of the blood-plasma in inflamed parts, and so acts on them locally as an antiseptic, but this is quite hypothetical. In the intestinal canal salicin is split up into glucose and saligenin, and this alkaloid in its turn into salicyluric, salicylic, and salicylous acids—the last of which is a local irritant.

A small portion of the salicyls passes away by the sweat, saliva, bile, and mucous secretions, but the principal part is removed by the kidneys. It can be detected in the secretion in about ten minutes after it has been taken, and the elimination goes on during from twenty-four to forty-eight hours. Thus we have a remedy which can remain in the system for a considerable period, a point of no little importance in reference to its antiseptic action. It might have been thought that it would not be necessary to give such frequent doses of a remedy which is so slow to disappear, but as the elimination begins early and is at first very rapid, the amount circulating in the system is speedily reduced, and, moreover, the liability to disagreeable and dangerous symptoms seems to forbid the attempt to meet the circumstance by very large quantities at long intervals.

Many believe that the antipyretic properties of the salicyls are due to their antiseptic power; if so, we can understand why the acid should be the most potent, for locally it is both antiseptic and disinfectant, and for this reason it would naturally be preferred in sarcina and in dyspepsia attended by putrefactive changes in the contents of the stomach. The effect on typhoid and other fevers may also be attributed to a destructive influence on organisms, and the very special value in acute rheumatism is held by some to be due to a fatal effect on a microzyme, to the presence of which they attribute this disease. And certainly the action is something more than merely antipyretic, for the remedy reduces the pain as well as the fever, though it has no claim to be called an anodyne; it also reduces the swelling of the joints, and, in fact, relieves the most urgent symptoms, so that it may be said to cure the disease. Yet, as we have said, relapses are apt to occur, as if the micro-organisms, if they exist, were only "scotched," not killed. As an antiperiodic or antimalarial, perhaps the effect is due to the prevention of the rise in temperature which forms so important a feature of every paroxysm of ague, but in these cases quinine is superior, so we may dismiss Dr. Maclagan's hypothesis as to the willow flourishing in damp situations, where rheumatism, to which it is antidotal, is also rife. Salicin is not only easier to take, but in rheumatic cases very much safer, for the acid and its salts exercise a particularly depressant action on the heart, and we know that in rheumatism this organ is very likely to suffer, and it is easy to understand that the additional depression caused by the remedy would aggravate the danger. Salicin appears to be comparatively devoid of the tendency to produce this depression, and can be given in full doses frequently without fear; in fact, small doses are of very little use.

If pneumonia be regarded as a general fever, a view which has obtained, among other able observers, the support of Austin Flint, who calls it "pneumonic fever," we might expect that the salicyls would prove valuable antipyretics in this disease, and as the antiseptic action is most obvious in the acid, it would be natural to prefer it. In diphtheria we can obtain a local effect on the throat, as we can also in scarlatina; but then it would interfere with the use of iron, which is sometimes so urgent. The mouth and throat, however, may be cleansed by the acid, locally applied, in either of these diseases, and the internal use will reduce the temperature. But in diphtheria the antipyretic effect is not usually urgent. Other forms of sore throat are often relieved by the salicyls, as are also hay-fever and influenza; in these also it may be employed locally as well as internally, or salicylic acid may be used topically while salicin is given internally. In the recurrent pyrexial attacks in the course of phthisis salicin may also be used, but the acid is too depressant and quinine is almost always to be preferred.

Salicylic acid is rather insoluble, and therefore mucilage is often employed to suspend it in water. Glycerine is a convenient solvent and will take up thirty or forty grains of the acid in the ounce, and when hot sixty grains; the glycerine can be diluted with warm water at the time of use. A solution of from one to two grains in the ounce will keep pretty well, and is also effective locally. In giving it internally it is well for the acid to be pretty freely diluted, which can be accomplished by using the glycerine mixed with water, a drachm to the ounce, instructing the patient to drink some water after it. The salts are freely soluble and possess a sweetish taste, which some do not object to; salicin is decidedly bitter, not very soluble, but can be given in water or in wafer paper.

KAIRIN.

During the last two years much has been written respecting this substance, which appears to be both a certain and potent antipyretic, and, perhaps, in proper cases, when cautiously given, it may be considered as safe, but its action certainly varies in different individuals, so that the proper dose for each has to be ascertained by careful observation at the commencement. Moreover, as some of the concomitant effects are such that the patient should be constantly watched while under its influence, it seems to be most adapted for hospital practice, where there is no difficulty in taking the temperature every hour or two. Kairin is described as the hydrochloride of oxyethyl—quinolin-hydride, or as a methyl hydrate of oxyquinolin, having the formula $C_{10}H_{13}NO$. It is a derivative of quinolin, and is said to exist under two forms, viz.: *kairin*, which is, perhaps, more durable and more agreeable in its effect, and *kairin M.*, which appears to have been the more extensively used.

Kairin was discovered by Dr. O. Fischer, of Munich, about two years ago, and was introduced as an antipyretic by Dr. Filehne, Professor of Physiology at Erlangen, who has employed it in all febrile conditions, including pneumonia and phthisis. He gives it in doses of 4 grains in wafer paper, directing water to be freely drank after it, but for the first three or four doses double the quantity, 8 grains. The temperature is reduced about one degree by each full dose, and as soon as it falls to 100° F. half doses only are given, because if altogether discontinued the temperature again rises rapidly, with perhaps a rigor, but when the smaller doses are continued only a gradual rise occurs. Should a rigor take place, a dose of 8 grains is immediately given. In feeble subjects smaller quantities suffice, say two or three grains, but in some persons 8 grains may fail to produce satisfactory reduction, and then more may be given, 12 or even 16 grains per hour for three or four doses, but to be discontinued as soon as the temperature falls to 100° F. By careful observation during the first day the proper dose for any individual may be determined, for the system does not get accustomed to the remedy, neither has any cumulative effect been observed.

Dr. Girat has experimented largely on animals, and finds that kairin lowers the temperature very constantly, while it also retards the pulse and respiration; he injected it hypodermically, and found that the member operated upon was paralyzed, the sensibility of the part was blunted, and sometimes complete anæsthesia took place. The kairin is eliminated chiefly by the kidneys, and its presence in the secretion may be detected in twenty-five minutes after the dose, and under its influence the urine acquires a dark green color. This effect is also produced upon man by its clinical use; discoloration sets in in about twelve hours and lasts twenty-four hours. Dr. Girat determined the toxic dose to range between one and two grains per pound of the animal's weight.

Dr. Paul Guttmann (*Berlin. klin. Woch.*, 1884) has employed kairin in 86 cases, in doses of 7 to 15 grains. It speedily reduced the temperature, which in many cases fell to normal within four hours, the reduction being accompanied usually by profuse sweating. Sometimes vomiting was produced, and occasionally singing in the ears. His cases included pneumonia, phthisis, pleurisy, measles, scarlet fever, and other febrile diseases. The sweating was particularly observable in cases of phthisis. The unpleasant symptoms were less noticed when the kairin was freshly prepared, but some specimens which had been kept produced not only disagreeable but even alarming symptoms, such as cyanosis and collapse. We may remark, however, that such accidents have occurred when the quality of the drug could be in no way impugned. For example, Freymuth found in several cases severe gastric disturbance and prostration, compelling him to discontinue the remedy. So did Seifert, as well as Riegel, and even Filehne has met with similar inconveniences.

Dr. Gottlieb Merkel treated nineteen cases of phthisis, pleurisy, carditis, scarlet fever, and typhoid with kairin exclusively, the patients being from sixteen to fifty-two years of age. These cases confirm the antipyretic power, and he seems so to have regulated the use as to have avoided disagreeable or dangerous symptoms, but he finds the greatest benefit when neither the heart nor the lungs are affected.

Dr. Hallopeau has employed it in pneumonia and phthisis, and considers the antipyretic effect to be certain, but the question remains whether such an effect is desirable in these cases. Dr. Menche cannot recommend it, for he several times observed collapse to be brought on in phthisis and pneumonia, though he had better results in rheumatism, in which he thinks it might take the place of salicylate of soda. So Dr. Korach (*Deut. Med. Zeit.*) met with severe collapse in pneumonia and typhoid, though he admits the antipyretic influence.

Professor Riegel reported (*Allg. Med. Central Zeit.*, July, 1883) unsatisfactory results in pneumonia; he obtained but slight reduction of temperature by four half-gramme doses (seven and one-half grs.) at intervals of one hour, and even larger quantities were comparatively powerless. When the temperature fell to normal it often rose again, in spite of repeated doses, and the pulse, though reduced in power, was not in frequency. Instead of the patient feeling better, as described by Filehne, severe depression, almost amounting to collapse, often rendered it necessary to resort to stimulants, and he therefore concludes that it is a dangerous remedy, especially in asthenic cases, chiefly on account of its depressing action on the heart. In a later communication (*Centralblatt f. Klin. Med.*, November, 1883) he repeats these objections, and states that, in consequence of the frequently threatened collapse, he has given up the use of kairin in pneumonia, in which he finds the antipyretic effect so slight, and in which, moreover, the danger does not arise from the high temperature. His statements are corroborated by Seifert and L. von Hoffer (*Centralblatt*). The latter gave muriate of kairin in pleurisy, pericarditis, pneumonia, tuberculosis, and fevers, and found it uncertain. In pneumonia, rigors and profuse sweats disturbed his patients; in tuberculosis he considers it contra-indicated, as the temperature only fell for a short time and afterward rose higher than before, this rise being accompanied by heavy perspirations, rigors, and collapse. He also mentions cyanosis and burning of the forehead as disagreeable effects sometimes noticed, but he never saw the digestive system interfered with. He concludes that in feeble persons it should only be given with caution.

We have but few observations by English physicians at present, but no doubt they will soon accumulate. Dr. Ashby, of Manchester, reports (*Brit. Med. J.*, December, 1883) that in a case of typhoid in a child of ten years two two-grain doses always reduced the temperature two degrees. Dr. Carter (*Liverpool Med.-Chir. Jour.*, January, 1884) mentions a case of hyperpyrexia in rheumatic fever in which kairin gave favorable results. Dr. Archer, of Liverpool, has recorded (*Brit. Med. J.*, April 12, 1884) a case of enteric fever treated with kairin, with the observations on temperature, pulse, and respiration taken at the time by his resident officer, Dr. Oldham. His case is said to have been "manifestly a desperate one" on admission to hospital, and although the patient died the temperature was greatly influenced by the remedy, and "on two or three occasions, when he was about to die, kairin apparently seemed to have a great influence in assisting to resuscitate the flagging vital powers, and we had some hope of rescuing him."

Kairin appears to have no claim to be an antiperiodic, even larger doses fail to cut short ague, and Cohn and Zadek (*Deut. Med. Woch.*) were unable to prevent the rise of temperature in remittents by giving the dose at the beginning of the exacerbation and repeating it hourly. In relapsing fever, however, full doses appear to prevent the relapses, but gastric disturbances and prostration are apt to follow. A very interesting observation has been made by Freymuth, who found that kairin had no effect upon the spirilli of relapsing fever, although it reduced the temperature, showing apparently that this last effect is not dependent on an antiseptic or germicide action.

Chinolin or Quinolin.

This derivative of quinine, chinchonin, or aniline, is a mobile, refracting liquid, lately introduced as antipyretic and antiseptic. It forms definite salts and the tartrate and salicylate have been employed in doses of five to ten, and even fifteen grains. Externally quinolin is a disinfectant, and a five per cent. solution has been used to paint the fauces, or as a gargle in diphtheria. Internally the dose is three to ten minims.

Resorcin,

named from being first derived from a gum resin, galbanum, and isomeric with orcin, which is derived from archil, is another recent introduction, possessing similar properties to quinolin, being a disinfectant externally, two to ten per cent. solutions being used, and these are not irritant, although resorcin itself is a caustic. It may be applied to the fauces in diphtheria either by painting with a solution or as a spray. Resorcin is a meta-dihydroxyl-benzol, a neutral crystalline, white body. It is now obtained from carbolic acid. It is freely soluble in water, alcohol, and ether. Internally it reduces the temperature and pulse in pyrexia, but does not affect the normal temperature. Excessive doses give rise to singing in the ears, deafness, mental disturbance, tremors, and other nervous symptoms. Above one drachm it is poisonous, but that amount has been given as a single dose; usually from seven to ten grains every two hours suffices. The taste being somewhat pungent, it should be well diluted.

Veratria

is a very powerful alkaloid possessing exceedingly irritating properties, and more adapted for external than internal use; nevertheless, it has been employed as an antipyretic, for it reduces the temperature, weakens the circulation, and retards respiration. It is a powerful poison to the cardiac muscle, as well as to its ganglia, and it depresses both the cardiac and respiratory centres. A minute dose is said to quicken the pulse and raise the blood-pressure, but after a full dose there is an immediate fall in the number of the heart-beats, as well as in the arterial pressure. In pyrexia the fall of the temperature may amount to several degrees Fahrenheit, but this is apt to be accompanied by vomiting, prostration, syncope, and collapse. Seeing that minute doses retard the respiration, even after section of the pneumogastrics, we must admit with Bezold and Hirt that it is a direct depressant of the respiratory centre in the medulla, eventually killing it. It seems to have been first employed as an antipyretic by W. Vogt, followed by Liebermeister, who in typhoid prescribes one-twelfth of a grain in a pill every two hours, and says that from four to six such doses suffice. He does not consider the collapse which so often ensues as dangerous, and says that it is easily controlled by restoratives and stimulants. Jürgensen avoids this alkaloid in pneumonia, because he regards the collapse as due to the effect on the heart, the failure of which is so often the great danger in this disease. Veratria is so much inferior to aconite and other remedies less dangerous and less disagreeable that I altogether exclude its internal use from my own practice. Taylor mentions ("On Poisons," 1875) a case in which one-sixteenth of a grain had been taken in a pill, and the patient was soon afterward "found insensible, the surface cold, the pulse failing, and there was every symptom of approaching dissolution." The patient ultimately recovered, but remained some hours in a doubtful condition. Yet larger doses than this have often been prescribed, but the case certainly justifies the opinion that it is wiser to reject this alkaloid as an internal remedy.

DIGITALIS.

Wunderlich seems to have introduced the use of digitalis as an antipyretic. Thomas, Ferber, and Liebermeister soon followed. They give as much as ten to twenty grains in the course of thirty-six hours, and follow it up by a full dose of quinine, thirty to forty-five grains, and this has sometimes succeeded when quinine alone failed in reducing the temperature. The indication for its use as an antipyretic is directly opposite to that in heart disease. The more frequent the pulse the less is digitalis indicated. The German physicians, who have chiefly used it, consider that in patients with powerful action of the heart no harm is done if it cause nausea and vomiting, but as soon as it does so the medicine should be stopped. In pneumonia it has also been used, but the effect upon the heart is late, and we therefore fail to get it just when we want it, while if we give large doses—and these are necessary to reduce the temperature—collapse is apt to come on just at the period of defervescence. Any one who has met with that, says Jürgensen, will not be anxious to repeat the experiment, though he thinks that when the heart is vigorous it may be safely employed to reduce the temperature.

As a cardiac regulator, not as a febrifuge, small doses at considerable intervals may be useful in pneumonia, as well as in other diseases, but while in this case we avoid the paralyzing influence on the heart we lose the effect on the temperature. For either purpose I regard it as vastly inferior to aconite, which is more certain and less dangerous. There seems to me to be little if any physiological basis for the use of digitalis as an antipyretic, while the clinical evidence is not fully satisfactory. A careful study of Wunderlich's[1] cases will show that the medicine was often given in such a manner that its action was produced about the period of the normal decline, and so far they fail to carry conviction that the fall was produced by the treatment. The various effects of digitalis in different doses—the fact that it is in one case a cardiac tonic or stimulant and in another a depressant, that its action is confessedly slow, and that at the best it is uncertain, leads me to prefer either aconite or quinine, according to the cases, to the almost total exclusion of digitalis.

In *bronchitis* it has been used to support the heart, but other things are quicker in their action as well as more certain, so that here again we may regard it as either unnecessary or dangerous. In *hæmoptysis* arising from heart disease the use of digitalis can be understood, and perhaps it may be admissible when this symptom is due to the congestion of incipient phthisis, provided always that the circulation is languid and there are other indications for its use. Some venture to give it in fully developed phthisis, particularly in combination with quinine, but here again the small quantities in which it is employed are incapable of affecting the temperature, and it must be regarded rather as a cardiac remedy. Further, it is very liable to derange the digestion, unless in minute quantities, and certainly in phthisis it is of the highest importance to dispense with anything which is likely to interfere with nutrition. The value of digitalis in phthisis was urged at the beginning of this century by Beddoes,[2] Kinglake,[3] and others. It has continued to be employed more or less down to the present time. Quite recently Dr. Jacobi stated, at the Medical Society of the State of New York (*Medical Record*, February 23, 1884), that he seldom treats a case of phthisis without it. But it is not as an antipyretic he employs it, but rather for the effect on the heart and indirectly on nutrition. It is therefore in small doses combined with other remedies, such as iron, nux vomica, arsenic, etc., that it has been most generally used and is most likely to be of service in phthisis. Thus it will be seen that in all respiratory diseases the administration of digitalis must be regulated by the condition of the heart. In fact, we regard it as a cardiac regulator, and consider the benefits which appear to have followed its use as due to this action, directly or indirectly.

---

[1] Wunderlich: Manual of Medical Thermometry.

[2] Beddoes, T.: Observations on the Management of the Consumptive, on the Powers of Digitalis Purpurea, and on the Cure of Scrofula. London, 1801.

[3] Kinglake, R.: Cases and Observations on the Medical Efficiency of Digitalis Purpurea in Phthisis Pulmonalis. 1801.

Aconite.

Greek writers often speak of a poison named aconite (ἀκόνιτον), and Dioscorides describes (lib. iv., cap. 77, 78) two kinds, but not in such a way as to enable us to identify the plants to which he refers. Latin poets seem to have used the word as a synonym for poison. The genus to which aconite belongs is very numerous. Decandolle included one hundred and seven varieties in twenty-two species. The aconitum paniculatum was once adopted by the London and Dublin pharmacopœias, but it is probably inert, for Christison having raised it at Edinburgh, from seeds supplied to him by Decandolle, found this plant destitute of medicinal properties. It is doubtful, indeed, how far aconitum paniculatum was ever employed, for Pereira could not find a sample in commerce, nor was he able to obtain a single specimen except such as had been grown in botanical gardens.

The aconitum napellus is now the only official plant.

Baron Stoerck,[1] in 1762, wrote an essay on this and other vegetable medicines, from which may perhaps be dated the introduction of aconite into use. It is true that some of his statements were much exaggerated and were even disputed at the time, especially in reference to conium, but many of his observations are worthy of consideration, though we must admit that no value attaches to his cases of cancer, which he reported as cured and which his contemporaries, four or five years after, tell us all died of the disease. Manghin[2] sums up some of the properties of aconite in a letter to Stoerck. Andreas[3] proposed it, in 1768, as a remedy in arthritic diseases, in which he was soon after supported by Spielmann,[4] and a little later by the Abbé Mann.[5] Several other publications on the subject appeared in the same century, after which there was a lull in the literature and the medicine seems to have fallen into comparative disuse. But in 1835, Dr. Turnbull[6] again called attention to the value of aconite in painful nervous affections, with an enthusiasm which seems to have caused distrust rather than confidence. In 1844, the Ed-

---

[1] Stoerck, Baron A.: Libellus quo demonstratur Stramonium, Hyoscyamum, Aconitum tuto exhiberi usu interno, et remedia esse in multis morbis maxime salutifera. Vienna, 1762.

[2] Manghin: Epistola ad Stoerck de Aconito. Vienna, 1766.

[3] Andreas, J. F.: De usu salutari Extracti Aconiti in Arthritide, observationibus comprobato. 1768.

[4] Spielmann, S. A.: De Aconito. 1769.

[5] Mann, M. L'Abbé: The Extraordinary Cure of the Gout by Hemlock and Wolfsbane. Trans. by P. Thicknesse. 1784.

[6] Turnbull, A.: On the Preparation and Medicinal Employment of Aconitine by the Endermic Method in the Treatment of Tic Douloureux and other Painful Affections. London, 1835.

inburgh Senatus Academicus awarded a gold medal to Dr. Alexander Fleming[1] for an inquiry into the properties of aconite. This essay was published the following year, and has continued to be regarded as a valuable contribution to therapeutics down to the present time. He recommended aconite in neuralgia of all kinds, in diseases of the heart, in rheumatism, in erysipelas, carcinoma, etc. It is curious that both before and since a similar wide field for its use has been claimed. Thus, as we have seen, Stoerck recommended it in cancer, Andreas in gout, others employed it in inflammation, and Marcello,[2] somewhat later, as a substitute for depletion in fevers, both continued and intermittent. No doubt, as we shall see, its anodyne influence when employed locally led to its use in painful diseases, while internally its febrifuge power caused it to be employed not only in pyrexia, but whenever the pulse was frequent, whence, perhaps, its repute in some cardiac diseases. My own experience of its use dates from my entrance into the profession, since as soon as I became a pupil I found that my revered teacher, the late William Huxtable, had been employing it in a large practice for many years, and continued daily to prescribe it.

When given to animals aconite produces muscular weakness, causing them to stagger, vascular depression, sometimes, but not often, vomiting, perhaps convulsions, and eventually death. There is also general anæsthesia, as was often demonstrated on dogs by the late Dr. Pereira, at the London Hospital. If, however, too large a dose be given, the animal may be killed so quickly that the anæsthesia is not observed. Death may occur, apparently from paralysis of the heart, in less than a minute after a hypodermic injection of aconitia. On opening the thorax immediately after death Pereira found that there was no pulsation of the heart visible, but Fleming states that he has seen it continue to beat for some time. Similar effects are produced on most animals.

In cases of accidental poisoning in man, the first symptom has usually been burning or tingling in the mouth and throat, followed by a similar sensation in the extremities, which rapidly spreads to the trunk. The pulse is greatly depressed and later becomes weak, irregular, or intermittent, and at last cannot be felt; respiration is also slow, feeble, shallow, irregular, and at last arrested. There is great prostration from loss of muscular power. The general sensibility is also reduced, so that with the tingling sensations more or less cutaneous anæsthesia is noticed. There is great anxiety, but the intellect usually remains clear to the last. The voice is reduced to a whisper and sometimes the special senses are lost.

---

[1] Fleming, A.: An Inquiry into the Physiological and Medical Properties of Aconitum Napellus, to which are added Observations on several other Species of Aconitum. London, 1845.

[2] Marcello, C.: Dell' uso dell' Aconito Napello nelle Febbri Continue ed Intermittenti irritative, come surrogato al Salasso ed al sanguisugio. 1856.

All this time the temperature steadily falls. The muscular weakness seems to be due to depression of the motor nerve-endings which gradually extends to the cord. The sensory nerve-endings are also affected and the paralysis extends inward, as it does also when the poison is locally applied. Sometimes the symptoms of poisoning come on almost immediately, but in other cases a considerable delay has been observed. Such variations are probably due to differences of dose or of the preparation and to the condition of the stomach with regard to food. A number of cases have occurred in which the root has been taken in mistake for horse-radish. As some time would naturally elapse before absorption when the poison was thus taken with a meal, it is not surprising that symptoms should not appear for three-quarters of an hour; though when the tincture has been taken on an empty stomach the effect has been almost immediate.

Many years ago I was called to a young lady aged twenty years, who had swallowed a dose of aconite liniment in mistake for some drops which she had been accustomed to take for hysteria. A very violent paroxysm of that malady had occurred in the evening, and a dose of the supposed remedy was administered. The stomach was empty at the time, and she complained immediately of a burning in the throat; in a very few minutes tingling of the lips, mouth, and throat was felt, and she said that her hands and feet had both gone to sleep. Within an hour this characteristic symptom had spread over the whole surface, and there was distinct loss of sensation; the pulse was slow and gradually became weak, and later still irregular and scarcely perceptible; the respiration was feeble, infrequent, and gradually became shallow and irregular; the diminution in the general sensibility gradually increased, the voice was reduced almost to a whisper, a cold clammy sweat stood upon the surface; there was extreme anxiety, great muscular weakness, no convulsion, no spasm, no pain, no vomiting. The intellect was perfectly clear. The horizontal position was strictly enforced and stimulants freely employed, but she continued in a state of peril for nine or ten hours.

It will be observed that in many respects the symptoms produced by aconite are the opposite to those caused by digitalis, which has therefore been employed with a certain amount of success as an antagonist. Dr. Milner Fothergill's experiments on this subject seem to show that digitalis may be useful as a protective "if given in sufficient dose and a sufficient time (from five to nine hours) before the dose of aconitine is administered" (*Brit. Med. Journ.*, August, 1877).

The effects of medicinal doses closely resemble those produced by larger quantities; thus locally aconite causes numbness or tingling, or a sensation often called "pins and needles." It lowers the sensibility to touch and temperature, no doubt through its powerful depressing influence on the sensory nerve-endings. If a portion of the root be chewed, or a drop of the tincture placed on the tongue, a similar sensation extends over the lips, mouth and throat. Full therapeutical doses reduce the pulse in frequency, force, and tension; they also retard the respiration and cause a feeling of weakness with the characteristic tingling of the extremities. As the pulse falls, the skin flushes and becomes moist. Besides this diaphoresis the medicine may also cause some diuresis. If the dose be repeated or have been at first very large the effect on the heart is more serious, and appears to be due to a direct action on the organ; the pulse, instead of having fallen about ten beats, may have been reduced fifteen or twenty, and become smaller, weaker, quicker, and eventually irregular. Muscular debility is now extreme, respiration shallow and hurried, or sometimes sighing; chilliness is complained of, the extremities are cold, prostration profound; if the patient attempt to sit up, giddiness, confusion of sight, or syncope occur, and the heart ceases in diastole. In all cases of an overdose, the horizontal position must be maintained, as there is great danger that on attempting to sit up, fatal syncope may occur. All these symptoms may occur from what have been sometimes described as full therapeutical doses, but it is most desirable that such quantities should very rarely be used, for a very large experience of this remedy convinces me that all its good effects may be obtained without running the risk of producing grave physiological effects.

The symptoms produced by full doses of aconite begin early and continue for a considerable time. A full medicinal dose will affect the pulse and produce other symptoms in from fifteen to thirty minutes, sometimes even earlier. The influence reaches its height in from one to two hours; it diminishes slowly, so that there is but little decrease for three or four hours more, but after that the symptoms gradually subside, though a degree of lassitude remains for several hours longer. If the dose be very large, or if a second dose should be administered before the effect of the first passes off, the effects will be still more marked, the pulse may fall to forty or even lower, but it soon rises and becomes at the same time small, weak, and irregular; the respiration also becomes shallow and hurried; sometimes sickness appears, with intense prostration, headache, giddiness, and interferences with or loss of special senses. These effects may last thirty-six or forty-eight hours. It may now be well to consider the mode of its action on different systems.

On the *circulation*, as we have seen, the effect is very decided. It depresses the heart by a direct action on its structure or on its ganglia; for the action has been shown to be produced on the frog's heart after its removal from the body, and on the application of the drug to the surface of the heart the usual phenomena are observed; after action of the vagi and in atropinized animals the symptoms of poisoning also occur. Perhaps the retardation of the pulse may depend on two influences, one in the heart itself, the other in the medulla. Dr. J. B. Nunneley was unable to detect any alterations in the vessels of the frog's web ("Proc. Royal Soc.," 1870), and Achscharumow (*Reichert's Archiv*, 1866) found, after dividing the sympathetic in the neck, that galvanizing the peripheral end produced the usual phenomena even in the most advanced stages of poisoning by aconite. Thus the drug seems not to affect the vasomotor nerves—a conclusion confirmed by other experiments. Nevertheless, impulses cannot be transmitted from the periphery to the vasomotor centres, as the poison destroys the conducting power either of the afferent nerves or of the cord.

This brings us to a consideration of the effects on the *nervous system*, on which very divergent views have been held. It is not improbable that great differences in the alkaloid or other preparations employed have vitiated some of the experiments. Even in toxic quantities loss of sensation occurs before serious disturbance of respiration, and long before loss of voluntary motion dogs will walk about, follow their master, answer to their names, and wag their tails when anæsthesia of the surface is so complete that they are insensible to pinching or pricking with needles. Liégeois and Hottot (*Journ. de Physiol.*, 1861), from very complete experiments on frogs, conclude that the anæsthesia is produced by paralyzing first of all a sensory perceptive centre above the cord, and secondly, the peripheral extremities of the nerves, which last effect afterward extends to the nerve-trunks. Loss of sensibility in frog's legs was simultaneous with or even preceded disturbance of respiration, and occurred not only long before the loss of voluntary motion, but even while reflex activity remained intact. The application of aconitia to a nerve-trunk paralyzes its sensibility, but when the veins of a frog's leg are tied and the drug injected into the artery, the skin loses its sensibility long before the nerve is affected. When voluntary movements are at length prevented Liégeois and Hottot ascribe this effect to the centre rather than to paralysis of the motor trunk, because irritation of the nerves induces contractions of their muscles; but the motor trunks must be somewhat affected, for the local application of aconite suspends the conducting powers, and after death their sensitiveness is lost earlier than normally.

In the *respiratory* system the effect of aconite is well marked and comes on early. The action appears to be on the centre of which it is a direct depressant, for arrest occurs in the frog before the motor nerves become affected, and previous section of the vagi does not prevent the effect of the poison on the respiration. The fall in the temperature in a fatal case recorded by Achscharumow amounted to three degrees Cent.

The effect on the *muscular* system sometimes amounts to complete prostration. Tremors and nervous twitching are rare; partial or complete loss of voice is not uncommon; muscular debility may pass off in a few hours or may last several days.

Alimentary System.—A sensation of warmth in the epigastric region is common enough, and nausea and even vomiting have been observed. Christison mentions ("On Poisons") that he had observed griping and diarrhœa, and others have spoken of an irritant action on the alimentary canal, but this must be very rare even when toxic doses have been taken, and I have never known it to occur from medicinal doses.

On *secretion* the effect is only moderate. Stoerck and others ascribe to it diaphoretic and diuretic qualities, and these it certainly possesses. Most authors admit that it is somewhat sudorific. As soon as the circulation is depressed, perspiration is likely to break out. The diuretic action is much less evident, and cannot be turned to any practical account, elimination probably occurs chiefly through the kidneys, and is rather slow, as may be supposed from the long time which the effects continue. Perhaps some portion may be carried off by the skin, for an eruption has occasionally been observed to make its appearance. It may lay claim to be in some circumstances an emmenagogue, as under its influence, given for other purposes, the menstrual flow is often restored.

If we call to mind the manner in which diuresis and diaphoresis occur, on the fall of the pulse and temperature, it will not be surprising that secretion generally may be promoted by any medicine which produces such fall.

The chief use of aconite is as a febrifuge to reduce arterial action and lower temperature; a subsidiary use is to allay pain and spasm, as well as to abate excitation of sensory nerves. H. C. Wood says that he has never used it "in those cases, such as pneumonia, in which a sudden and very powerful effect is desired, simply because veratrum viride seemed to him safer, more readily controlled, and equally effective." This statement is just the reverse of what I should make respecting these two drugs. Perhaps our experience differs, because each has learned to trust to one drug almost to the exclusion of the other. He admits as to veratrum that "in practice it should simply be used to lessen the force of the circulation," and further on he says, "When true sthenic arterial excitement is to be combated in any disease, except it be gastritis, veratrum viride may be employed as a prompt, thoroughly efficient, and at the same time very safe remedy—very safe, since it is almost incapable of producing death in the robust adult, unless used with great recklessness and in repeated doses. In the early stages of sthenic pneumonia, it offers, I believe, the best known method of reducing temperature and the pulse rate, and of ameliorating the disease." This is almost the kind of statement that I could indorse with regard to aconite, and then it would not be necessary to exclude gastritis; but one point should be constantly remembered, viz.: that it is quite unnecessary to give what have been called full therapeutical doses. Much better results are obtained from repeated small doses, and as soon as distinct effects are produced, the medicine, as I have long taught, should be suspended or given at longer intervals.

I have been in the habit of prescribing aconite in numerous diseases for the last thirty years. In the first edition of my work on Sore Throat, 1861, from the result of many hundred cases, I strongly recommended it in febrile and inflammatory diseases; at that date I entered a caution as to the dose, which in all the text-books was overstated. Admitting it to be true that a single dose of five minims or more of the tincture might be administered with benefit, I urged that repeated doses of much smaller amount are more advisable. The strength of the British officinal tincture is only about a third of that of the former London Pharmacopœia, but still the officinal dose is too large, except under rare circumstances; two minims can be repeated three or four times a day, but as a rule a single minim will suffice. In rare cases three may be required, but the effect should be watched. Aconite is not adapted for use in single large doses, after the manner in which some antipyretics are now employed, and any attempt to use it in this way is likely to lead to serious consequences. But in full doses, at moderate or short intervals, after the fashion of saline febrifuges, there is no more valuable remedy. A cautious use of this medicine is devoid of danger, though if recklessly pushed it may undoubtedly destroy life—rapidly and unexpectedly, for toxic symptoms develop themselves rather suddenly, if the first indications of its action be unheeded. In fact, it is a cumulative poison, and, consequently, is not to be prescribed in increasing doses. All the good effects may be obtained by small quantities, repeated at longer or shorter intervals, according to the rapidity of the action desired. I stated in 1861 that, although I had used it in thousands of cases, I had never produced alarming symptoms of poisoning, and knew no medicine which less frequently disappointed my expectations. Further, I said I had given it once, twice, and thrice a day for a considerable time, every four, three, two hours for a shorter time; and sometimes repeated a dose every half hour, carefully watching the patient. Others have since claimed great credit for the *discovery* that a dose may be given every *quarter* of an hour!

After a few doses, sometimes after a single one, the action of the drug is observed, the pulse is reduced in frequency and power; in some cases the power is increased, the frequency diminished. The skin becomes relaxed and bedewed with a gentle perspiration, nervous irritability and excitement are allayed, a calm comes over the patient, and often a sound sleep returns after a long absence. The pain is relieved almost as certainly as when it is locally applied for neuralgia. Clearly, then, it is a valuable sedative, exercising a marked influence over the heart. It was some appreciation of these properties which caused it to be recommended in heart disease and acute rheumatism. If a drop of the tincture be placed on the tongue, it is found to be acrid and bitter, and this taste is soon followed by a numbness or tingling in the mouth and fauces. Now, when the full action is produced, in giving this medicine, a similar sensation to this is perceived in other parts. The patient will declare or complain that he feels "numbed," or that he has the "pins and needles," or that he feels "just as if his feet had been asleep." This sensation may be very local—confined to the toes, fingers, or eyelids; or may extend up the extremities, almost over the whole body, according to the susceptibility of the individual. The remedy must now be discontinued, or the dose diminished, and only given just often enough to keep the system under its influence. This sensation is to aconite what salivation is to mercury. By it we shall not be misled. Like salivation, it may sometimes seem to fail, at others the effects may follow a single dose, but on the whole it is a certain measure of the patient's tolerance of the drug. The reader may easily produce the sensation on himself by taking a drop or two in water two, three, or four times a day.

The medicine is useful in all febrile ailments; in nervous excitement; indeed, whenever the heart's action is quickened or the temperature increased. There are few such diseases in which I have not at some period exhibited it. I often make it serve the place of salines, and in many cases it is an excellent substitute for digitalis. When sufficiently diluted tincture of aconite is tasteless—a recommendation of no small value to some people. For children a little syrup makes it as palatable as sweetmeats. If not diluted enough, it produces numbness or tingling of the mouth and throat by its contact. This property of acting locally on the membrane may be utilized for medical purposes. Thus in tonsillitis and other inflammations in the fauces and mouth it may be given in powders, made by rubbing up the tincture with a few grains of sugar. In 1854 I began to use it in the form of lozenges, a method I was employing with many other medicines. Of late years I have substituted jujubes under the name of pastilles for local remedies to the throat. A gargle may also be cautiously used, but should not be too strong. Sometimes we may paint the faucial membrane with the tincture diluted with glycerine and water. This arrests pain and often puts a stop to the inflammatory process, but if carelessly done may paralyze the soft palate for hours—a result the patient is not likely to forget, as if the uvula falls on the epiglottis it will cause a suffocative cough or a choking sensation which will distress and alarm him.

In catarrh of the upper air-passages a few doses of aconite will often be all the medicine required, and when there is febrile excitement it is specially indicated. In acute bronchitis, particularly in the early stage and as long as pyrexia is present, small doses may be administered rather frequently in place of the older salines and antimonials. Its febrifuge properties are as marked, its diaphoretic action often sufficient, and it seems also to possess some expectorant virtue, the membrane being soothed but not stimulated under its influence. In chronic bronchitis the propriety of giving it will be determined by the condition of the heart. The powerful depressant action on the circulation must never be forgotten, and we know that in chronic bronchitis and many other respiratory diseases all cardiac depressants are not unfrequently contra-indicated. In capillary bronchitis and catarrhal pneumonia aconite may be employed as a febrifuge with the same reservation respecting its action on the heart. In the early stage it may often replace the ordinary saline or diaphoretic mixture and preclude the necessity of small doses of antimony. At the commencement of acute pneumonia it will restrain the febrile excitement, relieve pain, soothe the nervous system, reduce the temperature, and promote perspiration. It does not preclude the use of quinine as an antipyretic; indeed, I have often given these medicines in combination or concurrently, and I have conducted pneumonia through its several stages with little medicine except aconite and quinine, and may remark that the employment of these in moderate doses is by no means to be regarded as simple expectancy. In pleurisy, aconite fulfils a twofold purpose, it reduces the febrile excitement and assuages the pain. In cases of slight extent and moderate intensity little other internal medication is called for so long as the pulse indicates its use. Of course, no one will suppose that it can arrest effusion into the pleura or to any extent promote reabsorption. It is for its restraining influence on the febrile excitement that in this and other diseases it is to be employed. In some cases of asthma it has occasionally given relief. In phthisis it is useful in the intercurrent attacks of inflammation and also to restrain pyrexia. It is when the pulse runs high, and when there is excessive susceptibility to pain, that it seems to be most useful. Here, of course, it is only employed as a palliative to meet temporary indications, but as such it may prove a comfort to the patient, though without in any way influencing the general progress of his disease. When a febrifuge is temporarily called for it is efficient, and has the advantage over many others that it does not derange digestion.

The *tincture* of the root is the best preparation for internal use. It is unfortunate that there have been such differences in the strength of tinctures in use. Fleming's tincture is too strong and should be discarded, as accidents have occurred from its use. The London Pharmacopœia tincture was three times the strength of our British Pharmacopœia, which is one in eight. The tincture of the United States Pharmacopœia corresponds with the former London Pharmacopœia, and is therefore three times the strength of our present British Pharmacopœia. The latter authority gives the dose as five to fifteen minims, but I have already said this is much too high. For general use I would recommend only one to three as safer and equally efficient. Only occasionally need five be given at a dose. The United States Pharmacopœia tincture would be efficient in from one-fourth of a minim to one minim. The *extract* is not nearly so uniform in power. The pastilles I use contain one-half minim of tincture in each. The alkaloid, aconitia, is so powerful when pure that it is only adapted for external use.

CHAPTER XVII.

NEUROTICS.

We pass now to remedies the chief value of which depends on their influence over the nervous system. The most striking group of these is that to which the term narcotics has been applied. In its widest sense this word may be used concerning anything that causes sleep or stupor. It is perhaps as well restricted to those which positively stupefy, and then other words may be used to express other qualities which may be more or less possessed by narcotics. Thus we have soporifics and hypnotics which induce sleep; sedatives which reduce excitement, and anodynes which relieve pain. Acro-narcotics have sometimes been grouped separately, cardiac and vascular sedatives and depresso-motors are terms which explain themselves, and sometimes neurotics used to produce a particular condition are named accordingly, as mydriatics, which dilate the pupils, and myotics, which contract them, and many other groups. A true sedative ought not to produce a previous stage of excitement, but the term is also loosely employed with reference to both soporific and anodyne qualities.

NARCOTICS.

Narcotics first cause excitement and then stupefy, just as we have seen alcohol does, but the period of excitement is shorter. They are, however, distinctly excitants, that is during the early stage of their operation, and they are therefore sometimes called stimulants; indeed, we sometimes employ them for this very quality. The late Dr. Anstie held that the relief of pain was due to this stimulant action, and that in neuralgia stimulant not narcotic doses would suffice. So in prostration from fever and other acute diseases the stimulating influence of small doses of opium has sometimes been sought, and I think that the same effect may be obtained from other narcotics. We use narcotics for various purposes: first of all, to produce sleep or for their hypnotic quality; then to abate pain, for their anodyne influence, or to overcome spasm, when they become anti-spasmodics; then to restrain secretion and sometimes hemorrhage, when they are astringents, or on the other hand to promote secretion, as in the case of some sudorfics. Further, they are used to control irritation and inflammation, when they may be called antiphlogistics; and lastly, as we have said, they are often valuable stimulants. In reference to respiration, some of the narcotics greatly retard it, and they are apt to interfere with expectoration and augment pulmonary congestion. Others, however, stimulate the respiratiory centre, instead of depressing it, and so quicken the breathing. The effect of these remedies varies greatly, according to the age, the idiosyncrasy and the habits of the patient. The last point is of the highest importance, inasmuch as persons may so accustom themselves to the use of narcotics that ordinary doses take no effect.

OPIUM.—Opium is the type of narcotics, and by far the most important of the class on account of the great range of its action. It has, indeed, often been regarded as the most important of all medicines, and spoken of as "God's great gift to man." The effect of a single small dose is usually mild excitement together with a willingness for quietude and a tendency to dose, or to a dreamy state of semi-sleep, but with an effort this gives way to a readiness to work. Movement is easy, fatigue is not felt, but in repose drowsiness comes on. The ideas flow fast, chasing each other in rapid succession, but are not so easily sustained. Pain, if present, is numbed or even subdued. Secretion is generally lessened, except that of the skin. Hence there is dryness of the mouth, nose, and fauces, no doubt of the entire respiratory mucous membrane, though later a secretion, but more viscid than usual, sets in. Sleep, when obtained, is often disturbed and dreamy, and the patient wakes up unrested, perhaps with nausea or at least a want of appetite, and a disposition to headache which sometimes becomes distressing.

With a larger medicinal dose the excitement is more marked, but of briefer duration. At first the pulse and respiration are both quickened, but they soon become slower, and the first stage gives way to the second, drowsiness is succeeded by sleep more or less profound, and on waking the uncomfortable condition is often still more marked. When a toxic dose is taken the preliminary excitement is scarcely perceived, but the patient is plunged rapidly into a comatose state, and lies unconscious, resembling a person intoxicated by alcohol or one suffering from congestion of the brain. The respiration is slow, deep, and stertorous, the pulse full and slow, though perhaps strong for a time—the pupils are contracted; if roused the patient soon relapses into his unconscious state. At this stage death does not often take place, but unless relief be afforded, prostration soon comes on, the coma becomes complete, so that the patient cannot be roused, and the pupils are closely contracted, though on the approach of death they may dilate widely. The respiratory effects are feeble, slow, and imperfect, the pauses between them long, and at length they stop altogether, death occurring through failure of respiration; the pulse is no longer full and slow, but has become feeble and rapid, and it gets weaker and quicker; the skin is cold and covered with clammy sweat; the countenance cyanosed. Even yet it is possible for the patient to be saved, but recovery will be gradual. Sometimes toxic symptoms are set up by moderate doses, and occasionally after small medicinal doses in consequence of the idiosyncrasy of the individual, serious symptoms occur; thus, within three minutes after a hypodermic injection of morphia profound coma has appeared. In other cases intense depression has followed a moderate dose; in others vomiting takes the place of sleep or follows it, and very often severe dull headache with giddiness and a general aggravation of the unpleasant symptoms previously mentioned are produced. When these symptoms are not very distressing rest in the recumbent posture renders them bearable, but when they are severe antidotes should be administered. Cases have been recorded in which a single dose has produced struggling for breath or syncope and even death. Other cases have occurred in which opium has set up delirium or convulsions. A more frequent source of anxiety is very prolonged sleep. A case has been communicated to me in which after a hypodermic injection of morphia the patient slept on soundly for about thirty hours. More rarely there is either delirium or convulsion.

Death mostly occurs from failure of respiration. This is due to a direct effect on the respiratory centre in the medulla, for in animals whose pneumogastrics had been cut Gscheidlen found the respiration was as much affected as when this had not been done. After death the bronchi present a highly congested appearance, and this congestion extends through the bronchial walls, the inner fibrous coat being specially affected. On the circulation opium produces a primary but transient acceleration of the pulse, which is followed by decrease in the rate with increase of its force and fulness, and to this succeeds either a gradual return to the natural condition or else as the patient loses strength the pulse again rises in rapidity. The slow full pulse of the second stage of poisoning appears to be due to the effect upon the inhibitory cardiac nerves, to which we may also ascribe the increase of the arterial pressure, for Gscheidlen found that morphia did not reduce the pulse after division of the vagi and that if they were cut during the second stage of poisoning the operation produced an extraordinary rise in the rate of the pulse. Other experiments seem to show that the peripheral ends of the vagi are stimulated, as are also the inhibitory cerebral centres. The quick, feeble pulse of the third stage seems to be due to paralysis of the peripheral vagi, for at that stage stimulation of the peripheral end of the divided nerve does not affect the heart.

It is unnecessary for our present purpose to trace the physiological effect of opium in all the organs, nor need we pass in review the numerous uses to which it has been put in general therapeutics, but presuming these to be familiar to the reader we shall content ourselves with mentioning those points in relation to the respiratory passages which we are desirous of bringing before them.[1]

Opium will often arrest an ordinary catarrh, so will a dose of morphia. Christison believed that it would cut short tonsillitis in the early stage, and certainly many a sore throat in persons subject to quinsy has thus been arrested. In a common cold the coryza is often stopped by a single dose, and so it is in a feverish catarrh or influenza, though in the latter case the symptoms are apt to return as the influence of the medicine passes away, but a second dose will again hold them in check. If properly combined with other remedies this will hasten recovery, or rather reduce the length of the attack. There is another form of coryza which is also checked and very often completely stopped by a dose of opium or morphia, viz., that produced by iodide of potassium. Moreover, I have often witnessed the immediate relief of all the symptoms of iodism by a single dose of morphia. Dover's powder as a diaphoretic is a favorite form to employ in catarrhal cases and is certainly one of the most efficient. It is, however, rather bulky, and for a full dose a single pill is not sufficient; generally I prefer morphia. Some add quinine to their opiate, but it is useless, unless a large dose be given, and a few small doses after the narcotic has done its work may be given with more advantage.

Opiates are often useful in asthma. In some cases a hypodermic injection of morphia will at once cut short a most painful paroxysm; on the other hand there are cases in which a single dose will bring on an attack. In the paroxysms of dyspnœa produced by emphysema considerable circumspection should be exercised in prescribing opiates, though they often afford great relief. In other neuroses of the respiratory organs opium may claim a place. In pertussis it is often used, and Copland thought that tinct. camph. comp. was the best form in which to prescribe it.

In bronchitis opiates allay irritation and restrain secretion, hence they enter into so many cough medicines. It is often taught that they are contra-indicated in severe bronchial affections, and much care must certainly be taken in prescribing them for old or weakly patients when the expectoration is copious and duskiness of the skin shows that oxidation is deficient. In such a condition to produce narcosis is certainly dangerous; expectoration accumulates during the profound sleep and the patient may be, so to say, drowned in his own bronchial secretion. Nev-

---

[1] In this chapter the history is passed by in consequence of its extent. There lies before me a list of upward of two hundred separate treatises on opium, to say nothing of morphia and other neurotics. It is impossible to analyze such a literature in the space at my disposal, and therefore scarcely any references will be given. It is hoped that the reader will be satisfied to have the subject treated as it presents itself from my own experience.

ertheless, even with rather free expectoration, with a violent cough and only slight râles and no sign of deficient oxidation, opiates may relieve the cough, and properly given they tend to diminish secretion. To narcotize by a large dose is one thing; to give a few small doses, watching their effect, is another thing; besides we can combine them with ammonium or other stimulants. It is the depressant influence on the respiratory centre which is to be feared, and to counteract this we must administer respiratory stimulants. The doses, too, must be carefully regulated if we would obtain the benefits of opium without its dangers. Even in old age and in infancy it may be made to yield some benefit if employed with skill and caution. I have given it to a lady of eighty-four in severe bronchitis, but with alternate doses of ammonia, atropia, and strychnia, and a friend over ninety years always took it for every cold he caught. Hypodermic injection is not desirable in these cases, since the full effect of the dose is produced immediately. What is wanted is rather to obtain the effect of repeated minute doses, so as to stimulate as well as to compose, and the distress of the patient needs something frequently. This is one of the great uses of ether; a few drops can be taken often and usually relieves the breathing.

It is sometimes feared to give opium in the dry, hacking cough of pulmonary disease, but I find it often useful. When these distressing coughs depend on morbid conditions of the throat, that part should be attended to; even then morphia lozenges may be useful, but other more efficient topical remedies ought to be employed, such as vapors, sprays, paintings, and other applications. The larynx should be inspected and if necessary laryngoscopal medication brought into play. When a diaphoretic is needed Dover's powder is suitable, but when the reverse is rather required atropia may be combined with morphia. This last combination is often of the utmost value, and by a variation in the proportions we may accomplish very different results; thus we may relieve the cough, perhaps the expectoration, and almost certainly the perspiration without at all depressing the respiratory centre. So, too, in bronchitis the secretion may be restrained and with it the cough when the respiration is not shallow nor the face blue, and in this case the relief thus produced may be permanent.

In paroxysmal dyspnœa due to cardiac disease and preventing sleep morphia given hypodermically is often valuable. It is strongly recommended in such cases by Dr. Clifford Allbutt, who also employed it to assuage the pain of angina pectoris or that caused by intra-thoracic tumors. Of course cardiac dyspnœa arising from dropsy, etc., will not be relieved by this treatment, which is only adapted for paroxysmal, not permanent dyspnœa. It may be added that it need not be given every night, but once, twice, or thrice a week, according to the urgency of the dyspnœa.

In chronic phthisis it is difficult to exaggerate the relief that may be afforded by this remedy or to summarize the indications for its use.

Sometimes the stimulant effect of opiates is a disadvantage. Graves taught us how this may be restrained by means of antimony, and the combination he suggested certainly enables us to extend the use of opium, though the discovery of other hypnotics has lessened the necessity for his combination. Aconite may be employed with a view of preventing the stimulant action, though it materially assists some of the other properties. Frequently during a course of aconite I administer an occasional dose of morphia or opium, and find that it not only accomplishes the purpose for which it is given but renders the aconite more efficacious, while the latter so reacts that a smaller dose of the opiate suffices. We may also employ the bromides in conjunction with opium in such a way as to modify its action. Not that the two medicines should be taken together but administered at suitable intervals. The headache, sickness, and other disagreeable symptoms which are apt to follow a full opiate may be greatly restrained, if not prevented, by bromide. Some persons for this purpose give a full dose with the opium; I prefer a smaller quantity —a couple of moderate doses, preceding the opium, so that the patient may be under the influence of the bromide. When it is taken it will often prevent or shorten the stage of excitement, a good night's rest is thus secured, and the subsequent headache, etc., prevented. Should it not succeed in doing this, another small dose can then be taken. The hypnotic effect of the opium is certainly increased by the bromide. My attention was first drawn to these reactions of the remedies on each other by observing the effects of morphia on patients who were taking a course of bromide. Chloral hydrate may also be made to modify the effects of opium, *e.g.*, a person under the influence of opium in the stage of excitement will drop off to sleep at once after a very small dose of chloral, and so several hours of repose may be gained. Very often, when persons accustomed to opiates were known to be kept awake by them for many hours together, have I cut short this distressing sleeplessness by five to ten grains of chloral hydrate. The action of belladonna in conjunction with opium is so important that it will be treated of further on, after treating of belladonna.

Children bear opium very badly, so do the very old. A single drop of laudanum has proved fatal to a young child; it should, therefore, only be given to children in exceptional cases and with the greatest caution, only in the liquid form.

Morphia salts are distinguished therapeutically by what may be termed negative properties, as compared with opium, thus they are less constipating, less sudorific, less stimulating. Moreover, they produce less discomfort afterward. Of course we are speaking of ordinary doses, since naturally, weight for weight, the alkaloid is much more powerful than the crude drug. The morphia salts do constipate, they are diaphoretic, and a stimulant stage usually precedes their narcotic action. These points may be observed after hypodermic injections, but in this case the effect on the alimentary canal may be less marked; still, generally it may be observed. The neutral tartrate of morphia is *sufficiently soluble* in distilled water, and keeps sufficiently well to afford us a non-irritating and non-acid solution (introduced by Messrs. T. & A. Smith, of Edinburgh).

Codeia seems to be only a feeble hypnotic, but it is sometimes useful, when morphia disagrees, to allay cough. With regard to the other opium alkaloids the statements made are very conflicting.

BELLADONNA AND ATROPIA produce effects quite contrary to those caused by opium and morphia, nevertheless they are often used in combination, and sometimes are said to reinforce each other. So we may with advantage take them up here. Belladonna has, more or less, been known as a counter-poison to opium for some three centuries, but special attention has only been drawn to them since 1862, when Dr. W. J. Norris wrote his paper in the *American Journal of the Medical Sciences*. Dr. Hughes Bennett (*Brit. Med. J.*, 1874) experimented carefully with atropia and morphia, and concluded that the one is antagonistic to the other within a limited area, since which time these conclusions have been confirmed and frequently acted upon. Modern experience with the alkaloid shows the accuracy of earlier observations with the crude drugs, and atropia is undoubtedly a valuable counter-poison to morphia or opium, and, moreover, may be used, as already stated, to prevent some of the disagreeable symptoms produced by opium. How, then, does it act? By its effect on the respiratory centre of the medulla, for belladonna stimulates this important point while, as we have seen, opium depresses it. In this respect, therefore, the one is the physiological antagonist of the other. It is often urged that the antagonism is not complete, that in some respects the one reinforces the other. This may be admitted, but does not alter the fact that they largely oppose each other, and just as we understand more fully the exact area of their antagonism, so much the more certainly shall we be able to avail ourselves of this property. No one will now dispute that animals poisoned by morphia have been saved from death by the administration of atropia, and several patients have recovered from accidental opium poisoning under the free use of hypodermic injections of atropia. We say the free use, but let it not be given recklessly, every dose should be watched. All depends on the respiration; a single large dose has been recommended, but this plan is not prudent; it is easy to repeat a small one if necessary. It is true that a poisonous dose has been given at once and the patient has recovered from both poisons, but it would be safer to give less and repeat according to the effect; we may thus keep always on the safe side. When the respiration begins to fail, an injection of a fortieth or a twentieth of a grain of atropia is enough *pro tem.* if the function responds, if not it can be repeated in a quarter of an hour. So, if it acts well but a relapse comes on another dose can be given. When the respiration has fallen to four, it may be doubled at once by an injection, which need not be repeated until the breathing again grows less frequent, or unless there has been no improvement for a long period. In this cautious way a series of relapses may be met, and so the patient tided over the time while the morphia is being eliminated. Dr. Fothergill injected a grain of atropia at once, in a case of accidental poisoning—this was certainly not without risk. He would advise generally, of course after emptying the stomach, one-fourth or one-third of a grain "before respiration is gravely affected." He warns us that there is a difference of susceptibility in patients, which is true as far as doses of a seventy-fifth to a twenty-fifth are concerned, but in massive quantities this can scarcely count, and if it could, a patient is quite as likely to be unusually susceptible as the reverse. Whenever we come to doses that might prove fatal or dangerous, difference in the degree of susceptibility is inappreciable. It is rather the amount of opium still in the system that accounts for the tolerance of the counter-poison. The cardiac and vaso-motor actions of belladonna are no doubt of use in opium narcosis, but it is the respiratory stimulus on which we must depend. But we have anticipated, and may as well go back to the general effects of belladonna.

In moderate doses, dryness of the mouth and throat with perhaps some disorder of the vision and dilated pupils may be produced. A larger quantity—a physiological dose will very decidedly disorder the vision and perhaps produce diplopia, fully dilate the pupils, and cause not only intense dryness in the throat, but distinct redness of the fauces; the pulse rises rather rapidly to 120 or more, and sometimes a scarlet flush comes over the face and neck and may travel over the whole body; it does not present the punctuations of scarlet fever and is not followed by desquamation, except very rarely. It is sometimes said that the pulse falls before the rise, if so, the fall must be very transient and is certainly not generally detectable; the medicine seems rather to produce a febrile state with very quick pulse, the beat of the heart being felt, and that unpleasantly, the temples throbbing, and the hot dry skin and dry mouth and throat being very disagreeable. Often the intellect is not affected until quite late even by poisonous quantities, but confusion and giddiness are common symptoms, as are extreme restlessness and a staggering gait. Neither is drowsiness produced, but rather the reverse, although, of course, it may happen that a patient falls asleep because the poison has counteracted a cause of previous wakefulness. Later, delirium comes on, and this is often furious; sometimes, too, convulsions may ensue; when the delirium is not furious it is always wakeful—the patient sees visions and dreams dreams, in which he is entirely absorbed, so that his attention cannot be drawn to the things around him; the delirium may persist for a long time, but in the end it subsides into stupor, just as the convulsions when present pass into paralysis. If these late symptoms appear early, it may generally be concluded that a very large quantity of the poison has been taken. We do not see cyanosis or other sign of respiratory failure except at the very last, then no doubt the heart and respiration both usually fail. But we must consider these and some other systems separately.

Circulatory System.—On the heart itself atropia acts as a direct depressant poison, but a large quantity must be present for such an effect to be apparent. On the cardiac-accelerator centres, or possibly on the nerves themselves, it acts as a stimulant, and unless in fatal doses it fails to destroy the excitability of these nerves. The ends of the vagus may be stimulated for a moment, but are quickly depressed, the pulse rapidly rising, and its rate cannot be reduced by faradization of the vagus. The vaso-motor system is also stimulated and the blood-pressure raised—at any rate so long as the doses are not excessive. Very large quantities depress the ganglia and even the cardiac muscle and the ventricle is found in diastole post-mortem.

Respiratory System.—Belladonna increases the number and depth of the respiration and this it does by directly stimulating the centre in the medulla, for the acceleration takes place even though the vagi have been previously divided. Loss of power in the respiratory nerves may occur at the close, and the patient may die from asphyxia; but this is probably not from failure of the centre, though some have conjectured that such failure occurs. It may be as well to mention that coincidentally with the stimulation of the centre there is perhaps depression of the bronchial ends of the vagus, lessening the tension of the muscular coat of the tubes and so facilitating the air-current, while at the same time there is depression of the afferent branches of the vagus which tends to relieve cough and dyspnœa by abating sensibility and reflex action.

Nervous System.—Belladonna or its alkaloid is an excitant, a deliriant, but not a hypnotic; so, as to the brain, it opposes opium. True it sometimes is given to procure sleep, but it can only do so indirectly; therefore, though it has been called a narcotic, and though coma comes on at the end in cases of poisoning, it scarcely deserves the name. In fact, in medicinal doses, it may be employed to remove or prevent the effects on the brain of morphia. By a careful graduation of the doses we may precisely antagonize the cerebral action of one of these medicines by the other. Dr. Fraser's experiments show that the action on the spinal centre is stimulant, although at first there is paralysis, because the function of the motor nerves is suspended. Lemattre, Meuriot, Bezold, and Bloebaum have shown that in large doses atropia can suspend the excitability of the efferent or motor nerve fibres. Still the dose required to produce total suspension is so large that it may cause death before excitability is entirely lost. Both the nerve-trunk and the peripheral intra-muscular terminations are affected, and no stage of preliminary excitement has ever been observed to precede the depression. This decided influence over the spinal nerves is important with regard to the therapeutical application of the remedy.

Muscular System.—The voluntary muscles appear to be unaffected and after death their contractility is unimpaired. On non-striated muscular fibre it is believed to exercise a paralyzing influence, to which is often referred its effect on the intestine. It may, however, be doubted whether this is not rather due to depression of the splanchnics permitting increased peristalsis, and to increase secretion. It is also believed by many to act directly on the unstriped muscular fibres surrounding the arterioles, and perhaps also in other situations, as, *e.g.*, in the bladder.

Glandular System.—The secretion of the salivary and mucous glands is arrested in a most remarkable way by belladonna, which also suspends the secretion of the skin. As first shown by Schiff, section of the chorda tympani arrests the secretion of the submaxillary gland, after which galvanization of the peripheral end produces an increased flow of saliva. Keuchel found that when atropia was administered to an animal before cutting the chorda-tympani galvanization of the peripheral end was unable to excite secretion, thus showing that the end of the nerve was paralyzed by the poison. Belladonna checks the secretion of milk, and this property is occasionally taken advantage of in therapeutics. In small doses it is also diuretic, though there is a good deal of variation in this respect. Dr. J. Harley estimated that the water of the urine was doubled in amount, and his experiments indicate an increase in the solid constituents, particularly phosphates and sulphates. After poisonous doses the secretion, though it may be increased at first, soon diminishes and eventually may be suppressed. Meuriot found that the secretion rises and falls with the arterial pressure. Perhaps the diuretic effect of small doses may be produced by raising the tension in the glomeruli of the Malpighian bodies.

Temperature.—Moderate doses produce a rise, toxic doses a fall; probably these changes correspond with the rise and fall in the blood-pressure. No doubt the increased metabolism tends to increase the production of heat.

The Eye.—Belladonna produces characteristic effects on this organ. Applied locally or given internally it dilates the pupil. Locally it paralyzes the peripheral ends of the ocular motor nerves, and perhaps also stimulates those of the sympathetic. When given internally it is carried in the blood to the eye, and there apparently acts in the same manner as if applied locally.

Atropia is almost entirely eliminated by the kidney. It passes into the urine quickly, and its effects therefore begin to decrease early. The production of a rash now and then suggests that possibly a portion may escape by the skin, but we have no proof that it does so. The urine of an atropinized animal dilates the pupil of another when applied locally, and there is little doubt that as a rule all medicinal doses pass out in the urine. Perhaps this accounts for some of the symptoms on the bladder which have been observed.

Therapeutically belladonna or atropine has been employed (*a*) to relieve pain, (*b*) to relax spasm, (*c*) to stimulate the heart and respiration, (*d*) to arrest secretion.

a. To relieve pain it is so vastly inferior to opium that we may say it is almost useless, though it has often been tried, and some still retain faith in its asserted anodyne qualities. We are speaking now of its internal use. There is a good deal of evidence in favor of its local application. Trousseau administered in neuralgia one-fifth of a grain every hour until giddiness came on, and then lessened the dose, but kept up the medication for three days. Anstie, Mr. Ch. Hunter, and Bartholow say that when it does act in relieving pain the effect is more permanent than that of morphia. But all advocates of its use lament its uncertainty, which Mr. C. Hunter finds very perplexing. It is the local application that is efficacious in this direction. No doubt internally and hypodermically in large doses it may affect the afferent nerves and so tend to assuage pain, while it may also modify the circulation in the part, and thus contribute to the relief of pain. This is why hypodermic injections for the relief of pain should be made as near as possible to the suffering point. But after all opium is the great anodyne and belladonna only exceptionally or indirectly useful for the purpose.

b. As an antispasmodic belladonna is more effectual, and this to some extent accounts for the reputation it acquired as an anodyne. When colic, cramp, and other spasmodic affections are relieved, of course, the pain is removed. In spasm of the voluntary muscles produced by injury to a nerve atropine is effectual provided it be injected directly into the affected muscle, as practised by Dr. Weir Mitchell; and then it doubtless acts by paralyzing the end organs of the nerves. In this way, too, it may relieve rheumatic spasms. By the mouth it is ineffectual for these purposes, probably because we cannot administer large enough doses to enable a sufficient amount to circulate in the part. In the involuntary muscles it seems to be more powerful, but even here, when possible, the local use is the most effectual, but when we cannot avail ourselves of this, we may administer it by the mouth. The action on non-striated muscular fibres accounts for its repute in colic, laryngismus stridulus, asthma, pertussis, some forms of constipation, spasm of the sphincter, and, indeed, a number of spasmodic affections, in which it has been more or less successfully employed.

c. As a respiratory and cardiac stimulant, belladonna is of much more importance. In full therapeutical doses atropine increases the frequency and force of the pulse, raises the temperature, and otherwise acts so as to produce a febrile condition, while, as we have already shown, it stimulates respiration. It may be employed simply as a cardiac stimulus. Graves recognized its value in typhus fever; it has often been employed in scarlet fever, erysipelas, etc. It has been used to rouse the heart in the collapse of cholera. Professor Schüfer advises a dose to be administered before the administration of chloroform as a preventive of cardiac failure. But we should remember that later on large doses exhaust the irritability of the cardiac ganglia. It is, then, as a respiratory stimulus that it is most valuable, acting directly on the centre. Its greatest use is when respiration is failing, as, for instance, in opium-poisoning. The stimulating effect on the respiratory centre may be compared to that of ammonia, but we must remember that the latter promotes bronchial secretion, quite an opposite effect to that of belladonna. The two, however, may often be given with advantage at the same time; thus, in bronchitis, when respiration is failing and the pulse feeble, even though rapid, with deficient oxidation, we may often tide over the danger by persistently stimulating the centre. Our predecessors relied on carbonate of ammonia, calling it a stimulating expectorant. We may give it just as they did, but reinforce it by a dose of atropine whenever its effect seems insufficient. Dr. Fraser found atropine at once restored the respiration in poisoning by calabar bean, and the antagonism to opium in this respect is well established. Besides the stimulating effect on the centre, there is a depressant influence on the periphery, which is doubtless of importance when the remedy is used in respiratory diseases; thus the depression of the afferent branches of the vagus in the bronchi would diminish reflex action as well as sensibility, and so tend to relieve cough and dyspnœa, while the depression of the bronchial termini of the vagus would reduce the tension of the muscular coat and so facilitate the air-current.

d. Atropia paralyzes the extreme branches of the chorda tympani; possibly, also, it acts on the gland-cells and also on a centre for sweat, if such a centre exist. We find the secretion of the mammary, sudoriparous, and salivary glands are all diminished or suspended under its influence. It will arrest excessive salivation from almost any cause; thus it will suspend mercurial ptyalism, and it will neutralize the effect of pilocarpine. It will check local sweats when used locally, and administered by the stomach or hypodermically it will control general sweats, even those produced by exercise or the Turkish bath. In the colliquative sweats of phthisis, Dr. Costa stated (*Phil. Med. Times*) in 1871 that a seventy-fifth of a grain at bedtime would prove an efficient remedy, and his observation has since been abundantly confirmed.

In diseases of the respiratory passages the uses of belladonna are numerous and important; in inflammation or congestion of the mucous lining attended with abundant secretion the indication for its use is distinct. Whether the nose, the fauces, the larynx, or bronchi be affected chiefly, a few doses will restore the normal circulation, and restrain the excessive secretion. In influenza or in catarrhal fever a large portion of the respiratory mucous tract is involved, and the general depression is often very marked; here, then, we have indications for the use of belladonna, which it is to be feared are often overlooked. It will relieve the intense depression by restoring the circulation, while it acts favorably on the inflamed membrane, in the stage, that is, of secretion: when the membrane is dry it will not be appropriate. When catarrh affects the nose only, or the nose and fauces, or perhaps the conjunctivæ, belladonna should be useful, for it dries the Schneiderian membrane and conjunctivæ, as well as the mouth, the fauces, and tonsils. In laryngeal catarrh, it is not so useful, perhaps, because here secretion is not so abundant, and yet in some cases of functional aphonia it exercises a favorable influence. In a couple of hours after a full dose the sensation of dryness passes away; a viscid secretion now appears and renders the mouth and throat clammy, while the tongue is covered with a white fur. If the secretion should be still too abundant, the dose may be repeated. It will be observed that in these cases, when not too severe, a single dose may suffice, but many prefer to give minute doses, say a drop or two of tinct. belladonnæ every hour until the patient is atropinized. When this plan is adopted it is desirable that the first dose or two should be larger than the succeeding ones. Here I would also remark that the indications from the state of the membrane point also to opium, which is an invaluable remedy in these cases, and may be advantageously combined with atropine, which reinforces the effect on the membrane, while it counteracts the depressing influence on the centre, and the opium in its turn frequently prevents the drying action of the belladonna on the skin and even secures a gentle perspiration. The modifying influence of one drug over another may often be secured in this way. A further example is afforded by aconite, which may be given at the same time as belladonna, whenever there is much feverishness; the aconite allays the fever, restrains or prevents the stimulating effects of the atropine on the circulation, and promotes perspiration, while the atropia may be thus compelled, as it were, to expend its energies on the mucous membrane.

In *asthma* belladonna relieves both cough and dyspnœa, as it does also the paroxysms of dyspnœa which occur in emphysema, but for these purposes full doses have to be given; Hyde Salter found this and gave ten minims of tinct. belladonnæ every two or three hours until a distinct effect was produced, and if the patient prove insusceptible still larger doses may be required, *e.g.*, half a drachm as a single dose to arrest or avert a paroxysm. As soon as distinct symptoms are produced the dose may be diminished, and if the paroxysms usually last long it is better to give divided doses. When a full dose is needed, I prefer the hypodermic method. Possibly some of the larger doses that have been required were in consequence of the tincture not being good; of course the alkaloid is more certain, but a well-prepared tincture is also efficient, or the succus may be preferred. The remedy is only useful where the symptoms indicate the necessity of its effects. Thus a moist skin, abundant expectoration, and quiet pulse are in its favor, but feverishness, with scanty expectoration, contra-indicate it. Fumigation by belladonna leaves dipped in a solution of nitre and dried, is sometimes preferred; these prepared leaves are burned in a close room and the patient breathes the fumes. Trousseau used cigarettes made of the leaves of belladonna, hyoscyamus, and stramonium moistened with a solution of opium. When the paroxysms of asthma come on pretty regularly in the morning a dose of atropia taken at bedtime will sometimes prevent the expected attack.

In *laryngismus*, in pertussis, and in any neurosis of the respiratory system, we may be glad to avail ourselves of the properties of belladonna. Its antispasmodic effect is here again to be considered, and as in asthma full doses are required, that is to say, to obtain good results, we must produce the physiological effects, though as soon as they are manifested the dose should be lessened. Children are more insusceptible to the action than adults, so that as much as ten minims of the tincture may have to be given every hour to a child of one or two years old, and even that may not dilate the pupil. The cases in which it is most effectual are those in which the bronchial secretion is considerable. It is disfavored by some when bronchitis is present, but this is probably because it has been given during the febrile stage; it is most effectual when pyrexia has completely subsided, about the third week, when it will be found to exercise considerable influence over the paroxysms. It is true that their violence should be abating about this period, but with every allowance for that it must be admitted that belladonna exercises a favorable influence, and so it will at an earlier date when the cough is even more convulsive, provided the bronchial secretion be free and the fever not considerable, or be restrained by other remedies given concurrently.

In *diphtheria* belladonna may be administered, either with a view of supporting the respiration or of restraining exudation. As it suspends secretion in the fauces, it has been assumed that if given in time it may prevent the formation of the false membranes. As diphtheria is an exceedingly depressing disease, it may be safely employed in the hope that it may do this, as if not, its influence will be favorable in supporting the circulation. At a late stage it may be used to stimulate the respiratory centre, and can be given hypodermically when ammonia cannot be taken.

In the night sweats of phthisis atropia is almost certain to maintain its reputation. It seldom fails to arrest this distressing symptom, and the dose required for this purpose is usually so small that no inconvenience is to be apprehended. Moreover, it does not require to be long continued. One small dose at bedtime will often suspend the sweats for several nights. Sometimes it requires to be given for two or three nights in succession, and then no more may be needed for some time. The $\frac{1}{120}$ grain will often suffice, sometimes even less, but at others $\frac{1}{16}$ or $\frac{1}{60}$ may be required.

BELLADONNA WITH OPIUM.—We have seen that in some instances opium and belladonna neutralize each other, as also do their alkaloids, morphia and atropia. These opposing actions are so important that it is desirable to consider them further. We may utilize the two remedies at the same time, giving them in combination or alternately with each other, especially in painful neuroses. Opium will cut short asthma or catarrh; so will belladonna. Opium produces certain unpleasant symptoms, belladonna prevents most of these. From an early period the opposite effects upon the pupil could not be overlooked. Morphia and atropia are often combined in hypodermic injections; these same alkaloids may also be administered by the mouth. Morphia produces less disagreeable effects than opium, and we may further reduce the unpleasant action by adding atropine. When we only want the good effect of opium or morphia, we can often neutralize the evil consequences by the addition of a suitable dose of atropia. The antagonism between the two agents is not universal, that is, does not cover the whole field of action, but the area over which it exists is exceedingly favorable for our therapeutical efforts. In some points there is no antagonism; for instance, atropia does not prevent the relief of pain by opium; indeed, many have supposed that it rather assists the anodyne action. Erlenmeyer says that the antagonism is complete in regard to the action on the brain, so that when the two remedies are given together no coma results; but on the sensory nerves there is no antagonism, and accordingly he combines them for the relief of pain. Bartholow holds that the anodyne effect of opium is even assisted by atropia. On the other hand, the existence of the antagonism has been denied by Brown-Séquard and Dr. John Harley. The latter has criticised the recorded cases of opium poisoning which have been treated by atropia with considerable ingenuity, but has scarcely succeeded in reversing the general judgment. The Edinburgh committee presided over by Dr. Hughes Bennett reported that: (1) sulphate of atropia is within a limited range physiologically antagonistic to meconate of morphia; (2) meconate of morphia does not act antidotally after a large dose of atropia, thus, while atropia is an antidote to morphia, morphia is not an antidote to atropia; (3) meconate of morphia does not antagonize the effect of atropia on the branches of the vagi supplying the heart. From this it may be concluded that atropia is of more value to correct the action of morphia than the reverse, and this coincides with our clinical experience; but there are cases in which morphia may with advantage be made the corrective of a full medicinal dose of atropia.

Therapeutically the question of dose is of most importance, and it varies with regard to each remedy in different individuals; considerable experience is therefore required to obtain the best results. If one-fiftieth grain of atropia will stop certain effects of a quarter grain of morphia it is by no means certain that one-hundredth grain will serve to correct one-eighth. Indeed the presumption might rather be the other way. Both alkaloids affect the brain, but differently; it has been said that when they are combined the sleep is longer, and some assert that atropia prevents the insomnia of morphia, that is, shortens the stage of excitement. My experience is rather the reverse of this, which certainly might be anticipated from the physiological effect of atropia, which itself produces excitement, and so, *a priori*, might be expected to increase the excitement of opium. It is possible that the atropia may tend to prolong the sleep induced by morphia, but it is very difficult to judge. It may reinforce the anodyne property, but in the small dose required the amount which could obtain local access to the painful part must be infinitely small, and we have seen that any anodyne quality depends on this local access. Morphia is our sure anodyne, and we only add the atropia to enable the patient to tolerate it, or to prevent the inconveniences to which it may give rise. Nausea, vomiting, headache, constipation, syncope, all the train of disagreeable symptoms which so often follow the opiate may be prevented by a properly graduated quantity of atropia added to the morphia, and this whether taken by the stomach or injected under the skin. It is said that the illusions produced by belladonna are prevented by opium and the sleep is more rational, but we hardly give doses of atropia which produce illusions. Morphia depresses, atropia stimulates the heart. A very minute amount of atropia will prevent morphia from contracting the pupil. Morphia lessens, atropia augments the secretion of the kidneys; on the skin the action is just the reverse, morphia promoting, atropia arresting perspiration; both remedies arrest the secretion along the respiratory tract; morphia retards, atropia increases peristaltic movements, and the same may be said of their action on the intestinal secretion. Above all, opiates depress, but belladonna stimulates the respiratory centre. It will now be seen how often the one may be made to supplement the other, and that their counter-influences are of special value in respiratory diseases, particularly catarrhal affections and neuroses.

As an average it takes from $\frac{1}{50}$ to $\frac{1}{75}$ of a grain of atropia to neutralize the effect of a quarter grain of morphia, but much less doses may suffice to secure the freedom from inconvenience which is desired. Bartholow recommends $\frac{1}{150}$ or $\frac{1}{100}$ grain of atropia to a quarter and half a grain of morphia. I have usually found a larger proportion of atropia necessary. In giving atropia by the mouth, the dose being so small and the drug powerful, it seems best to give it in liquid form, the liquor atropiae sulphatis affords a convenient dose of one to two minims ($\frac{1}{120}$ to $\frac{1}{60}$), and this combines well with liquor morphiae. Sometimes it is desirable to administer aconite at the same time, for the sake of its febrifuge property and to restrain the action of the atropia on the heart.

OTHER ANTAGONISMS.—Medicines may be partially antidotal or antagonistic, that is they may counteract each other in some respects but not in others, and we may call them antagonistic, well knowing that the area over which they thus act is limited. Still their counteraction is extremely interesting and important, and may even suggest to us how medicines may antagonize diseases. It seems idle to deny the antagonism because it is not complete; should we not relieve some of the effects of a poison, even although others might be beyond our control?

Atropia and Physostigma.—Kleinwächter and Bourneville recorded cases in which they observed an antagonism between these drugs. Bartholow received the prize of the American Medical Association for an essay on the subject in 1868, and the following year Dr. Fraser brought before the Edinburgh Royal Society some account of his researches, in which he has shown the area over which this antagonism exists. He showed that atropia averted the effects of a lethal dose and therefore is the antidote for calabar-bean. From one-fiftieth to one-thirtieth grain should be injected under the skin, and repeated at intervals until the pupils dilate and the bronchial secretion is checked. This leads us to observe that calabar-bean immensely increases the secretion of the lachrymal, salivary, bronchial, and intestinal glands, while atropia controls these effects and establishes its own, just as it counteracts the contraction of the pupils and dilates them. Some other opposite effects have been demonstrated chiefly by Dr. Fraser, thus " physostigma increases the excitability of the vagi nerves, while atropia diminishes or suspends this excitability; physostigma diminishes the arterial blood-pressure, while atropia increases it." It is remarkable that a minute dose of atropia which is insufficient to produce an appreciable effect will suffice to avert many of the effects of physostigma. Although atropia may prevent death from calabar-bean, the converse of this fact has not yet been demonstrated. The committee of the British Medical Association reports that sulphate of atropia antagonizes to a slight extent the fatal action of calabar-bean, but that the area is more limited than indicated by Dr. Fraser.

Atropia and Pilocarpine.—Jaborandi excites perspiration and salivation. Applied to the eye it contracts the pupil, it retards the heart and afterward arrests it in diastole. These effects are the opposite of those of atropia, which is able to control them so that a hypodermic injection of one-hundredth of a grain will at once check the action of a dose of jaborandi or pilocarpine.

Atropia and Bromal.—In reference to the bronchial membrane it is interesting to observe that these two oppose each other. Bromal kills by producing extreme excess of the bronchial and salivary secretions, by which the animal is choked; atropia arrests th esesecretions, and so far is antidotal, but of course bromal is no antidote to atropia, as this latter does not destroy life by its action on these glands.

Atropia and Muscarin.—Schmiedeberg showed the antagonism of these substances on the heart. Dr. Lauder Brunton has enlarged this area by showing that the dyspnœa caused by muscarin appears to be due to powerful contraction of the pulmonary vessels, blanching the lungs. The right heart is distended owing to the condition of the pulmonary vessels. Now atropia at once removes the spasm of the vessels and sets the loaded right heart free, thus completely removing the dyspnœa. Muscarin also stimulates the termini of the chorda tympani and so salivates, it also excites perspiration, in both these respects being opposed by atropia. It appears, however, to dilate the pupil, though in most other respects it antagonizes atropia. Muscarin seems to act on the heart by stimulating the intracardiac inhibitory apparatus, much in the same way as pilocarpine probably acts.

Atropia and Prussic Acid.—Preyer says that atropia paralyzes the peripheral branches of the vagus and in this way prevents hydrocyanic acid from arresting the contractions of the heart. If so, it would be an antidote to that poison, which, however, produces death so rapidly that there is seldom time for treatment of any kind.

Atropia and Aconite.—Atropia given with aconite, or a little before it, antagonizes the action on the heart, but when delayed for about a quarter of an hour after a lethal dose is unable to prevent death. As before remarked, we may avail ourselves of this antagonism when employing therapeutical doses; thus it often happens that the general effects of aconite are most desirable and may be obtained by its regular administration every few hours, while an occasional dose of atropia may be advisable for the sake of its effects on the mucous membrane. On the other hand, when regular doses of atropia are required, it may be desirable to counteract some of its effects by aconite.

Some of these antagonisms are not easy to understand and show that the action is more complex than has been supposed. It is not a single effect which expresses the properties of a medicine, although its therapeutical value may depend on that one. Atropia, as we have seen, antagonizes the effect on the heart of pilocarpine, muscarin, and aconite, though both poison and counter-poison seem to paralyze the excito-motor and muscular substance. Further, pilocarpine antagonizes the effect on the heart of muscarin, and yet both seem to act in precisely the same manner. Moreover, atropia, which we have seen antagonizes so many poisons, does not prevent the effect of digitalis nor of veratria on the frog's heart. Though digitalin antagonizes muscarin, pilocarpine, aconite, and atropia, Dr. Ringer has suggested "that these antagonisms may be due to chemical displacements."

Before leaving this subject it may be remarked that some other antagonisms are of special interest, and progress is being made in their study. Caffeine and morphia are distinctly antagonistic, and guaranine modifies the action of morphia, but not very markedly. Calabar-bean, as we have seen, provokes the bronchial secretion, and its action is greatly modified by chloral hydrate, which in some cases has prevented death after a lethal dose, but it must be given before the full action of the physostigma is produced. The chloral depresses the respiration and paralyzes the centre, as well as the circulation, in which it appears rather likely to reinforce the effect of the calabar-bean, for that certainly depresses the respiratory centre. Physostigma has also been given in strychnia poisoning, as well as in tetanus, but has scarcely answered the expectations formed of it, and the same may be said as to its antidotal power toward atropia and chloral ; in fact, the chief use at present of calabar-bean or its alkaloid, eserin, is for local use in diseases of the eye. Strychnia prevents some of the effects of aconite, but the reverse does not seem to hold good, strychnia is also to some extent counteracted by the bromides and by chloral hydrate, which have been employed with some success as antidotes. It is obvious that the more completely we understand the mode of action of any of these powerful remedies, the more likely we are to be able to influence that action whether by counteracting or preventing it, or by increasing or reinforcing it, or otherwise modifying it. A knowledge, too, of the area of the antagonisms between them helps us in arranging suitable combinations ; thus, though a simple prescription may often be best and it is certainly desirable to cultivate simplicity in prescribing, it very often occurs that great benefit may be obtained by suitable additions to the chief remedy.

Further, we may modify the action of our remedies by giving others before, with, or after them, we may administer variously acting agents alternately between each other. To modify opium we give with it belladonna ; to stimulate the respiratory centre we give belladonna, or atropia alone, and we may reinforce its action by alternating it with ammonia or ether ; and furthermore, while pushing these remedies we may obtain the effect of morphia on the mucous membrane by a small dose at bedtime, its depressant effect on the centre being neutralized by the other medicines. These delicate alkaloids are easily destroyed, caustic alkalies decompose them, even lime-water destroys atropia so much that Dr. J. Harley proposed it as an antidote in belladonna poisoning. It is usually better to give these powerful alkaloids in a fluid form unless great reliance can be placed on the preparation.

Atropia in combination with morphia is often given as a pill. Atropia alone has the disadvantage that a solution does not keep well, and should therefore be freshly prepared. The gelatine disks, however, prepared by Savory & Moore, keep perfectly; they are always ready for hypodermic injection, and are equally available for administration by the stomach, as a disk can be washed down with a little water without tasting.

HYOSCYAMUS AND STRAMONIUM.—These two remedies may be regarded as allies of belladonna, in fact, daturia and hyoscyamia were for some time regarded as identical with atropia, but some differences have been established, and other very similar alkaloids have been obtained from the solanaceæ. The alkaloids have been used for subcutaneous injection as anodynes, but are not superior to atropia, and very inferior to morphia. Hyoscyamine has also been employed for the secondary sedative effect on the cerebrum in maniacal excitement. The herbs hyoscyamus and stramonium seem both to be more decidedly hypnotic than belladonna, and hyoscyamus has a special repute as an anodyne, and its secondary or soothing effect comes on earlier and is more marked; hence, perhaps, its reputation for producing sleep and relieving pain. Moreover, hyoscyamus seems to exercise a special influence on the mucous membranes. Thus the respiratory, the gastric and the intestinal lining are all soothed by it, as is also that of the bladder. In these cases it has also been given with alkalies, and even with liquor potassæ, but it has been shown that this decomposes it. Its somewhat laxative and anodyne influence on the bowels makes it a valuable adjuvant to aperients.

All the preparations of stramonium have obtained considerable repute as antispasmodics in asthma, and their use has been extended to cases of laryngeal cough. A favorite method of employing stramonium is by fumigation or inhalation. A popular plan is to smoke it like tobacco, and smokers mix tobacco with it; this, however, is not to be recommended, as it is desirable to draw the fumes of the stramonium into the bronchi, where its local action is believed to be considerable in relaxing spasm, and so relieving the dyspnœa, whereas the fumes of tobacco irritate rather than soothe. Twenty grains of the dried leaves or ten of the powdered root may be smoked at a time, the fumes being inhaled, or any other convenient method of inhaling the fumes may be employed. It will be seen that it is in pure nervous asthma that it is indicated; in the dyspnœa of heart disease, or that caused by structural changes in the lungs, it is useless. Sometimes it fails in asthma, but in other cases it is very successful, the cause of this difference not having been ascertained. Stramonium seems to be more directly depressant to the nerves of the bronchi than belladonna, and thus, perhaps, may be considered as to some extent a respiratory sedative, especially as regards its local action. Given internally, the extract is more powerful and the dose accordingly smaller. Datura tatula, an allied plant of more robust growth belonging to the stramonium genus, has been introduced of late years, as a substitute for the older remedy, and as often succeeding when that fails; it may be used in the form of cigars, cigarettes, fumigations, etc., in the same manner as stramonium. An extract and tincture are also made. Daturia extracted from it is more powerful than atropia, from which its salts differ somewhat in solubility and in crystalline form. The dose is $\frac{1}{120}$ to $\frac{1}{60}$ of a grain, but it should be employed with great caution; the five-thousandth of a grain applied locally affects the pupils.

CAMPHOR.—Camphor excites the cerebrum and produces a kind of intoxication, evidently exercising a considerable influence on the nervous system. It is eliminated by the skin and the bronchial mucous membrane. It is, therefore, natural to employ it in respiratory diseases, in which, indeed, it has long enjoyed a popular reputation. In *acute catarrh*, inhaled or used as a snuff, it is a popular remedy, and in *hay fever* as much may be said. Some authorities have recommended it in whooping-cough and other spasmodic affections, others look upon it rather as a stimulant or perhaps expectorant, giving it in combination with ammonia in chronic bronchitis, capillary bronchitis, and in emphysema. Its action in respiratory diseases seems to be not dissimilar to that of turpentine, to which it presents other analogies; for instance, it is antiseptic, antispasmodic, etc. The late Dr. Copland attributed to it special value in bronchitis and asthma; he combined it with ammonia as an appropriate stimulant when expectoration was arrested from want of power; when expectorants were admissible he often added it to them, as he did to diuretics, opium, and other remedies. He declared that "in nearly all stages of bronchitis, camphor is a most valuable medicine," and added, "its virtues have been singlarly overlooked by the writers on this disease," and further pointed out that when exhaustion and difficulty of expectoration become urgent, "it is one of the most valuable remedies we possess." But in spite of this opinion of a most able observer, the remedy has not been extensively used for these purposes, except by a small number of physicians, who have satisfied themselves of its value. Though camphor has been in tolerably common use for some two hundred and fifty years, during which it has given rise to a very considerable literature, and been recommended for all sorts of diseases, from a common cold to cholera, its exact medicinal value has scarcely yet been fully ascertained, and it is perhaps most frequently employed, rather as an adjuvant than for other purposes. As an antispasmodic it is not unfrequently combined with musk and other powerful nervines.

Camphor forms some very curious compounds with chloral, thymol, phenol, etc., most of which possess distinctly anodyne properties. Camphor-monobromide is sedative and antispasmodic, as well as hypnotic, and from Bouneville's researches appears also to depress the circulation and lower temperature.

CHAPTER XVIII.

PNEUMATICS.

This term (πνευματικά, from πνεώ, I breathe) was employed by Pereira to signify therapeutical agents which acted by their influence over respiration and calorification. He included those which affected the respiratory muscles, the mucous membrane, the breathing, and the calorific function. In reference to the muscles, something has already been said in the chapter on Neurotics; thus the efficacy of stramonium and other medicines in relieving spasm of the bronchial muscular fibres has been pointed out, and it may be stated that other medicines possess similar or opposite properties. As to those medicines which Pereira considered diminished want of breath, and which he termed "torporifics," these also are for the most part neurotics. Some of them have been supposed to produce a condition analogous to the physiological states of hibernation, ordinary sleep, asphyxia, or syncope, but most of them may be classed as narcotics, and several are distinct depressants of the respiratory centre. So again, substances which influence animal heat have been considered amongst refrigerants and antipyretics on the one hand and stimulants and neurotics on the other; the calefacients of Pereira being in reality excitants or respiratory stimulants, *i.e.*, accelerators of circulation and respiration. This leaves us only those substances which act on the mucous membrane and an immense number of which are commonly spoken of as expectorants. It will, however, be necessary to extend our view beyond this, and we must therefore rearrange our pneumatics.

If whatever affects the respiratory system directly or indirectly were to be included among pneumatics the word might as well have been employed as the title to this volume, since it only professedly includes respiratory therapeutics. Many, perhaps, would object as it is to some of the agents we have included, but surely those which only indirectly affect the respiration ought not to be overlooked as remedies for disorders of that function, while other functions are so closely related to it as constantly to claim attention. Nutrition, therefore, and everything affecting it are of equal importance in diseases of the respiratory and other systems, while in consequence of the intimate association between respiration and circulation derangement of the one almost invariably produces disorder of the other. We may, then, employ the word pneumatics for all those substances which influence directly or indirectly the respiratory system; or rather, we apply it to these substances whenever they are used for this purpose, since almost all of them possess other properties for which they may be even more frequently employed. Some of these pneumatics have consequently been considered elsewhere, and it will not be essential to repeat what has been said concerning them, it being a matter of convenience to consider the various actions of a remedy in connection with each other. Plenty of remedies enter into more than one group in every classification, and so in this; our pneumatics, therefore, will include those which act more or less directly on some portion of the respiratory organs or influence distinctly their function.

Regarding respiration as a provision for interchange between the atmosphere and the blood, we find that the function may be greatly affected by changes in the quality of the air. Thus variations in its temperature and in the amount of moisture it contains affect every portion of the membrane over which it passes in its passage toward the blood, and may even affect the pulmonary cells and the circulating fluid itself. So well known is the effect of unusual cold that catarrh is continually ascribed to it, but dampness is equally injurious, and the two combined are still more likely to give rise to affections of the respiratory tract. So disease of any portion of the mucous membrane extending from the lips and nose to the extreme ramifications of the air-passages, and even catarrhal pneumonia may be produced by cold and damp. The opposite condition of warmth and moisture is soothing and sometimes we endeavor to maintain it by regulating the atmosphere of the room or by providing a warm stream of air through inhalers or respirators.

Variations in the pressure of the atmosphere also exercise considerable influence. As previously stated, the diminished pressure on lofty mountains increases the frequency and depth of the respirations as well as the vascularity of the lung, and thus accounts for the tendency to hæmoptysis observed under such circumstances. In descending mines an opposite condition obtains, and sometimes we may produce a somewhat similar influence by causing the patient to breathe compressed air.

Deficiency in the quantity of air available for respiration may also take place, or the supply may be interrupted for a brief period, or it may be replaced by a respirable gas like nitrous oxide. Increase of the supply may be produced by resorting to pressure, or an excessive proportion of oxygen may be provided. Then, again, the amount of air at the disposal of a patient remaining the same, more or less may be actually used according to the activity of the respiration, and this we may sometimes vary by regulation of his exercise and rest. So in disease, when one portion of the lung is rendered useless we see the remainder working harder in order to make up for the deficiency thus occasioned.

The chemical quality of the air may be altered. This may occur from defect of oxygen as well as from the presence of impurities. Sometimes the attempt has been made to increase the amount of oxygen, but without much success; at other times inhalations of oxygen, or of this gas mixed in various proportions with air for short periods at a time have been prescribed. A resort to mountain air, or to the seaside, or an ocean voyage is a more usual method of improving the quality of the air breathed. The variations in the air, which we have already noticed, go far to make up the complex influence of climate. But this subject, important as it is in reference to respiratory therapeutics, is too extensive to be included in the present chapter and will therefore be postponed.

The circumstances relating to the blood and general circulation have already been considered, but we must not quite pass over the vascular supply of the bronchi and lungs. The circulation through the bronchial vessels may be stimulated or depressed by agents which act upon the general circulation, as well as by remedies which possess a topical action, thus exercise, evacuants, some respiratory stimulants and expectorants may notably stimulate the bronchial circulation, while depressants and some other remedies produce an opposite effect. So the pulmonary circulation may be modified by the systemic, and therefore whatever acts on the latter affects the former; besides which a direct influence on the lesser circulation may be produced by these agents. Muscarin appears to possess a very direct action, powerfully contracting the pulmonary vessels.

The muscular system must not be overlooked. The muscles of respiration, as we have seen, may be affected by various agencies within our control; but it is the muscular coat of the bronchi and perhaps also the diaphragm which we most frequently seek to affect by medicines. Inasmuch as spasm of the bronchial fibres gives rise to most distressing symptoms, substances which control this condition, whether acting directly or indirectly, are sometimes grouped together as antispasmodics. Many of these are depressants of the respiratory nervous apparatus; others, acting perhaps through the same channels, are grouped as narcotics. Sometimes expectorants become antispasmodics by relieving the vessels and thereby causing a free flow into the tubes, when after the secretion is coughed up the spasm subsides. In the same way the removal of mucus by emetics produces subsidence of the spasm, and the act of vomiting, as is well known, greatly relaxes muscular fibres.

With regard to the nervous system, we have already considered several stimulants and sedatives of the respiratory and cardiac centres, and we may add that other agents of this kind are within our grasp. Of these, ammonia and strychnia are important as stimulants, while chloral, the bromides, conium, etc., may be added to the depressants we have had occasion to describe. Afferent sensory nerves from all parts convey impulses to the medulla and produce impressions on the centre which influence reflexly the respiratory movements. But impulses from the respiratory organs themselves are more important, inasmuch as they are direct. Now the entire surface of the air-passages is abundantly supplied by the vagi, which are the special afferent respiratory nerves. By them impressions are constantly being collected and conveyed to the centre, and when such impressions are unusually powerful they may overflow, so to say, and thus affect other centres. They may even reach the convolutions, when they will be perceived as sensations referred more or less distinctly to the respiratory organs. We may thus have undefined, uneasy sensations, or a degree of irritation, or oppression and distress, or, again, distinct pain. Further, motor filaments of the vagi rise in the centre and are distributed to the bronchial muscles and so regulate their contraction, thus controlling the calibre of the tubes. In this way the bronchi are completely under the control of the medulla, and moreover the very impulses which originate in the air-passages conveyed to the centre react through it on the passages themselves. Now the afferent respiratory nerves may be stimulated on the one hand, as, $e.g.$, by cold air, irritating gases, or certain medicines which determine to the respiratory tract. On the other hand, they may be depressed, as by warmth and moisture as well as by remedies acting more or less directly upon them, and some of which deserve to be called respiratory anodynes, from the soothing influence which they exercise and the manner in which they restrain uneasy sensations and relieve pain.

Indirect influences transmitted through other nerves also readily affect the respiratory centre, and may therefore be employed to act upon it. Everyone knows how readily it is stimulated through the fifth by irritating the nostrils, or by the sudden application of cold to the forehead. In the same way it may be influenced through the olfactory, optic, and auditory nerves by strong odors, intense light, and loud sounds, as it also may through the cutaneous nerves, as from counter-irritants. On the other hand, sedative influences may be transmitted from the surface, as in warm baths, general or local, fomentations, poultices, and anodyne local applications. We have now to add that efferent nervous impulses may be also more or less modified by our remedies, $e.g.$, strychnia stimulates the spinal centres of the respiratory muscles as well as the peripheral nerves, perhaps both directly and indirectly, and electricity may be called in as an immediate stimulant. On the other hand, opium seems to be a depressant of the entire efferent tract, while calabar-bean depresses the spinal centres and conium expends its energy chiefly on the motor nerves.

The glandular system of the respiratory tract is also within the reach of our remedies, and that not only as a whole but in some of its parts. So that the secretion from the nose, throat, larynx, trachea, and bronchi may be changed in quantity or quality, and this alteration may affect a small portion or the whole of the membrane. The glands may be stimulated by ammonia, sulphur, iodine, most of the expectorants, and by some remedies which perhaps act topically, being excreted through the membrane. Their action may be restrained by belladonna, opium, and perhaps astringents. Cold and heat, warmth and moisture, or other local remedies also influence the secretion directly, while other remedies act upon it either through the circulation or the nervous system, some of these being irritating, others soothing.

We are now prepared to consider the various effects produced upon the air-passages by the remedies at our command. It will be observed that their action is often very complex, though sometimes it is simple and perhaps direct. So many substances and of such opposite qualities have been supposed to act upon the secretion, or to affect its production that no little confusion has been introduced. Excess of secretion is so obvious a symptom that it must have attracted attention from the earliest times. Many substances were believed to be capable of increasing it, but only a few were supposed to possess the power of restraining it, and accordingly medicines of most opposite qualities have been grouped together as expectorants. If the origin of the word (*ex pectore*) justified the usage, we might perhaps apply it to whatever affects the sputa, or that which is brought up from the chest, though originally it was no doubt intended for whatever increased the secretion. Such increase might, however, only be apparent, the removal of a larger quantity not necessarily implying greater activity of production. Like other functions, that of the respiratory membrane may be increased, decreased, or altered. No doubt it is most frequently increased, and as this constitutes the most obvious and the most frequent symptom of disease, agents which could promote it seem to have received the most attention. It is perhaps due to the notion which so unfortunately prevailed for a long time, as to the value of evacuants, that this idea of promoting secretion was allowed to put into the shade the equally important subject of the possibility of restraining it. Still, it is curious that those who could see the desirability of restraining mucous diarrhœa should consider that an analogous discharge from the respiratory tract should be promoted rather than restrained. It was, perhaps, a happy thing that some so-called expectorants really acted by diminishing and altering the sputa, rather than by stimulating its production. Here, perhaps, it is but fair to admit that some early writers [1] were conscious of the evils that might arise from the indiscriminate use of expectorants.

---

[1] Ludolf, J.: De usu et abusu medicamentorum Expectorantium. 1723.
Buechner, A. E.: De incongruo Expectorantium usu frequenti morborum pectoralium causa. 1756.

Here we may pause to refer to certain remedies which act locally on small portions of the aërian membranes. These are first of all *errhines* (from ἐν, in, ῥίν, nose), which are introduced into the nostrils for the purpose of increasing the nasal discharge. Many of them are taken as snuffs; some act merely mechanically, but others are distinct irritants to the mucous membrane. Various acrid powders have been used for this purpose, but liquids and vapors may be also employed. It will be observed that all are stimulants to the secretion and we have no name for substances possessing opposite qualities. But we perhaps oftener use such as tend to restrain secretion, and which possess emollient, soothing, or astringent properties. Some internal remedies promote the nasal secretion, but are not called errhines, as they are not applied locally. Other internal medicines restrain this secretion. Absorption readily takes place from the healthy pituitary membrane. *Sternutatories* (*sternutatoria*, from *sternuo*, I sneeze), or ptarmics as they are also called (from πταίρω, I sneeze), also stimulate this membrane, but these terms are employed for medicines which are applied locally in order to produce sneezing but not to bring on a discharge, though it is obvious that such a stimulant will temporarily increase both the nasal mucus and the tears. Cough medicines, sometimes called *bechics* (βηχικά, from βήξ, a cough), are not used to stimulate, but rather to soothe the throat and larynx: they comprise a number of emollients and demulcents supposed to act locally during the act of swallowing and to sheathe the surface from irritants. Various forms of linctus, lozenges, and similar remedies are given with this intention, and sometimes small quantities of expectorants or narcotics are added, to which is due, perhaps, the most important part of the effect. Direct local applications can be made to the larynx by means of the laryngoscope, information concerning which will be found in the author's "Laryngoscopy." Passing further down the respiratory tree we come to medicines acting upon the lower portion of the membrane, and first of all

EXPECTORANTS.

In a wide sense this term is frequently applied to whatever facilitates the evacuation of the bronchial secretion, as well as to whatever increases its flow, and this whether directly or indirectly. It will be observed that there is a great difference in these two actions. In each case more phlegm is raised, but in the one that is only because the removal is effected of that which is already there, while in the other case there is the additional outpouring of fresh secretion. Now the natural method of evacuating bronchial mucus is by coughing, so that whatever excites this action assists the process. An irritating gas or any other local stimulant to the respiratory membrane may thus be said to be expectorant. In health, although moisture is continually exhaling from the pulmonary surface, very little is condensed into liquid and retained in the passages; only enough to maintain the moisture of the membrane, and in combination with the products of the gland to form a sufficiently tenacious secretion to cover the surface and entangle such particles as may be inspired and thus enable them to be coughed up. Small pellets or larger glutinous masses are thus frequently brought up by persons who have been exposed to dust, etc. In perfect health a resident in a pure atmosphere, *i.e.*, in a good climate, might perhaps scarcely ever cough up such masses, but in towns it is not uncommon for these pellets to be extruded nearly every morning, and in densely populated cities like London and Manchester, where much coal is burnt, the sooty particles taken in give rise to a black phlegm, which is coughed up more or less regularly. When the membrane has become relaxed, perhaps from the recurrence of catarrh, a secretion of this kind may become habitual. In disease the cough may be weak and insufficient to accomplish the increased labor, and it may then happen that it is desirable to excite the muscular action by appropriate stimuli, or it may be necessary to act upon the membrane and the muscles engaged indirectly through the nervous system; stimulants of the respiratory centre may be called for. But in other cases it may be possible to so sustain and strengthen the system as to give the patient more power to cough; nutrients and tonics may perhaps do this.

But a cough may be incessant and yet ineffectual, and in such cases it may be necessary to employ remedies to restrain it. These act on the mucous membrane or on the nervous system, and perhaps on the process of secretion. The irritability of the pulmonary surface needs to be soothed while the vagi are controlled, and it may be that depressants of the centre are also required. When there is no secretion to be raised, it may even be desirable to employ such expectorants as provoke the secernent action, concurrently with sedatives. When the troublesome but useless cough is kept up by disease at some small point in the larynx, that must be treated by local applications, not by expectorants, which are equally useless in the cough caused by elongated uvula.

Other hindrances may occur to the evacuation of bronchial secretion, the tubes may be in a state of spasm, preventing the removal and interfering with the respiration. When the accumulation is considerable and the spasm intense an emetic will sometimes relieve the distress. The act of vomiting produces relaxation of the muscular system and at the same time greatly aids mechanically in clearing the tubes. In other cases antispasmodics will bring about the relaxation of the bronchial muscles. Depressants of the nervous system also do this, whether of the centre or periphery. Narcotics are sometimes employed for this purpose, but in full doses are not without danger.

We may next direct our attention to the accumulated secretion. We may act upon this with a view to vary its quality or quantity. As to its consistency, when, as often happens, it is thick and tenacious, and therefore difficult to raise, we may endeavor to liquefy it in order to make it easier to get up, *i.e.*, we may try to dilute it. This we may do first of all by inhalations of steam, which may also be laden with other vapors or by atomized liquids containing chemical substances in solution, such as certain salines which are believed to assist in attenuating the sputa. But we should not forget that these vapors and sprays act upon the mucous membrane as well as its secretions, and absorption may even take place. We may also act upon the secretion through the system, by the internal administration of medicines. Iodide of potassium is a powerful attenuant, which also increases the quantity of the secretion, as does, perhaps, chloride of ammonium, but it is not unlikely that they do not merely dilute it but distinctly alter its character. Alkalies used to be considered as liquefying the sputa, but recent experiments throw some doubt on their action. Of course, whatever increases the quantity of the secretion tends to attenuate it. Another group of remedies which facilitate expectoration may produce an opposite effect, thickening the secretion. When there is a large quantity of thin, watery mucus it may be more difficult to raise than if it possessed more tenaciousness. In such cases whatever thickens it in a moderate degree may expedite its exit. Pure, dry air, even if cold, will do this, and a dry climate may be most desirable. Certain vapors and sprays may also be employed to act on the membrane and internally the administration of astringents or of opium may accomplish the end. Of course, whatever checks the act of secretion tends to thicken the sputa. Other changes in quality may also be brought about. Essential oils and other substances which impart their odor to the breath alter the secretion. Some not only add their own odor but deodorize the sputa, are, in fact, disinfectant or antiseptic expectorants, and they are the more entitled to the name, inasmuch as most of them stimulate excretion.

We come now to those expectorants which affect the quantity of sputa secreted. These also are of two kinds, those which stimulate the secernent function and those which restrain it. The first are sometimes classed with eliminants, and, indeed, have been employed with an idea of relieving the system by increasing the discharge. Such a notion is, however, obsolete; they may act by directly irritating the mucous membrane, as when ammonia and other stimulants are inhaled. Others are introduced through the system, and being brought to the pulmonary surface may act upon the vessels there, causing them to unload themselves; or they may stimulate the glands to increased action; or, again, they may stimulate the nerves; or, once more, circulating in the blood may directly influence the centre. Many of these true expectorants are unquestionably eliminated through the respiratory mucous membrane, and it is natural to seek there the explanation of their action. Many of them are very volatile, and it would appear that this quality facilitates their excretion through the aërian membrane. It is, indeed, a general rule that the more volatile a substance the more easily it escapes from the lungs. The respiratory membrane is in truth especially adapted for the interchange between the air and gaseous substances of the blood, so that we might anticipate that volatile substances would here find a natural outlet, and this, in fact, they do. When they are present in too great a quantity to escape by this route the excess not unfrequently passes off by the skin, this being also adapted for the removal of vaporous substances which may be compared with the insensible perspiration. It is true that some medicines of this kind seem to be attracted to the lungs almost exclusively, while others are determined toward the skin; but often the one appears to supplement the other, though in other cases there is rather a determination toward other mucous membranes than toward the cutaneous surface. We have said that these volatile substances are excreted through the lungs, and there exercise their influence where they not only affect the quantity but alter the quality of the secretion, and may, perhaps, replace the deteriorated phlegm by a more healthy flow, thus being, as is sometimes said, alterative. Some have doubted whether the effect is produced in consequence of elimination through the bronchial membrane or glands. In "Neligan's Medicines," edited by Macnamara, this question is raised, and it is said that it is perhaps only the odorous principle which escapes through this channel. To this it may be replied, very likely, but in such case it is perhaps this very principle to which the medicinal virtues belong. Further, the elimination by this channel, not only of volatile but of other expectorants, has at length been ascertained by experiment.

Remedies which diminish the quantity of the secretion are directly antagonistic to the true expectorant which we have just considered, but some of the former have often been grouped with the latter. When the expectoration is very profuse, its evacuation may be rendered easier by restraining secretion, and so we have remedies which might be termed "paradoxical expectorants," if we admit them to be expectorants at all. When they diminish the quantity they mostly render the phlegm thicker and more tenacious. They may act directly on the mucous membrane or its glands; others act only through the nervous system, either on the periphery, as some neurotics, or on the centre, as narcotics; others contract the vessels as astringents; and it may happen that some first of all stimulate the vessels to pour out an increased quantity of liquid, by which they are relieved, and afterward on their contracting the secondary effect is produced. Finally, it is not impossible that some may act indirectly by altering the quality of the blood, for in recent experiments a solution of soda injected into the circulation was followed by arrest of bronchial secretion.

Here it may be convenient to consider the most recent researches into the phenomena of the secretion of mucus and the changes which may be effected by medicines. Professor Rossbach reported (*Berl. klin. Woch.*), in 1882, the results of a series of experiments, mostly on cats, which have since been confirmed by Petronne (*Lo Sperimentale*, 1883). The method of procedure was to open the trachea and watch the appearance of the mucus which in the normal state covers the membrane. This they found was only a thin layer, which persisted even when large quantities of air passed over it. When the mucus was gently removed with blotting-paper the layer reformed within a minute or two, but the collection was never so great as to run into drops or to flow down. From this it would appear that the secretion is not continually going on, but that whenever it is reduced in quantity by evaporation a fresh outflow is produced to compensate for the loss. It is probable that the constitution of the mucus is not uniform, but that the portion last exuded is more watery and alkaline and therefore adapted to hold the mucin in solution. According to Rossbach's experiments no nervous current, or only a feeble one, is transmitted from the centre to the glands of the air-passages, for the secretion continues in its usual manner after section of the laryngo-tracheal nerves. But vaso-motor fibres seem to reach the laryngeal lining, as the secretion appears to have some connection with the vascularity of the membrane.

Rossbach next observed the effects of temperature, by applying hot poultices and ice-bags to the animal's body. Within half a minute of the application of ice the whole mucous surface of the larynx and trachea was blanched from the complete contraction of the blood-vessels, but in from one to two minutes this passed away and was succeeded by a bluish red tinge, while such quantities of mucus were secreted as to freely flow down. On the removal of the ice and replacing the poultice the color instantly changed to deep red. Fresh applications of ice again set up vascular spasm, but more slowly and to a lesser degree. These phenomena Rossbach considered direct reflexes.

Equally interesting results followed the application of medicinal substances. No effect was observed from a weak solution of carbonate of soda when inhaled, but the application of weak ammonia or vinegar, by painting the membrane, caused intense hyperæmia and increased secretion. Astringents applied in the same way produced an opacity of the epithelium, so that deeper changes could not be observed, together with complete arrest of the secretion. Air saturated with turpentine blown upon the membrane caused gradual diminution, and finally arrest of the secretion, which, however, returned when the irritant was removed. A watery solution of turpentine increased the secretion, but at the same time diminished the vascularity of the membrane. From this it would appear that turpentine is a true expectorant, though at the same time it removes congestion and would account for the value which has been placed upon it by some clinical observers. Other true expectorants, so far as these experiments go, are emetia, apomorphia, and pilocarpine; but neither of these affected the vascularity of the membrane. It will be observed that clinical experience confirms these results, especially as regards emetia; pilocarpine is not well adapted as an expectorant on account of its effect on the salivary glands, nor unless a powerful sudorific should be at the same time desired, but apomorphia is easily tolerated and does not often disturb the appetite or cause nausea, points which my clinical experience confirm.

The antagonists of these expectorants, atropia and its allies, were also shown to produce the effects which had been ascertained by observation; thus, the membrane could be completely dried by the application of atropia, but at the same time the vascularity was increased. In the same way morphia diminishes but does not arrest the bronchial secretion as atropia does. The combinations of morphia and atropia recommended in our chapter on Neurotics are thus shown by experiment to possess the qualities there stated, and, as we have said, these experiments have been repeated with similar results.

Petronne agrees with Rossbach that anæmia of the membrane and atony of the glands is brought about by the effect of soda on the vasomotor and secretory nerves, when that alkali is injected into the blood, but a considerable quantity of the carbonate had to be employed. An experiment of this kind is certainly not so satisfactory as the others we have named, it seems to show, indeed, that great excess of soda in the blood may hinder, or even arrest the tracheal secretion, but those who have been accustomed to employ it with an opposite intention may reply that they only use small doses, or else local applications, by means of an atomized solution. Perhaps the use of such sprays by attenuating the secretion may render it more easy to expel. Inhalations of ammonia or of vinegar are considered by Petronne as rational in chronic dry catarrh, but, of course, as inadmissible in acute cases. When the membrane was bathed with these liquids it caused extensive hyperæmia and much secretion. Astringents within four to six minutes he found produced great pallor of the membrane. His observations with nitrate of silver on the tracheal membrane of rabbits, guinea-pigs, and dogs are interesting: he used a solution up to four per cent., which dried the epithelium and completely arrested secretion, but he was able to satisfy himself that there was no diminution in the vascularity. On the other hand, in the pharynx and the nasal cavity the four per cent. solution produced a considerable excess in the secretion of the mucous glands in these regions, but in the larynx the same solution in the course of half an hour produced dryness; hence, inhalations of tannin or alum may be used as astringents, but applications of nitrate of silver will differ in their effects according to the part of the membrane to which they are applied. A spray of spirits of turpentine on a small portion of membrane diminished and soon arrested secretion, but a watery solution, one or two per cent., gave rise to hyper-secretion and vascular depletion, thus completely confirming Rossbach's statements. This property of calling forth secretion and at the same time producing anæmia of the membrane is as important as it is interesting, and enables us to understand the favorable influence of terebinthinate sprays and vapors in chronic catarrh and thickening of the membrane. With apomorphia, pilocarpine, and emetia, experimenting upon dogs and cats, he produced swelling of the tracheal membrane with a glandular condition, and so much secretion in the respiratory tract that crepitant râles were heard all over the chest, and section of the nerves did not prevent these phenomena. He considers that pilocarpine produces secretory troubles through the whole system and should therefore be used with caution, but like the other expectorant alkaloids may be useful in dry catarrh and croup. As to their antagonists, atropia and morphia, he entirely confirms Rossbach. Atropia completely checks the secretion in forty or fifty minutes, but increases the vascularity, and as section of the laryngo-tracheal nerve does not prevent the phenomena the action of the alkaloid may be exerted upon the mucus-forming cells and the terminal nerve-filaments.

The preceding account of Rossbach's and Petronne's experiments upon animals may perhaps be appropriately followed by a briefer statement of some of my researches on the subject of expectorants. It may be stated at the outset that my experience very largely corroborates their statements, though some differences may yet have to be accounted for. My observations have been made upon patients chiefly with the aid of the laryngoscope, but in the pharynx and nostrils this is unnecessary. In all cases, however, reflected light has been employed. The parts have been painted with solutions in some cases, in others they have been treated with vapors and atomized sprays. Comparing the results, I have not been able to satisfy myself as to the effect of soda, partly, perhaps, because the solution was not of the same strength. I have never regarded this bicarbonate as specially expectorant, whether employed locally or administered internally. Potash, however, promotes secretion, and is therefore a useful ingredient in expectorant mixtures; not only the bicarbonate, but the citrate, and I think also the nitrate, possesses this property. Some of the salts of lithia are also endowed with similar virtues: under the influence of these salts the secretion becomes thinner and more abundant, while congestion abates. But the most remarkable salt in this respect is iodide of potassium, the action of which, when taken internally, is obvious in the nasal passages, the pharynx, and larynx. Further, by auscultation, it may be demonstrated to act also on the bronchial membrane. When the system is saturated and severe iodism produced, it is perfectly true that a state of congestion may be set up, but at an earlier stage, or rather under the influence of small quantities, or when solutions are painted on the part, or sprays inhaled, there is seen rather the exudation of a considerable quantity of thin, watery mucus, with no increase of vascularity, and sometimes even a little pallor. This salt is therefore expectorant, promoting the bronchial secretion, as also may be incidentally remarked it does that of the salivary glands. As to such irritants as ammonia and acetic acid, I can entirely confirm the statements quoted. So I can for the most part what has been said of astringents, but these agents differ somewhat among themselves, some do not appear to produce the opacity described by Rossbach, unless the solutions are very concentrated, and all of them may be so diluted as to prevent this result and yet retain their power to diminish secretion and vascular fulness. I have applied tannin to the membrane of the larynx, in both weak and highly concentrated solutions, with the result of reducing vascularity and secretion. In the pharynx the effect is much less obvious; in the nasal passages it is more marked, and this and other weak astringents may be applied in the same manner and the effect observed with the rhinoscope. As a rule it may be stated that only weak solutions should be experimented with in this region. The result of applying nitrate of silver with a brush differs very greatly in different parts, as well as in different proportions. I have employed the nitrate in solutions varying from one grain to sixty in the ounce, in a few instances one hundred and twenty grains, and various proportions between. These concentrated solutions are of course caustic. Milder ones produce opacity of the membrane and arrest secretion; still weaker solutions, and sometimes very weak ones, produce in the larynx a sensation of dryness, which arises from diminished secretions as well as a diminution of vascular fulness. In the nasal cavity and in the pharynx, perhaps as a secondary effect, congestion follows the application of ten to twenty grain solutions, accompanied by hyper-secretion, and this even when the application is confined to a small area.

Pilocarpine and emetia internally are decided expectorants, but the former is not well adapted to be used for this purpose in ordinary cases, though the latter may be made to replace ipecacuanha. I have no conclusive experiments as to the local application of pilocarpine, but as to ipecacuanha and its alkaloid can confirm their expectorant qualities. On the other hand, apomorphia I have tried locally with very decided results, as well as administered it internally. It produces a considerable increase in the secretion, with little if any diminution of vascularity, and is, therefore, an excellent simple expectorant, and may also be given as such internally in very small doses. For the last two or three years, in consequence of the result of experiments, I have been using apomorphia as a spray to relieve preternatural dryness and irritability of the larynx, and can recommend it for this purpose. In a case in which incessant cough had been a most serious complication for many weeks, exhausting the patient—a phthisical one—and defeating all the efforts of the medical attendants, a spray of apomorphia produced more relief than any other remedy, local or general, and in distinctly laryngeal cough, with a preternaturally dry membrane, even when congestion was present, the local use of this alkaloid has been of great service, apparently by provoking a fresh flow of thin, unirritating mucus. The power of apomorphia to produce considerable bronchial secretion, when given internally, has been confirmed by Jürgensen and Wertner, who have given it in bronchitis, pneumonia, and œdema of the lungs, with good results.

With atropia my results are exactly antagonistic to those observed with apomorphia. It is very easy to check, and even to arrest the secretion by means of atropia ; it acts as promptly and certainly on the laryngeal membrane as it does in arresting salivation. In from half to three-quarters of an hour the membrane may be dried and rendered hyperæmic. This may be effected by a hypodermic injection or by a local application ; it is obvious that the latter should be made with considerable circumspection, on account of the potency of this alkaloid. It would not do to paint the ordinary solution with a laryngeal brush. Morphia, though less powerful and perhaps less rapid, also diminishes the secretion, but it does not in ordinary quantities cause hyperæmia ; it is difficult to completely arrest secretion by opiates. With regard to the combined action of morphia and atropia, it is scarcely necessary to add anything to the statements which I have advanced in the sections on opium and belladonna in a former chapter. (*Vide* Neurotics.)

It is commonly said that expectorants are rather uncertain in their action. This is not surprising if we are to include among them all the groups we have described, and then regard them all as acting in the same way, or look upon the respiratory surface as a medium for the action of evacuants. This membrane ought not to be regarded as an emunctory, and it is fortunate that the notion of evacuating by all channels has pretty well passed away. Nevertheless, we observe that gases which are ordinarily eliminated by the bowels do sometimes escape through the respiratory membrane and taint the breath, and so it is not surprising that other substances finding their way into the blood should be removed by this channel, though at other times they are eliminated by the skin or kidneys. The uncertainty complained of will, to a considerable extent, disappear if we are sufficiently exact in the selection of our expectorants and other pneumatics according to the precise effects we are desirous of producing. Even then, however, it is not to be denied that a great element of uncertainty does exist, which is chiefly due to the difficulty we experience in coping with the many grave results which are produced by diseases of the respiratory organs.

Expectorants, and indeed all the pneumatics, may be conveniently divided into two groups: 1, those which are applied locally to the mucous membrane, that is, which are inhaled as gases, fumes, vapors, or sprays, or which are brought into contact with accessible portions of the membrane as in the case of gargles, lozenges, and topical medication by manipulative proceedings; 2, those which are administered through the system, and which may be called, therefore, general expectorants. No doubt, as we have seen, many of these act topically, being brought to the part through the circulation and there excreted, at any rate in part. But being present in the blood, other effects in the system may also be produced. Moreover, the several members of the group act through different channels, some on blood-vessels, others on the nerve-supply, others on the glands of the membrane, others circulating in the blood find access to the medulla and act upon that, and others may perhaps produce their effect by acting on the blood-plasma or the red corpuscles. The influence of some narcotics is very general; thus opium not only restrains secretion and affects the entire respiratory nervous apparatus, but has so great an influence on the nervous system of the whole body as to make its influence exceedingly complex and, so to say, universal. So we find other general expectorants exercising such important influence on the circulation that their effects are felt throughout the system, whether as stimulants or depressants.

We propose now to rearrange in a few useful groups the principal respiratory remedies—pneumatics—with brief comments on some of the more important, a method which it is hoped will economize space. We will begin with

GENERAL EXPECTORANTS.

These augment the amount of secretion poured into the respiratory passages by stimulating the activity of the secernent function. Their effects upon the general system are so diverse and so considerable, especially upon the circulation, that they may be divided into groups according to their general effects, which, as it happens, to a large extent correspond with their local action. Three groups may thus be arranged: 1, Depressants; 2, stimulants; 3, alterants.

1. DEPRESSANT EXPECTORANTS.—The most important of these are the nauseants, which have already been described in connection with emetics, as in large doses they produce vomiting. Antimony and ipecacuanha act directly and indirectly as expectorants, as we have seen they do as emetics and nauseants. They hold an almost unique place as depressant expectorants, greatly reducing the circulation, and so are of more use in febrile cases. It has sometimes been thought that the expectorant action is only produced as a result of the nausea, but emetia has been

detected in the secretion, and as well as antimony is doubtless partly eliminated through the respiratory membrane and may therefore act upon it locally, at the same time the centre is depressed as well as the circulation. Moreover, an expectorant action can be produced in doses which are insufficient to excite nausea. Anyone with large experience of these medicines, and especially of antimony, can confirm this statement. (See Nauseants.)

Apomorphia, though a prompt emetic, and chiefly employed for that purpose, also possesses expectorant properties. It is, however, scarcely a depressant, except through the vomiting it produces. It does not seem to depress the circulation as a necessary part of its own influence, for in emetic doses the pulse may rise on the approach of nausea and reach its highest point just before vomiting begins, falling between the acts of vomiting. The blood-pressure, too, is not lessened by its action. Further, it stimulates the respiratory centre, and thus, as shown by Dr. Gee [1] and by Siebert,[2] greatly accelerates the respiration. From these observations, which have been fully confirmed by Quehl,[3] Riegel and Boehm,[4] Bourgeois,[5] Dujardin-Beaumetz,[6] Budin and Coyne,[7] Brunton (*Practitioner*), and others, it would seem that in itself it is hardly entitled to be called a depressant expectorant, but neither can it be, perhaps, considered stimulant, at any rate in expectorant doses. Large quantities or repeated doses may set up prostration with depression of the respiratory centre, causing slow and shallow respirations, and, as Harnack [8] proved, death from their gradual secession; there may then be a rise in the pulse-rate and a fall of temperature of, according to Ziolkowski,[9] one-tenth to half a degree C., though Bourgeois maintains that it has no effect on the temperature, and Moerz [10] observed a rise of one-fifth during the act of vomiting. After section of the vagi the respiration is more accelerated, so the effect appears to be due to the action on the centre. Carville mentions a case in which three-tenths of a grain caused prostration and collapse in an adult (*Gaz. Heb.*, 1874); another is reported by Prevost (*Medical Record*, 1875), in which a smaller dose produced serious collapse. In children, Harnack [a] found this condition easily produced, but Loeb (*Schmidt's Jahrb.*)

---

[1] Gee, S.: Clinical Society's Transactions. 1869.
[2] Siebert, V.: Untersuchungen über die physiologischen Wirkungen des Apomorphins. 1871.
[3] Quehl, Max: Ueber die physiologischen Wirkungen des Apomorphins. 1872.
[4] Riegel and Boehm: Deutsches Archiv für klin. Med. 1872.
[5] Bourgeois, J. B. V. : De l'apomorphine Recherches cliniques sur un nouvel émétique. 1874.
[6] Dujardin-Beaumetz: Note sur l'action thérapentique de l'Apomorphine. 1874.
[7] Budin and Coyne: Sur certains effets de l'Apomorphine. 1875.
[8] Harnack: Archiv. exp. Path. u. Ther., Bd. ii.
[9] Ziolkowski: Apomorphin. 1872.
[10] Moerz. A.: Prager Vierteljahr. 1872.

injected 0.002 grain under the skin of an infant thirteen months old suffering from capillary bronchitis; it produced free vomiting, by which the child was much exhausted. Occasionally without causing vomiting apomorphia has produced unpleasant and even alarming symptoms. Sometimes these appear to have been due to the use of a solution which had become green, a change it soon undergoes. It is therefore desirable to use only fresh solutions.

Lobelia and *tobacco* are both powerful depressant expectorants. They kill by depressing the respiratory centre. The active principles, lobelin and nicotin, are very powerful depressing and irritating poisons. They are quickly eliminated, chiefly through the kidneys. Both medicines have been said to be expectorant, but are not much used as such. Lobelia is reputed to possess a considerable power over spasm of the bronchial muscles, and for this reason is esteemed by some physicians, who give from ten to twenty drops of the tincture every quarter of an hour until nausea is produced, in order to arrest paroxysms of asthma. Others give as much as a drachm, and repeat it in two hours if no relief follow. But these large doses sometimes produce alarming collapse, and their effect must be closely watched. Ten minims of the ethereal tincture may be added to expectorant mixtures when there is a tendency to bronchial spasm. Lobelia is a powerful irritant of the stomach and bowels, and has been sometimes popularly used, probably on this account, but it has given rise to a number of fatal results. Even under medical supervision grave symptoms have frequently arisen, especially in weakly or young patients. Indeed, it ought not to be given to children, and altogether is not a remedy in which I have much confidence.

2. STIMULANT EXPECTORANTS.—These include (*a*) substances which act directly on the membrane, those (*b*) which stimulate the centre, and (*c*) those which, though exercising an influence over the respiratory organs, excite the circulation. Sometimes general stimulants become expectorants, as by sustaining the system for a time they either promote the bronchial secretion, or, more important still, in the cases in which they are required, enable the patient to expel it. Hot beverages and liquid nutrients assist in fulfilling these indications. In other instances exercise and gymnastics are appropriate. Alcohol, ether, essential oils, terebinthinates, and oleo-resins find their place here. Stimulants of the centre, like belladonna and strychnia may also be used with a view to sustaining respiration and assisting the removal of the sputa.

Ammonia and its carbonate are powerful general stimulants with expectorant properties, operating on the central nervous system and especially stimulating the respiratory and cardiac centres. As a volatile stimulant ammonia may be supposed to determine toward the respiratory membrane, but it appears to be oxidized in the system, at least a portion of it, and to escape, as shown by Bence-Jones ("Phil. Trans.,"

1851) by the kidneys. A small quantity injected into the blood of an animal greatly accelerates the respiration. Not only does section of the vagi fail to prevent this, but the change from the slow, deep breathing of divided vagi to the extremely rapid breathing of ammonia-poisoning is most remarkable. The increase in the pulse is chiefly due to the effect on the cardiac centre, though it also exercises a powerful but evanescent stimulant action on the heart. On injection into the veins Lange found a momentary fall of arterial pressure, followed by an enormous sudden rise, corresponding with the increase of the pulse-rate, and which by other experiments (*Archiv f. exper. Path. u. Ther.*) he showed to be due to an effect upon the heart itself, or on the peripheral vaso-motor nerves, or on the muscular fibres in the coats of the arteries. In poisonous quantities Funke found the heart was quickly paralyzed (*Pflueger's Archiv*). Felz and Ritter (*Jahr. de l'Anatomie et de la Phys.*, 1874) observed that the red corpuscles were injured in a dog killed by ammonia. As an expectorant the carbonate is generally employed, and is appropriate in feeble patients when the secretion is tenacious and difficult to raise. By sustaining the centre it improves the coughing power, and enables the patient to get rid of the accumulation. It is therefore in the bronchitis of old and weakly people, in pneumonia when the heart is failing, or when it occurs in the course of typhoid, that it is most appropriate. In short, it is as a general stimulant, with a very special action on the respiratory centre, as well as on the circulation, that it is most valuable. That it also does affect the bronchial membrane seems to be shown by the fact that in some instances of death by poisoning this membrane has been found intensely congested. It is often added to other expectorants, such as senega, in order to increase their effects. Compare its action with other respiratory stimulants.

Scilla.—Hippocrates employed squill (σκίλλα) for various purposes, and Pliny describes (lib. xix., cap. 30) two sorts, and mentions that Pythagoras had written a treatise on the medicinal virtues of squill and invented the *acetum*. Durastantes[1] extolled this preparation in 1567, Alberti[2] and Schulze[3] recommend it in asthma, and Brichenden[4] says that in his time it was used in both asthma and dropsy. Its action as a diuretic was discussed by Cullen[5] and Home,[6] while Vogt[7] contrasted it

---

[1] Durastantes, J. M.: Libri duo Medici, i., De Aceti Scillini compositione mirificis, ob sanitatem ac vitam duitissime producendam viribus, ac congruo usu. 1567.

[2] Alberti, M.: De Squilla. 1722.

[3] Schulze, J. H.: Disp. sistens ægrotum Asthmaticum usu radicis Scillæ sublevatum. 1737. Also, Examen chemicum radicis Scillæ marinæ. 1739.

[4] Brichenden, J.: De radice Scillæ. 1759.

[5] Cullen, W.: Treatise Materia Medica. 1789.

[6] Home: Clinical Experiments. 1783.

[7] Vogt: Lehrb. d. Pharmakodyn. 1828.

with digitalis. The diuretic effect was naturally noticed before the expectorant, but the latter has been well established. It was also early recognized that scilla is very irritant, full doses causing nausea, vomiting, and perhaps purging. In toxic quantities inflammation and even gangrene of the stomach and intestines have been seen and strangury and convulsions have occurred.

Scilla is a powerful stimulant of the mucous membrane, we may perhaps say of all the mucous membranes, but it exercises a special influence on the bronchial surface, where scillain has been shown to be eliminated. It is further entitled to a place among the stimulant expectorants as it possesses a very decided action on the heart, in respect to which, as well as to its diuretic qualities, it may be compared to digitalis, with which, indeed, it is very often advantageously combined. The force of the cardiac contraction is increased by this medicine, but not the frequency, which is rather retarded. At the same time contraction of the peripheral vessel is produced, with a rise in the blood-pressure. This is followed by relaxation, which mostly begins in the renal arterioles, and thus accounts for the diuretic action. Squill increases both the secretion and the vascularity of the bronchial membrane; it is therefore contra-indicated in acute cases. Active congestion of this membrane becomes acute inflammation under the influence of scilla. It is true that in strong persons it may seem to do good, for by increasing the amount of secretion and at the same time rendering it thinner this remedy may facilitate its expulsion and so give rise to a deceptive appearance of relief, while it is in reality aggravating the disease. But in chronic cases with some passive congestion it may be very appropriate. In relaxed conditions of the membrane, in leuco-phlegmatic constitutions, and in aged or debilitated patients, when the expectoration is viscid and raised with difficulty, provided the skin be cool, soft, and moist, and the pulse slow, soft, or weak, the best effect may be obtained. If the right heart should be at the same time secondarily affected, the remedy will be still more applicable, as it will act favorably in sustaining the circulation. But if the pulse be hard and quick, and the expectoration purulent, there is considerable risk of its doing mischief. It is quite possible that in such a case it may indirectly benefit a strong patient, by exciting a profuse secretion and thereby unloading the capillaries. In doubtful cases it may be combined with antimony or with refrigerants; but when these are required it is usually too soon to begin squills. The rule is, in acute inflammations it is not to be given—nor is any stimulant expectorant. Ipecacuanha, it is true, has a stimulating influence on the membrane, in increasing and liquefying the secretion, but in the doses usually prescribed the irritation is not great and the other qualities of this drug make it a depressant expectorant. Scilla is apt to derange digestion, and may therefore be inadmissible when otherwise indicated lest it should interfere with nutrition. In such cases combinations with ammonia, belladonna, or carminatives may be employed, or these and other remedies substituted, with perhaps minute doses of ipecacuanha, reserving the squill for a later period.

Senega was introduced as a remedy for the bites of snakes and other venomous creatures by Dr. Tennent,[1] a Scotch physician, residing in Pennsylvania in 1738, and soon after we find from C. Linnæus[2] that it had been employed in fevers and inflammations, and he considered it diuretic. By 1782 Hellmuth[3] was able to collect remarks from various writers as to its use in pleurisy, pneumonia, hydrothorax, asthma, rheumatism, dropsy, etc.; it was considered to promote most of the secretions, and therefore was employed in numerous diseases. But it has failed to maintain its reputation, except in bronchial affections, and in these it is still confidently prescribed. Sundelin[4] took twenty grains of the powder every two hours for three doses; he found it greatly irritated the back of the tongue and throat and increased the flow of saliva. A little later it caused burning at the epigastrium, nausea, and vomiting, as well as griping pains and watery purging. The gastric uneasiness with loss of appetite lasted three days. The skin was rendered warmer and moister, and diuresis, with a feeling of heat in the urinary passages was produced. Larger doses caused violent vomiting and purging, with giddiness and anxiety. It appears to excite the vascular system, and to stimulate all the mucous membranes; but its chief value is for its effect on the respiratory membrane. Its irritant quality has caused the powder to be used as an errhine, but it is internally that it is chiefly employed. It promotes the secretion of the bronchial membrane, and renders it less tenacious. It is believed to stimulate the circulation in that membrane, as well as the nerves, so that it assists the expulsion of the sputa, and is therefore a valuable stimulating expectorant, acting in both ways. Senega also acts upon the heart in the same way as squill and foxglove, diminishing the frequency, but not the force of the beat; probably, also, it affects the circulation, and through this may be diuretic. It contains saponin, which has some analogies with digitonin, and may be perhaps identical with it. This active principle is locally very irritating and affects the whole circulatory and nervous systems; it is excreted by the bronchial membrane (which perhaps explains the expectorant action of senega) as well as by the skin and kidneys, both of which it stimulates to increased action. As a diuretic it increases the amount of the solid constituents of the urine as well as the water.

So powerful a stimulant to the respiratory membrane is to be avoided

---

[1] Tennent, J.: On the Rattle-snake Root. 1738. Also, Epistle to Dr. Mead concerning the Efficacy of the Seneca Snake Root. 1742.

[2] Linnæus, C.: Radix Senega. 1749.

[3] Hellmuth, L. C.: De radice Senega. 1782.

[4] Sundelin: Handb. d. spec. Heilmittell.

in acute inflammatory conditions. It may even be capable of converting congestion into inflammation. In pneumonia and phthisis it is also as a rule too stimulating. As it irritates the stomach and bowels, it is desirable to avoid it where the digestion is easily disordered and when there is a danger of impairing nutrition. Moreover, most persons consider it nauseous, so much so that many decline to take it. Its greatest value is in chronic bronchitis, in debilitated constitutions, and when ammonia is called for. The infusion then forms a good vehicle for carbonate of ammonia. Some persons venture to give it before the febrile stage has completely passed by, but most observers admit that it is much better at a later period. Some have so much faith in its specific action on the membrane, that in order to obtain that at an earlier period, *i.e.*, before they would venture to give it alone, they give with it, or at the same time, small doses of antimony. This, of course, increases the tendency to vomiting which is produced by the senega, and it would be more rational and more in accordance with the knowledge we have obtained of its action to postpone its use until antimony becomes no longer appropriate.

Turpentine.—Several terebinthinates have been employed from an early period, though it is not easy to identify the particular plants spoken of by ancient writers. Probably the pistachia terebinthus is the terebinthus of the ancients (the Τερμίνθος of Dioscorides and Theophrastus). Oil of turpentine is a powerful stimulant to the various mucous membranes, and also to muscular fibre ; it is largely excreted by the lungs and kidneys, perhaps also to some extent by the intestines. Some of it passes out through the skin, and an eruption is occasionally produced by it ; it is extremely irritant to the kidneys, which fact prevents its being used when otherwise it is indicated ; but it has been employed in many very different diseases. It is a depressant of the nervous system, an effect which begins to manifest itself as soon as the reflex stimulant influence has been produced. A full dose produces, after preliminary excitement, dulness, drowsiness, and unsteady gait, and a larger quantity, coma. At the same time the heart is interfered with, and the blood-pressure falls. These facts may perhaps explain the remarkable power of turpentine to arrest hæmoptysis, and other forms of internal hemorrhage ; a full dose sometimes acting like a charm, when all astringents have failed. This effect may perhaps be partly due to an effect upon the vessels. In the bronchial walls it appears to affect the vascular supply, and also the muscular fibres. It must also be a disinfectant to the sputa, and for this reason is useful in chronic bronchitis, bronchiectasis, and pulmonary gangrene. Locally we know that the fumes of terebinthinates and allied substances have long been found to exercise an important influence on the respiratory membrane. This local influence is no doubt exercised upon the capillaries, glands, and nerves of the membrane during the exhalation of the remedy, which imparts its own odor to the breath.

The inconvenience attending the administration of oil of turpentine is derived not from its effect on the gastro-pulmonary membrane, but on the genito-urinary tract, to which it determines, and also on the kidneys, which it greatly irritates, even small quantities often giving rise to lumbar pain, diminution or sometimes suppression of urine, strangury, and hæmaturia. Although, therefore, it has been used as a diuretic, and is generally speaking antispasmodic, any excess in the dose produces opposite effects, and it is often necessary to omit its use on account of renal irritation. Besides these symptoms toxic doses often, but not invariably, occasion vomiting and purging. The giddiness and intoxication which succeed the preliminary, no doubt largely reflex, excitement rapidly pass into a state of extreme depression. Complete unconsciousness comes on in some cases with dilated pupils, the pulse becomes rapid and weak, and the respiratory centre fails. At a very early period the effect on the renal organs was observed, Ranchin,[1] in 1640, having given us an account of the violet odor imparted to the urine by the remedy, and Wilhelmi[2] gives an account of its use in diseases of this membrane and compares it with balsams. Durand[3] a century afterward recommended it in biliary calculi. During the present century the properties of the medicine have been more carefully studied and its position among therapeutic agents determined.

All the turpentines properly so called are odorous exudations from the stems of trees belonging to the coniferæ ; some, as frankincense, are solid, others, like Canada balsam, liquid ; all of them are mixtures in various proportions of volatile oils and resins, these can be separated by distillation. Thus pure oil of turpentine, sometimes improperly called spirit or essence of turpentine, is the volatile oil obtained from various species of pinus by distillation from the crude exudation, the solid resin being left. When freshly distilled the pure oil is colorless and limpid, but on exposure to the air it absorbs oxygen, gets yellowish and thicker, and a resin is produced. Moreover, it converts part of the oxygen into ozone, for which reason it is employed in the guaiacum process for detecting blood. The oil is very slightly soluble in water and less soluble in alcohol than most volatile oil, it is, however, readily taken up by ether ; in its turn it dissolves fixed oils and fats, resins, india-rubber, etc. Prepared from different sources it behaves differently toward polarized

---

[1] Ranchin, F. : Traité curieux sur l'odeur de la violette que les Terébinthines donnent aux urines. 1640.

[2] Wilhelmi, J. : De Terebinthina. 1699.

[3] Durand, J. F. : Observations sur l'efficacité d'un mélange d'Ether sulfurique et d'Huile volatile de Térébenthine dans les Coliques Hépatiques produites pars des pierres biliares. 1790.

light, but commercial samples often consist of a mixture of several oils. Pure oil = $C_{10}H_{16}$, and is therefore isomeric with many other essential oils, and is resolvable into terebine, cymene, a camphoraceous body, colophene, etc. We have dwelt thus particularly on its properties on account of its intimate relations with other essential oils and allied remedies, some of which we will now notice. Firwood oil, obtained from the pinus sylvestris, may be regarded as identical with common turpentine, but rather less disagreeable in flavor.

Essential oils are rather complex compounds, possessing powerful odors of very various kinds, many of which are greatly esteemed for their agreeable flavor and fragrance. They contain camphors, turpenes, resins, etc., and are allied to balsams and gum resins on the one side and to carbolic, benzoic, and cinnamic acids on the other. Many of them at a low temperature separate into a solid camphor-like body called stearoptene (στέαρ, fat, πηνός, volatile), and the pure fluid hydro-carbon eleeoptene (ἔλαιον, oil, πηνός). We may make three groups: (*a*) Pure hydrocarbons or non-oxygenated turpenes, isomeric with oil of turpentine, $C_{10}H_{16}$, examples of this group are oil of juniper, lavender, peppermint, cloves, and several others; (*b*) oxygenated, of which cinnamon, caraway, etc., furnish examples; (*c*) those which contain besides C and H a proportion of S, whether with or without N, these are marked by their powerful and often fetid odor, as in the case of oils of mustard, horseradish, and assafœtida.

Essential oils are partly excreted by the pulmonary mucous membrane and stimulate it in a manner resembling turpentine and camphor, and may therefore sometimes be substituted as more agreeable medicines. As we have seen, they differ considerably among themselves in fragrancy, flavor, and other qualities. And so they do in their therapeutical value, but they may all be regarded as weakly antiseptics, as stimulants, and antispasmodics, and inasmuch as they are eliminated to some extent through the bronchial membrane, claim to be regarded as expectorants. Many of them, however, exert their principal effect on the alimentary membrane, exciting the nerves of taste and smell, promoting the flow of saliva, stimulating the stomach, and increasing appetite; they are therefore commonly called carminatives and stomachics. The muscular coat is also stimulated, and the consequent contractions expel flatus and remove the pain caused by distention; they are therefore in a sense antispasmodic. This stimulating influence extends along the intestines and some of them are highly prized in colic and as corrigents of purgatives. They enter the blood and directly excite the whole nervous system, though perhaps some of their stimulating influence may be regarded as reflex. The pulse rises in force and frequency; under their influence the blood-pressure is also raised, but these remedies are not often given in doses sufficient to affect the circulation or to be much felt through the general nervous system. They are excreted by all the mucous membranes, chiefly through the kidneys and bronchial membrane, upon which, as we have said, they may exercise their special effect while being eliminated.

Oil of eucalyptus, which has lately been so largely used, is antiseptic, and being excreted by the kidneys and lungs imparts its odor to their excretions, disinfecting them and at the same time stimulating the mucous surface ; it has therefore been freely used in cystitis and pyelitis on the one hand, and on the other much more freely in bronchitis, asthma, and other respiratory affections. Oil of anise has generally been held to determine toward the respiratory membranes, while the mints, cloves, pigments, caraway, coriander, etc., most affect the alimentary membrane, and juniper is a diuretic and is identical with turpentine in its composition and specific gravity, and resembles it in imparting a violet odor to the urine. Garlic, onions, leeks, and other strong-smelling vegetable substances employed in cookery, owe their qualities to their essential oil, and have long had a popular reputation as expectorants in chronic catarrh. In some districts raw onions are eaten by elderly sufferers from bronchitis, under the belief that they help their winter cough, and with the same view leeks and garlic are freely used in their diet.

Camphor may be regarded as an oxidized product of an essential oil, or a stearopten ; its formula is $C_{10}H_{16}O$, and it has so distinct an effect on the bronchial mucous membrane, through which it is eliminated unchanged, that it has long enjoyed a popular reputation in catarrh and other affections, the powder being even used as snuff. In moderate quantities it does not give its odor to the breath, and it is not eliminated as such, but in a changed condition through the kidneys. It is somewhat antiseptic and generally regarded as a diffusible stimulant and antispasmodic. In ordinary doses it is rapidly absorbed and apparently rapidly decomposed, and does not produce a great effect, but in full doses its chief action is on the brain, where it may produce giddiness, drowsiness, and a species of intoxication. (*Cf.* Neurotics.) Oil of camphor is a hydrocarbon $C_{10}H_{16}$, isomeric with oil of turpentine, and possessing in a high degree the odor and other qualities of camphor.

Canada balsam is the turpentine which exudes from the balm of Gilead fir—abies balsamea. It has sometimes been given internally, for the same purposes as ordinary turpentine, and is believed to be less irritating. It is not a true balsam, but yields a volatile oil on distillation. Fine specimens are sometimes sold as balm of Gilead, which, however, is the product the balsamodendron opobalsamum. Other turpentines have also been used, and they all possess, though in different degrees, similar qualities.

Balsams.—The true balsams are the fragrant, resinous substances which contain benzoic or cinnamic acids, and all of them have some claim to be considered expectorants. In the Pharmacopœia we have four true balsams, viz. : benzoin, storax, and the balsams of Peru and Tolu. Copaiba is not a balsam, but an oleo-resin, as is Canada balsam. *Balsam of Peru*, though a stimulating tonic and expectorant, is not much used internally in the present day. Like its allies it may be considered antiseptic, it seems also to be a nervous sedative. *Balsam of Tolu* possesses similar qualities, but its flavor is much more agreeable, and consequently it retains its place as a valued ingredient of cough-mixtures, though in the quantities in which it is mostly employed it is more a flavoring ingredient than an active therapeutical agent. At one time the use of balsams and other somewhat stimulating substances was almost universal, but now they have become unduly neglected, and certainly in chronic bronchitis, winter cough, and any old standing affections of the respiratory membrane, and in bronchitic asthma they deserve attention. They began to be neglected from the time when Dr. Fothergill[1] pointed out the evils that might follow their indiscriminate use, about one hundred years ago. But Trousseau and Pidoux[2] have expressed considerable confidence in them as remedies in chronic bronchial catarrh, though in the present day it is their local use which is most trusted.

Benzoin is not named by ancient writers in such a way as to enable us to identify it, but it must have been known at an early period, and has long been used in incense. It seems to have been confounded with other balsams, all of which were believed to exercise considerable influence and were highly valued as medicines. In 1530 Perez[3] wrote a treatise upon their use. In 1592 Alpinus[4] distinguished some of them, and in 1597 Guibert[5] pointed out the distinction between benzoin and myrrh.

Benzoin is the most irritating of the balsams, therefore the most likely to derange the digestion; its effect on the respiratory mucous membrane resembles that of the others, but it is more stimulating and therefore should be avoided where there is any degree of irritation. In chronic cases, in leuco-phlegmatic constitutions it may be employed. Benzoic acid has also been used internally for this as well as other purposes, as well as topically. *Storax* was well known to the ancients; it is mentioned by Hippocrates, Theophrastus, Dioscorides, and later on by Galen, under the name στύραξ, and Pliny calls it styrax. The source of this and some other drugs related to it have been carefully investigated by the late Daniel Hanbury,[6] whose papers on such subjects are full of

---

[1] Fothergill, John: Works. edited by Lettsom. 1783.

[2] Trousseau et Pidoux: Traité de thérapeutique. Third edit. 1874.

[3] Perez. G.: De Balsamo, y de sus utilidades para les enfermedades del cuerpo humano. 1530.

[4] Alpinus, Prosper: De Balsamo Dialogus; in quo verissima balsami plantæ, Opobalsami, Carpobalsami et Xylobalsami, cognitio elucescit. 1592.

[5] Guibert, N.: De Balsamo ejusque lachryma quod Opobalsamum dicitur. 1603.

[6] Hanbury, Daniel, F.R.S.: Science Papers. 1876.

information and interest. The therapeutical use of styrax is the same as the other balsams.

Gum Resins.—Olibanum and myrrh are both of great interest on account of their long history and the immense value at one time attached to them. Olibanum, the Lebonah of the Jews, the Lubán of the Arabs, the λίβανος and λιβανωτὸς of the Greeks, was doubtless the frankincense of the ancients. It is mentioned in the book of Exodus (xxx., 34). It contains a volatile oil besides gum and resin. It was long regarded as most useful in chronic mucous discharges, particularly those of the respiratory membrane, and was also employed in other chest affections, even in hæmoptysis ; it may be regarded as a stimulating expectorant and is most efficacious as a fumigation. Myrrh was a most important article of commerce at the very dawn of history, the company of Ishmaelites who bought Joseph from his brethren " came from Gilead with their camels bearing spicery and balm and myrrh, going to carry it down to Egypt " (Genesis xxxvii., 25). Its Hebrew name is mahr (מֹר), most likely in allusion to its bitter taste ; the Greeks called it σμυρνα, and in the Æolic dialect μύρρα. Hippocrates valued it in several diseases and Dioscorides described several kinds. But, notwithstanding the early period at which the product was known and its importance as an article of commerce, we had no accurate account of the tree which produces it until quite modern times.

An interesting account of this investigation has also been given by Hanbury.[1] Guibert[2] distinguished it from benzoin, for Haller, speaking of his work, says, " Agit de myrrha, de vino myrrhato, de benzoino ; negat pocula myrrhina facta fuisse ex benzoino, et benzoina myrrha differre docet." In the next century Polisius[3] collected a curious account of the uses of this drug and the numerous formulæ in which it was employed, an example followed by an English writer[4] at the commencement of the present century.

Myrrh promotes the appetite and is considered as a good stomachic. Like the other remedies to which it is related it appears to be eliminated by the mucous membranes, upon which it acts as a moderate stimulus. In large doses it may irritate the stomach and cause a degree of pyrexia with a full pulse and a feeling of warmth in the respiratory passages, but it does not appear to possess to any extent the nervine and antispasmodic properties of the fetid gum-resins, such as assafœtida, galbanum, etc. It is, however, considered more distinctly tonic, in which it is

---

[1] Hanbury, D. : Op. cit.

[2] Guibert, N. : Assertio de murrhinis, sive de iis quæ murrhino nomine exprimuntur, adversus quosdam de iis minus recte disserentes. 1597.

[3] Polisius, G. S. : Myrrhologia seu myrrhæ disquisitio curiosa. 1688.

[4] Stackhouse, J. : Extracts from Modern Authors respecting the Balsam and Myrrh Trees. 1815.

distinguished from the balsams. In chronic bronchitis it may therefore be useful. It has been lauded in phthisis, in which it was supposed to check purulent expectoration, but it is more likely to prove injurious in this disease, at any rate in large doses. It has been much employed in gargles and mouth-washes; it may be considered somewhat antiseptic.

Balm of Gilead, the balm of the Old Testament, the Βάλσαμον of the Greeks, also called balm of Mecca, once so highly valued, is now seldom employed in Europe, but in Asia still retains its reputation both as a medicine and a perfume. It contains a volatile oil and its properties are closely allied to the terebinthinates.

Bdellium is the name applied to two gum-resins, one of them probably the bdellium of the Bible, also called Indian bdellium, or false myrrh, the other called African bdellium; the former is the more fragrant and resembles myrrh, the latter contains a volatile oil. Neither are now much used.

Ammoniacum still retains considerable repute as a stimulant expectorant, and in this respect is perhaps overrated, being scarcely superior to several of the almost obsolete gum-resins and inferior to the fetid ones. Two kinds have been described, and that used by the Greeks, ἀμμωνιακόν, differs from the article now in use. Pliny says it grew near the temple of Jupiter Ammon, to which the name has been traced, but others think it quite as likely that it is only a corruption of armeniacum, having been probably imported through Armenia. From Abu Mansur Mowafik ben Ali[1] we learn that it was found in Persia as early as the tenth century. It is mentioned by various writers, Greek and Roman, from the first to the thirteenth century, as an incense or fumigation, *thymiama* and suffimen.

Ammoniacum is decidedly a local irritant, and large doses are therefore apt to disturb the stomach. It appears to be eliminated by the mucous membranes, upon which it acts as a stimulant. It contains a small quantity of volatile oil, which probably escapes through the lungs and accounts for its stimulant action on the respiratory membrane. The irritation of the skin produced by the plaster would seem to be due to the resin. In the small doses usually given it can hardly be considered to have much action on the system, but full doses are apt to cause vomiting and purging by their local irritating action. In chronic bronchial catarrh and in asthma, and in some cases of emphysema if there is a good deal of secretion from a relaxed condition of the membrane, it sometimes appears to be useful, though Trousseau and Pidoux (*op. cit.*) took it in doses of two drachms without producing any effect, local or general. It is certainly less efficacious than the fetid gum resins. Cullen[2] preferred

---

[1] Liber Fundamentorum Pharmacologiæ. 1055.
[2] Cullen, W.: First Lines of the Practice of Physic. 1784. Fourth ed.

assafœtida as an expectorant, and when any antispasmodic influence is desired it would be well to combine it with one of its more disagreeable allies.

These *fetid gum resins* have held a distinct place in materia medica from the earliest times, and are still valued as possessing very distinct properties, dependent, probably, for the most part on the essential oil which can be obtained by distillation. In spite of what we consider their extremely disagreeable odor they have been and are prized in the East for this very property, and have sometimes been employed among us in cookery, though only in the most minute quantities. The most important and the strongest is *assafœtida*. Although undoubtedly used by the ancients the history is not very clear, on account of the confusion between it and the *succus cyreniacus*. It seems as if it had been introduced as a substitute for the cyrenian juice, probably from Persia, the word *assa* corresponding with *laser*, pointing to that source, but Myrepsus,[1] who lived about 1227, speaks of ἀσαφιτιδα.

Assafœtida is a nervine stimulant, antispasmodic, and may be regarded also as expectorant. In truth all these fetid gums might very well be grouped together as antispasmodic expectorants. Some have called it disinfectant, as such it would be a good illustration of a more powerful odor overpowering a weaker one, and so obtaining the credit of destroying it. It is readily absorbed and is eliminated through all the secretions, charging them with its disagreeable smell. It exercises a decidedly stimulant influence on the mucous membranes, particularly the bronchial, through which a considerable portion of the essential oil escapes. So completely does it pervade the system that the pus of superficial ulcers has been observed to smell strongly of the drug. Joerg[2] found that in doses of twenty grains assafœtida caused in healthy persons irritation of the stomach and bowels, with increase of the secretion of the alimentary membrane and of the fæces. The bronchial secretion was also increased, the respiration quickened, and together with this the pulse rose, the animal heat seemed augmented, and perspiration was promoted. These and other excitant effects were verified on nine healthy persons, the dose in no case exceeding twenty grains. Nevertheless Trousseau and Pidoux swallowed half an ounce for a dose, with no other effect, they tell us, than having to live for two days in an atmosphere more horribly fetid than the drug itself. It is said that the perspiration of Asiatics who use assafœtida daily is very fetid, a circumstance alluded to by Aristophanes,[3] but probably it is the breath which is more tainted and distributes the odor round the person, though there is little doubt that

---

[1] Myrepsus, N.: Antidotarius, cap. xxvii., cited by Alston, Materia Medica, *vide* Sprengel's Hist.

[2] Joerg: *Vide* Wibmer. Werk. d. Arzneim. u. Gifte.

[3] Aristophanes: Equites, act ii., scene 4.

the skin also excretes some. As a stimulant and antispasmodic expectorant assafœtida is well adapted to relieve chronic bronchial disease, attended with dyspnœa and wheezing, especially in old and weakly persons where the disease is of long standing, and where the dyspnœa is somewhat paroxysmal. Another class of cases in which it often produces benefit is recurrent bronchial catarrh, attended with wasting, in young women who seem predisposed to phthisis and in whom debility and amenorrhœa cause anxiety. It is scarcely as an expectorant, however, that in such cases it is most useful, for under its influence the bronchial secretion often diminishes; but there is no doubt that the effect on the bronchial membrane is favorable, it may almost be called soothing, and the other well-known properties of the drug indicate its use. Combinations with ammonia, where a more stimulating influence is needed, are most valuable and should be used when the circulation is languid. In pure spasmodic asthma, assafœtida is uncertain. Cullen,[1] who placed great reliance on the drug in other diseases, found it of no benefit in this, and his opinion is indorsed by Pereira.[2] Still it may often be advantageously taken in combination with other remedies between the paroxysms, with a view of giving tone to the respiratory system, especially when attacks are provoked by damp. It is more useful still in bronchitic asthma. In laryngismus and in whooping-cough it has been very strongly recommended both by Millar[3] and Kopp,[4] but I cannot report much in its favor, and it is difficult to persuade children to take it.

Galbanum.—The next most important of the fetid gum-resins is mentioned among sweet spices in the book of Exodus (xxx., 34). It was used by Hippocrates and other Greeks. Dioscorides spoke of $\gamma\alpha\lambda\beta\alpha\nu\eta$ as the produce of $\mu\epsilon\tau\omega\pi\iota\sigma\nu$, which grew in Syria. Its physiological and therapeutical properties resemble those of assafœtida, and it may be ranked between that drug and ammoniacum, but as it yields more essential oil than either it has been supposed to be more excitant to the circulation. Clinical observation does not support this idea, and it may be that the oil, though greater in quantity, is of a less powerful kind. It is decidedly less antispasmodic and probably less expectorant, and for the most part only employed in combination with its more powerful ally.

Similar observations may be made respecting *Sagapenum*, the $\sigma\alpha\gamma\dot{\alpha}\pi\eta\nu\sigma\nu$ of Hippocrates and Dioscorides, the *sacopenium* of Pliny, which, so far as the respiratory membrane is concerned, may be ranked between assafœtida and galbanum, though it may be remembered that it acts more decidedly on the alimentary mucous membrane, and has therefore been

---

[1] Cullen, W.: Treatise on Materia Medica. 1789.
[2] Pereira, J.: Elements of Materia Medica and Therapeutics. 1853.
[3] Millar, J.: Observations on the Asthma and the Whooping-cough. 1769.
[4] Kopp: London Medical Gazette, vol. i.

used as a warm aperient in constipation with flatulence. The essential oil has a strong odor resembling garlic, and a bitter garlicky taste, which is shared by the resin.

Opoponax was also employed by the Greek fathers of medicine, who mention, some three and others four, kinds under the names πανάκες and ὀποπάναξ. Dioscorides, who describes three kinds of the gum-resin, gives a good account of opoponax (ὀποπάναξ), which he tells us was procured from the πάνακες ἡράκλειον. In composition this resembles the preceding gum-resins and is most like ammoniacum in its properties; it may often be advantageously combined with this, and though excluded from the British Pharmacopœia, perhaps deserves a place as much as a number of other little used substances. The essential oil is eliminated through the lungs and acts on the bronchial membrane, stimulating it to increased secretion but not often appearing to irritate.

Elemi, apparently a word of Ethiopian origin, is probably derived from several of the terebinthaceæ, and its effects are similar to other terebinthinates, but it is abandoned to external use. The ointment is sometimes called the balm of Arcæus, being a substitute for an ointment introduced by that writer.[1]

Copaiba has sometimes been employed in bronchial affections, instead of some of the other oleo-resins, and Armstrong,[2] Halle, Bretonneau,[3] Bayle,[4] and others have recommended it, but it has no superiority to the other substances, is extremely disagreeable to take, sometimes gives rise to severe gastric irritation, and on account of its associations and of its affecting the bronchial less distinctly than the urinary passages is better abandoned to the treatment of diseases of the latter.

Mastich, obtained from the Island of Scio, also contains a volatile oil, and so possesses in some degree the properties of the terebinthinates. Highly prized for centuries, it was used by Hippocrates, who also employed the leaves of the tree σχῖνον, as well as the resin, and an oil obtained from the fruit. It is scarcely ever now used in medicine, though it so long enjoyed a high reputation, and should have some power in chronic affections of the mucous membranes.

Inula Helenium, the ἑλένιον of Hippocrates and Dioscorides, our elecampane, is an aromatic tonic as well as an expectorant, also somewhat diaphoretic. Thus it seems to promote the secretion of most mucous membranes and of the skin. In large doses it is sufficiently irritating to cause nausea and vomiting. Its action resembles that of senega, but it is milder, and as it has had some repute as a tonic in dyspepsia it may

---

[1] Arcæus: De recta Curand. Vulner. Ratione. 1658.

[2] Armstrong, J.: On the Efficacy of Copaiba in Inflammation of Mucous Membranes. 1823.

[3] Bretonneau: Mémoire sur la diphthérite.

[4] Bayle, A. L. J.: Bibl. de Thérapeutique. 1828.

be presumed that it is less likely to irritate the stomach than senega, and is certainly less nauseous. It is in rather chronic bronchial affections with profuse secretion, in the absence of pyrexia, that it is most likely to be useful. The name *inulin* has been sometimes applied to a substance resembling starch obtained from Helenium, but which its discoverer, Rose, called alantin. The active principle seems rather to reside in a camphoraceous body, which has been called alantol, or alant camphor, somewhat resembling menthol in taste and smell.

Larix.—This interesting tree has again been brought somewhat into notice after having suffered a long eclipse. From it is obtained Venice turpentine, to which wonderful virtues were long attributed, and which has lately acquired a sudden but ephemeral notoriety. Orenburgh gum and manna de Briançon were also the products of the larch. Dr. Frizel, of Dublin, introduced the inner bark, in 1858, as a somewhat stimulant medicine, possessing also astringent properties, with a special action on mucous membranes, especially the bronchial, and a tincture of this bark has found its way into the Pharmacopœia, and is often used to check profuse expectoration in chronic bronchitis. It has also been used in hæmoptysis and other hemorrhages. Dr. Stenhouse obtained from the bark a volatile active principle, which he could not find in other pines, and which he called larixinic acid, on account of its reaction, but which some have termed larixin, an inconvenient name, as it may be confused with another body to be named just now. The common larch is also the source of a fungus, which for a long period was highly valued, the agaricus laricis, or boletus laricis, or boletus officinalis, boletus purgans, polyporus laricis vel officinalis, as it was variously called. This fungus was known to the ancients, Dioscorides describing it under the name Ἀγαρικόν, while in the East it is now known as ἀγαρικόν τολευκόν and Κατρὰν μαντυρί, It may still be found at herbalists', under the name of agaric, female agaric, white agaric, or larch agaric. The active principle has been referred to the resin, but Martius[1] isolated a white amorphous bitter-tasting powder which he termed laricin. The resin certainly possesses irritant qualities and is purgative. The powder of agaric is a local irritant, causing watering of the eyes, sneezing, cough, etc., when applied to the nostrils. Swallowed, it acts as an emetic and a purgative, causing considerable nausea and griping, but its most important use was to arrest colliquative sweating in phthisis. De Haen recommended it for this purpose, and Bisson[2] and many others[3] reported favorably of it, includ-

---

[1] *Vide* Buchner's Repertorium, 1846.

[2] Bisson: Mémoire sur l'emploie de l'Agaric blanc contre les sueurs dans la phthisie pulmonaire. 1832.

[3] *Vide* Rubel, J.: De Agarico officinali. 1778.

Jacquin: Diss. de Agarico officinali. 1778.

Murray, J. A.: Apparatus Medicaminum, tam simplicium quam compositorum. 1776-90.

ing Andral, who, however, afterward abandoned it as of little use. The dose of agaric was from three grains to eight, at bedtime, to restrain sweating, and from one-half a drachm to a drachm as a cathartic.

Tar and its derivatives are all more or less stimulant expectorants, but as they are more frequently used in respiratory diseases for their antiseptic qualities, they will be considered with other antiseptics in a separate group.

3. ALTERANT EXPECTORANTS.—These are chiefly alkaline, saline, or antiseptic. They promote the secretion of the bronchial membrane, but with little or no irritant effect, and may even rather depress than excite the circulation.

Alkalies.—Something has already been said respecting recent experiments, which tend to throw doubt on the long-acknowledged power of alkaline carbonates to promote bronchial secretion and to render it less viscid. It had long been considered certain that these carbonates were partially eliminated by the bronchial membrane, but the experiments alluded to cast doubt upon this. It must be remembered, however, that the injection of a large quantity of soda into the blood is a very different thing from the administration of moderate medicinal quantities. Then it may be remarked that soda is a natural constituent of the blood, and a few grains more or less would scarcely make much difference by the mere fact of their presence. Clinical observation leads me to regard soda as very inferior to potash as an agent for attenuating the sputa, and might almost induce me to admit that it is powerless to do so. But the local effect on some mucous membranes resembles that of potash, though it is less marked, and we have Virchow's authority for stating that soda stimulates the movements of the cilia, and sometimes restores them after they have ceased. Then again it can scarcely be forgotten that mucin, which is readily precipitated by acids, dissolves freely in alkaline fluids. Even the precipitated form of mucin is easily redissolved by alkalies.

The application of a solution of a potassium salt to the mucous membrane of the mouth increases the flow of saliva and even causes an opacity in the fluid which has been conjectured to arise from alteration of the secreting cells; but irritation of the sympathetic nerve-filaments of the submaxillary gland will give rise to the same appearance. The question is, whether the bronchial membrane is affected in the same way. That some potash salts affect the mucous tissue is undeniable from their action on the bowels. Further, their effect in the mouth and pharynx confirms the general observation that they really increase the bronchial secretion and render it more fluid, while they perhaps increase the activity of the cilia as well as the glands. Gubler conjectured that potash antagonizes soda in the blood and in respiratory combustion, inasmuch as it is found normally in the red corpuscles.

Chlorate of potash has enjoyed a considerable reputation, though partly on grounds for which there is no foundation. It has been largely used in croup and diphtheria, but does not deserve confidence. It could not be expected to sufficiently liquefy the exudation. It will, however, sometimes cause a degree of salivation, and in bronchitis, when the expectoration is viscid and scanty, it seems to give relief by promoting the flow of the secretion and rendering it more watery, so that it may be termed, perhaps, an alterant expectorant. Dr. Laborde[1] says that it both modifies and dilutes the expectoration in acute and chronic bronchitis, and Dr. Sedgwick[2] considers it useful in catarrh. Koehler[3] had long before employed it in phthisis, but without obtaining any benefit, though Dr. Fountain[4] at a later period endeavored to revive its use under the notion that it would act as a liberator of oxygen in the system. This idea unfortunately seems to have prevailed for a considerable time after it was shown that the chlorate is eliminated unchanged by the kidneys. Wöhler[5] as early as 1824 reported that the chlorate among other salts passed off unchanged in the urine, and his observation has been fully confirmed; nevertheless, the idea continues from time to time to crop up. It was entertained as recently as 1871 by Baudrimont,[6] and has several times since been complacently stated by contributors to the medical journals, but though this hypothesis must be abandoned, it does not follow that in some way the chlorate may not be of some service, as we know that other salines are. That it may affect mucous membranes is natural to suppose, since, as shown by Isambert,[7] it escapes in all the secretions, but its chief value is undoubtedly as a local remedy. Dr. Harkin[8] reported that it improved the color and strength of some phthisical patients, and lessened cough and diarrhœa. The late Dr. Symonds[9] seems to have thought that it could promote the healing of a cavity. Dr. Spender[10] expresses a regret that its value in phthisis is not better

---

[1] Laborde, J. V.: De la valeur du Chlorate de Potasse dans le traitement des Gingivites Chroniques, etc. 1858. Also, Laborde, J. V.: Étude comparative de l'action physiologique des Chlorates de Potasse et de Soude, des Bromures de Potassium et de Sodium. Déductions relatives à l'emploi thérapeutique comparé de ces substances 1875.

[2] Sedgwick, L.: British Medical Journal, 1873.

[3] Koehler: Lancet, 1836-37.

[4] Fountain, E. J.: On the Treatment of Phthisis by the Chlorates of Potash, with Observations on Oxygen as a Therapeutic Agent. 1860.

[5] Wöhler: Zeitschrift f. Physiologie, 1824.

[6] Baudrimont, E.: Recherches sur l'action intime des substances qui aident à la décomposition du Chlorate de Potasse pour en dégager l'oxygène. 1871.

[7] Isambert, E.: Études chimiques, physiologiques, et cliniques, sur l'emploi thérapeutique du Chlorate de Potasse, spécialement dans les Affections Diphtheritiques. 1856.

[8] Harkin: Dublin Medical Quarterly Journal, 1861.

[9] Symonds: British Medical Journal, 1868.

[10] Spender: Brit. and For. Med.-Chirur. Rev., 1872.

known, and M. Gimbert[1] endeavors to show that in certain forms of this disease it is of considerable value. On the other hand Dr. Cotton, in his trials at the Brompton Hospital, could trace no definite effects to the remedy, though sometimes it improved the condition of cachetic patients, while Dr. Austin Flint[2] found benefit in only one out of fourteen cases. We may perhaps be permitted to infer from this discordant evidence that like other salines it is only useful in those conditions of the system which indicate this class of remedies, and this is the conclusion to which my own experience points. That in considerable doses it affects mucous membranes there can be little doubt, and the way in which it does this is probably by a local effect while being eliminated, and certainly its great power as a topical application in aphtha and other affections of these membranes is not to be denied, and gives it a place in therapeutics quite distinct from other potash salts. It may also be here observed that the chlorate is destitute of the powerful cardiac depressant action possessed by the nitrate. With regard to the other salts both of soda and potash the statements already made will suffice.

Ammonium Chloride.—The early history of this salt, as it is often told, may after all be only a fable, for sal ammoniac, which is said to derive its name from Ammonia, the name of a district of Libya where stood the oracle of Jupiter Ammon, though mentioned by various writers, may have been only rock-salt. Besides, this district has been said to have obtained its name from the nature of the soil ($ἄμμος$, sand), and after all the name may have been derived from $ἄμμωι$, a word of Egyptian origin. Herodotus, it is true, mentions sal ammoniac ($ἅλς\ ἀμμωνιακός$) as found in this locality, but perhaps the name was applied to some other salt. Dr. Royle[3] has remarked that the Hindoos must have been acquainted with sal ammoniac ever since they burnt bricks, as they now do with the manure of animals, for some of it usually crystallizes at one end of the kiln. Moreover, they seem to have found out how to make the carbonate from it, and from them the Arabians probably learned the process. Geber[4] knew the method of purifying it by sublimation. Muys[5] recommended it in doses of one to two drachms in intermittents, and it became rather largely used for various purposes.

The older writers thought that chloride of ammonium acted as a sedative to the heart, and at the same time quickened the capillary circulation. They also observed that it increased secretion, especially from the mucous membranes, and attributed to it the power of promoting absorp-

---

[1] Gimbert: De l'emploi du Chlorate du Potasse dans certaines formes de la Phthisie Pulmonaire. 1872.
[2] Austin Flint: American Journal Medical Sciences, 1861.
[3] Royle: Antiquity of Hindu Medicine. 1837.
[4] Geber: The Works of Geber. London, 1678.
[5] Muys, W. G. : De Salis Ammoniaci Prœclaro ad Febres Intermittentes usu. 1716.

tion, though in a less degree than mercury. It was also said to impoverish the blood, but in ordinary doses no impairment of the plasticity will be noticed, and no distinct effect on the circulation, though it has been shown that after poisonous doses administered to animals, the blood contained less solids than normally. Rabuteau[1] has shown that it probably passes off by all the secretions. He found it speedily in the saliva, but the major portion is removed through the kidneys, and he was able to recover from the urine nearly the whole amount which had been taken. It notably increased the amount of urea, and Böcker[2] found that it increased all the solids of the urine except uric acid, which was slightly diminished. In Germany sal ammoniac has been largely prescribed for gastric catarrh. In England it is chiefly used as a remedy for chronic bronchitis, and sometimes for acute catarrh; but it is scarcely appropriate in the early stage of the latter, or when there is decided pyrexia. Perhaps its most appropriate use is after the more active symptoms have subsided, but before a stimulant expectorant seems called for. Dr. Patton commends it (*Practitioner*, vol. vi.) in the later stages of pneumonia. Some writers seem to have supposed that it is decomposed in the system, ammonia being liberated and acting as a stimulant, but there is no foundation for the idea, and even large doses have no exciting effect. Dr. Copeland recommended it in passive hemorrhages, and Dr. Warburton Begbie has used it for this purpose with success (*Lancet*, 1875). The late Dr. Beigel (*Lancet*, 1867) arranged an apparatus for the inhalation of freshly formed chloride, and Liebermann has reported favorably of a similar apparatus of Loewin's (*Bull. de Thérap.*, 1873).

Sulphur has been known from the earliest ages, it is mentioned in the Book of Genesis (xvi., 24) and in Homer's Iliad (lib. xvi.) as well as other ancient writers. It passes into the secretions and may be considered a mild stimulus of the skin and mucous membranes, particularly the bronchial, but in large doses it acts on the intestinal lining. Among other uses it has accordingly been employed in chronic catarrh, asthma, bronchitis, croup, diphtheria, and pneumonia. Buchheim and some others doubt whether it really increases the bronchial and cutaneous secretions, but it is difficult to shut our eyes to the fact; there is a difference of opinion at to the renal secretion. There is a vast amount of experience as to its favorable influence in respiratory diseases. It was a recognized remedy among the ancients. Galen sent consumptives to breathe the sulphurous vapors at Etna. It acquired the name of Pectoral Balsam. Dr. Graves and many others expressed confidence in its favorable influence over the respiratory membrane, and Duclos[3] advocates the continued

---

[1] Rabuteau: L'Union Médicale, 1871.
[2] Böcker, F. W.: Beiträge zur Heilkunde. 1849.
[3] Duclos: Bull. de Thérap., 1861.

use in asthma of small doses. Binz[1] suggests that the bronchial nerves may be soothed by the direct action of sulphuretted hydrogen, as it is excreted through the membrane. Many mineral waters of considerable repute in the treatment of respiratory diseases are supposed to owe their virtues to the sulphides they contain.[2] The local use of sulphur will be considered further on.

Iodine is the most remarkable of all the alterative expectorants. As the element was only discovered in 1811 and introduced a few years later by Coindet[3] into therapeutics we may regard it as a conquest of the present century, but before this remedies containing iodine were in use; such, for instance, as burnt sponge, which for a long time enjoyed a considerable reputation. Iodine is also present in many mineral waters. Indeed, it is rather widely distributed in small quantities in both the organic and inorganic kingdoms, as may be seen by reference to Sarphati's[4] list of substances containing it. Of course a medicine so potent and, as shown by Coindet and later by Lugol,[5] of special value in scrofula, was sure to be employed in phthisis and other diseases of the respiratory organs and to give rise to a very considerable literature which we have not space to examine.[6]

Topically iodine is an irritant, and the vapor when inhaled strongly stimulates the respiratory membrane. It has been employed at times extensively for this purpose, either alone or in combination with other stimulants or with anodynes.

Camphor sprinkled with the tincture and inhaled is a popular remedy for coryza; so is a combination of iodine with carbolic acid. Combinations with kreasote, ether, chloroform, etc., have their uses, but it should be remembered whenever these inhalations are employed, the iodine may enter the system and produce its characteristic effects. Conversely we may administer it or the salts by the stomach for the purpose of influencing the respiratory membrane.

Iodine is quickly taken into the blood, probably as iodide of sodium or perhaps as an albuminate; it passes rapidly to all the tissues, particularly to the lymphatic glands and secreting surfaces; but only scantily to the nervous centres. It would appear, therefore, as if it must increase

---

[1] Binz: Elements of Therapeutics.

[2] Tichborne and Prosser James: Mineral Waters of Europe. 1883.

[3] Coindet, C. W.: Biblioth. Univ. de Genève, vol. xvi.; also Observations on the Remarkable Effects of Iodine. Trans. by Dr J. R. Johnson. 1821.

[4] Sarphati, S. E.: Commentatio de Iodio. 1835.

[5] Lugol, J. G. A.: Mémoires, sur l'iode, etc. 1829–1831. Also trans. by O'Shaughnessy, with appendix, 1831.

[6] For early literature see Bayle's Bibl. de Thér., 1828. For later references to 1854 see Titon, H. A.: Recherches sur l'absorption et la valeur thérapeutique des préparations Iodées. 1854.

tissue metamorphosis, but in medicinal quantities loss of weight is quite exceptional, and it does not appear to increase the excretion of urea. It has been conjectured that it may act on the blood-plasma, rather than the tissues, or that it may spare the liver, which would account for these results. They may, however, be partially due to the rapidity with which the medicine is eliminated. It is found very soon after it has been taken in the mucous secretions. It appears also, though later, in the perspiration, bile, and even the milk. It seems, in fact, rapidly to pervade the system and to be as rapidly excreted in all directions, but the major part is removed by the kidneys, though it can scarcely be termed a diuretic, and it is found in the secretions chiefly as iodide of sodium, but a portion of this salt is decomposed, and hence the free element exercises its local action. Pereira observed that the pocket-handkerchiefs of iodized patients had a distinct odor of the element.

The prominent symptoms of iodism show the effect on the respiratory membrane, on the whole of which it acts powerfully. The nasal passages are perhaps the first to be affected, but the conjunctivæ and the frontal sinuses speedily participate, and thus we have sneezing, profuse coryza, and watering of the eyes, with often distention of the brow and severe headache. Often the patient complains that he has caught a violent cold and the watery discharge from the eyes and nose saturates his handkerchiefs. The effect on the buccal and faucial membrane is also obvious in the coated tongue, flow of mucus, swelling and redness of the gums, palate, and throat. Salivation shows the stimulant action on the glands and even when this is not prominent, the continual taste of the medicine proves that it is being secreted in the saliva. The effect on the pharynx and larynx gives rise to irritation in the throat, a sensation of heat, and cough; the burning and rawness extend along the trachea and over the chest. Further, that the bronchial membrane is also engaged is shown by the coughing up of a frothy mucus or a larger quantity of a more liquid secretion, according to the degree of iodism, and on auscultation moist râles indicate the extra fluid in the tubes. Thus throughout the respiratory membrane a considerable degree of congestion is produced; but at an earlier stage, or in the milder cases, there is less active congestion, and the membrane may only pour out a quantity of thin, watery mucus, with scarcely any increase of vascularity.

Such effects would seem to indicate that the preparations of iodine in appropriate doses must be regarded as true expectorants. When the secretion is deficient, the membrane dry, and perhaps swollen, the cough painful, frequent, or constant, and yet useless, nothing, or very little being raised, iodide of potassium will bring on a secretion of thin mucus, and so re-establish a moist condition of the membrane, thereby relieving the irritation, and at the same time unload the vessels and remove the swelling. Why, then, should it not be more often employed in this condition?

Another kind of case in which it may be used may be named, when the tubes are not dry, but contain an increased quantity of mucus, but that so tenacious as to be with difficulty removed, being brought up as "stringy" phlegm with no little difficulty; sometimes it seems to plug some of the tubes. In these cases the cough is troublesome, but ineffectual, and whatever will liquefy or dilute the secretion will give relief. Iodide of potassium causes an extra flow of thin mucus, and is also believed to exercise an attenuant effect on the secretion already in the tubes, but, whether it does this or not, the extra flow dilutes what is already there, and, even if the thick phlegm will not mix with or dissolve in the new watery outflow, it will at least float in it and be loosened from the walls of the tubes by it, and so be more easily coughed up.

Iodine has also been said to be an antispasmodic expectorant, but if so, it would for the most part be by an indirect influence. Free secretion of itself naturally tends to remove tenacious phlegm, or foreign bodies, as well as to relieve engorged vessels, and so far to relax spasm set up by such influences. The inhalation of ethyl-iodide, however, will often rapidly relax the spasm of asthma, and recently Prof. Germain Sée has employed with considerable success sprays of iodide of potassium for the same purpose. It seems to me probable that the action in the last case is, as stated, indirect, but with regard to the vapor of ethyl-iodide the effect is too rapid to be thus accounted for.

We have no proof that iodine depresses the respiratory centre; its action on the mucous membrane is therefore not to be referred to the nervous system; and though there are many other uses for this remedy which have almost overshadowed its expectorant properties, these last can scarcely be considered as the least valuable.

Very curious differences in the susceptibility of patients to the influence of iodine may be constantly observed. Some are so extremely susceptible that minute doses at once produce iodism, and others are so easily affected by it that it is extremely disagreeable to them. In some this idiosyncrasy shows itself in salivation, in others in coryza. Sometimes there is only a moderate salivation, the fluid being impregnated with the taste of the iodide as it is being continually eliminated, so that all the food tastes of the salt, to the great disgust of the patient, and the digestion is apt to be deranged. To many persons this taste is exceedingly repugnant. When the respiratory membrane is affected, the patient seems to catch cold, sneezing being early followed by free secretion. Occasionally the catarrh is so easily induced that we cannot use the remedy, but usually it can be restrained or arrested. A dose of morphia will generally accomplish this, and sometimes very small quantities of an opiate given with the iodide enable the patient to tolerate it. In mild cases a single dose of morphia arrests the coryza, and prevents the development of worse symptoms. The antagonism of these two medicines, so far as the mucous membrane is concerned, is fairly complete, but we must not forget that opium affects the centre and it is often undesirable to give it. Belladonna is also to the same extent antagonistic to iodine. Possessing a very special influence on the salivary glands, it is perhaps preferable where they are chiefly excited, but experience satisfies me that it also antagonizes the effect of iodides on the pituitary, faucial, laryngeal, tracheal, and bronchial lining. I have used atropia to restrain and arrest the effects of the iodides in these situations with great success, especially where opium was contra-indicated or undesirable, but much more frequently I rely upon morphia.

There is an opposite condition to the extreme susceptibility noticed above. Some patients seem almost insusceptible to the action of iodine; and others, to whom it is given in large quantities with great benefit for various purposes, never experience any irritation of the respiratory passages. This seems to me specially the case when large doses of iodide are required. In syphilis I have administered very large doses for long periods, and that, too, when the fauces and larynx have been seriously diseased, without exciting irritation of the membrane. Such cases are, indeed, of daily occurrence, and yet out of numerous instances of iodism I have met with few in syphilitics. It would be interesting to learn whether this is the general experience.

The uses of iodine in respiratory diseases may be deduced from its action. In bronchitis it is only admissible as a promoter of secretion in the manner already explained. Where the expectoration is thin and easily coughed up, the iodides are useless, if not contra-indicated. In asthma, Trousseau and Jaccoud both found it useful, and long before it had been used by Horace Green, and it also constituted the chief ingredient of a once popular quack remedy. Dr. Salter sometimes obtained benefit from full doses. Dr. C. J. B. Williams employed it in combination with carbonate and stramonium (*Med. Times*, 1872). Dr. Reed employed liquor iodi in constitutional dry asthma, when the paroxysms came on without obvious cause (*Med. Record*, 1879), and during the last few years M. Sée has reported (*La France Med.*) a number of cases at all ages, in which he has given doses of twenty to forty-five grains, daily reducing the dose as the improvement continued. Often he found the breathing become easy in an hour or two after such a dose given during the paroxysm. He has also employed with great benefit sprays containing the iodide. There seems to me no doubt that the remedy is often very beneficial, especially in those cases in which the paroxysms are excited by cold, or whenever they are relieved on the appearance of secretion. It is the power to excite this secretion that is probably the key to the use of the iodide in asthma. It may also be added that in gouty and rheumatic patients its influence is favorable.

So potent a medicine was sure to be tried in phthisis, and at first there were not wanting observers who hoped that a curative agent had been discovered, among whom may be mentioned Bardsley,[1] Gairdner,[2] Scudamore,[3] Clarke,[4] and others. It must, however, be confessed that disappointment awaited them. At the Brompton Hospital the experience of Dr. Cotton (*Med. Times*, 1859) was not favorable; weight was seldom gained, dyspepsia was often produced, and a wasting increased. Others have observed irritation and even hæmoptysis follow the use of this remedy. But in certain cases of pneumonic phthisis, carefully used for a short time, it may possibly be beneficial, and in strumous constitutions, where there is a fear of the development of tubercle, it may perhaps be used with advantage, but any attempt to produce a distinct impression on the respiratory organs may excite local irritation and aggravate febrile excitement. Even the local use by inhalations of the vapor of iodine, valuable as it often is, requires to be prescribed with considerable circumspection, and the effect should always be watched. It is quite possible for iodism to be induced by such inhalation. I have known these phenomena brought on by the accidental inhalation of the vapor during the preparation of compounds of iodine, as well as by the exposure of the element with the intention of its becoming evaporated as a method of its administration. It is in laryngeal phthisis that the local effect of the vapor is most useful, indeed in extensive pulmonary disease it has not yet accomplished much, and Pereira declared he had never seen it do any good,

In pneumonia, when consolidation has continued for a considerable period, small doses of iodide of potassium will sometimes set up the process of absorption. Some physicians have employed the iodide at an earlier stage. Dr. Gualdi treated thirty-nine cases from the beginning with frequent doses of iodide of potassium, and obtained excellent results; two only died, and one of these from a complication at the outset. In all cases the expectoration from being tenacious and viscid became on the second day fluid, resembling bloody serum, the fever ceased and with it the exhaustion, though the state of the lungs was not improved. The appetite of the convalescents was greatly increased. Dr. Gualdi found the treatment succeed better in young persons than adults, and urges

---

[1] Bardsley, J. L.: Hospital Facts and Observations. On Iodine, etc. 1830.

[2] Gairdner, W.: Essay on the Effects of Iodine on the Human Constitution, with Practical Observations on its Use in the Cure of Bronchocele, Scrofula, and Tuberculous Diseases of the Chest and Abdomen. 1824.

[3] Scudamore, Sir C.: Cases illustrating the Remedial Power of the Inhalation of Iodine and Conium in Tubercular Phthisis and various Disordered States of the Lungs and Air-passages. 1834.

[4] Clarke, Sir A.: On the Exhibition of Iodine in Tubercular Consumption and other Diseases of the Chest, and in the Treatment of Scrofulous, Cancerous, and Cutaneous Diseases, etc. 1845.

that the remedy should be given at the commencement of the disease. In croup and diphtheria the iodides have been employed with a view of separating the false membrane by causing a profuse flow of watery secretion.

Iodoform differs so much from the other preparations as to require a few words to itself. It does not seem to irritate the mucous membranes like the other preparations, so that iodism very seldom follows its use, and it is well tolerated by the stomach. It is rapidly absorbed. Högyes[1] says the first step is its solution in such fatty matter as it may meet with, this solution being in turn decomposed by albumen, as a compound of which it enters the blood. The iodine is eliminated in combination with sodium in the same manner as when taken in other forms, but some iodoform seems to escape from the skin, or in the breath, for a person who has taken it for some time evolves the characteristic odor. Considering its chemical relations with chloroform it was conjectured that it might be anæsthetic, and locally it seems to possess this property. Maitre pointed out (*Bouchardat's Annales*, 1857) that its application relieved pain, and examined its action on the nervous system, respecting which experimenters are not agreed. He compared it to alcohol, and in some instances it seems to narcotize, as shown by McKendrick (*Edin. Med. Journ.*, 1874) and again by Högyes (*Med. Record*, 1879). I have employed it largely internally, though not as an expectorant, and the effect on the mucous membrane seems but slight, yet the other medicinal effects of iodine may be obtained from this preparation. I compared it with others in a paper read at the Medical Society of London, in 1871 (*Medical Press*, 1871–72), and have continued to use it largely ever since (*Brit. Med. Jour.*, 1878). Of late it has been employed with a view to an antiseptic action, especially in Italy.[2] Dr. Dreschfeld introduced it into this country in 1882, as likely to be of service in phthisis, and Dr. A. Ransome read a paper on it at the British Medical Association, in 1883. He gave it in doses of 1½ grain three times a day, and found that it disturbed the digestion, whereupon, on the suggestion of Dr. March, he added two grains of croton-chloral-hydrate. He used at the same time inhalations and had the patients weighed before and after treatment. He attributes to iodoform "some slight improvement, even in cases in which it was manifestly hopeless to expect a cure," and thinks that "in the earlier stages of the disease, it is decidedly worthy of further trial." I have very seldom found such doses interfere with digestion, I have often given three grains three times a day for weeks together without the slightest inconvenience. A combination with an equal quantity of storax greatly controls the disagreeable odor; but

---

[1] Högyes: Archiv f. exper. Path. und Pharm.
[2] Lo Sperimentale, 1883, and Annali Univers. di Med. e Chir., 1883.

when the system becomes saturated, so to say, with the medicine, the patient begins to feel the inconvenience of continually perceiving the smell.

Antiseptic and Disinfectant Pneumatics.

We have already had occasion to speak of antiseptics, when treating of antipyretics, and perhaps it would be as well to recur to the subject here, inasmuch as the pneumatics of which we are about to speak are such as are supposed to act through the system. The idea of disinfection is by no means modern. Homer speaks of sulphur as a disinfectant, and the preservative power of salt and of vinegar was known at an early age. Even the word antiseptic is by no means so modern as some have supposed, and it was preceded by antiloimic (αντι, and λοιμος, pestilence), a term which was applied to any substance supposed to prevent infection from the plague or other pestilence. We have seen that Pringle[1] used the word antiseptic in the title of his paper, and he was soon followed by MacBride.[2] In 1767 the Dijon Academy of Science offered a prize for an essay on antiseptics, and two years afterward three of the essays[3] submitted were published.

The same year Godwin[4] produced his "Septicologie," and soon after Alexander,[5] Brownrigg,[6] and Henry[7] published English essays on the subject. Cartheuser[8] and Callisen[9] followed in Latin, Bucholz[10] in German, and a little later Mugis[11] and Grewe[12] in Latin. Nearly fifty years then elapsed before Kaiser[13] published his experiments on the comparative antiseptic power of several agents. Nearly forty years more before Lister began what may be considered the modern antiseptic system in surgery, for an exposition of which we must refer to his contributions

---

[1] Pringle, J.: Op. cit.

[2] MacBride, D.: Experimental Essays. 1764.

[3] Boisseau, B. C., Bordenave, S., et Godart, J.: Dissertations sur les Antiseptiques, qui concoururent pour le Prix proposé par l'Académie des Sciences de Dijon en 1767–1769.

[4] Godwin, J.: Septicologie, ou Dissertation sur les Antiseptiques. 1769.

[5] Alexander, W. Experimental Essays: On Antiseptics, etc., 1770.

[6] Brownrigg, W.: On the Means of Preventing the Communication of Pestilential Contagion, and of Eradicating it in Infected Places. 1771.

[7] Henry. S.: On Antiseptic Substances. 1773.

[8] Cartheuser, J. F.: De remediis Antisepticis. 1774.

[9] Callisen, H.: De Antisepticis. 1775.

[10] Bucholz, C. F.: Chym Versuche über einige der neuesten einheimischen antisepischen Substanzen. 1776.

[11] Mugis, P. N.: De antisepticis proprie dictis. 1781.

[12] Grewe, T.: De putridine et Antisepticis. 1782.

[13] Kaiser, J.: Experimenta ad Comparandam vim Antisepticam Aceti, Nitri, Salis communis, et Chloreti calcis instituta. 1831.

and those of his followers in the various journals since 1869. Dr. Sansom's[1] able treatise refers to the medical as well as the surgical uses of antiseptics.

In diseases of the respiratory organs we may select a remedy which is believed to possess antiseptic powers for two purposes : first, that when it is excreted through the bronchial membrane it may disinfect the sputa ; second, in the hope that it may prove destructive to the organisms supposed to be present in the body. With regard to the first indication, we have seen that a number of expectorants may act in this manner, but the use of antiseptic inhalations would seem to be more efficacious. As to the second point, the whole theory of the parasitic origin of disease is involved. This theory is undoubtedly a fascinating one, admitting the presence of an organism to be the cause of a disease, the business of the therapeutist would be to search for a remedy which would be poisonous to the parasite but not to the patient. In quinine, as we have seen, these requisites appear to meet. Minute quantities are fatal to whole colonies of putrefactive organisms, and very large quantities can be taken into the human body, and will linger in the system for a considerable period. If, therefore, a disease be caused by an organism to which quinine is fatal, its administration will effect a cure, provided no irreparable injury has been inflicted. It is believed by many that the virtues of quinine really depend on this property. Unfortunately it is powerless over some other organisms, and we have to search for more potent poisons.

Buchner denies the value of antiseptic treatment, and seems to think that it is impossible to satisfactorily carry it out, at any rate with the agents now at our disposal. Koch has shown that corrosive sublimate is fatal to the bacteria of anthrax in the proportion of one part to two hundred thousand. To reach this proportion in the mass of a man's blood would require doses of two-fifths of a grain. Such proportionate doses have been injected into the veins of animals by Binz, without killing them, and he seems to think that it would not be impossible to employ this remedy in man, but Buchner (*Centralb. f. klin. Med.*, 1883) remarks that the sublimate combines with the albumen of the blood, and that the resultant albuminate, though remaining in solution, is less diffusible, less capable of osmosis, and therefore less likely to be equally distributed through the tissues, while it is at the same time less poisonous. He found double the quantity of albuminate of mercury to that of the sublimate was required to destroy the bacteria, and he calculates that for quinine to act as an antiseptic it would have to be given in

---

[1] Sansom, A. E. : The Antiseptic System : A Treatise on Carbolic Acid and its Compounds. A Theory and Practice of Disinfection, and the Practical Application of Antiseptics, especially in Medicine and Surgery. 1871.

three-ounce doses, but he admits that there is probably some relation between the value of a remedy in disease and its power of destroying micro-organisms outside the body, but what that relation is remains to be proved.

Different micro-organisms vary greatly in the effects they produce, and it has been stated rather confidently that perfectly healthy tissues harbor germs. The recent researches of Hauser and Zahn discountenance this idea. Hauser (*Centralb. f. die Med.-Wissensch.*, 1884) removed entire organs and parts of tissues by heated instruments from animals just killed to super-heated glass vessels, plugging the mouths with cotton-wool. He kept the vessels in a moist chamber, at a temperature of 30° C. Zahn (*Virchow's Archives*, 1884) employed the blood of healthy animals, collecting it with full precautions in tubes previously filled with oxygen, hydrogen, and carbonic acid, so as to exclude atmospheric air. The tubes were kept sealed at a temperature of 37° to 38° C. for months without any sign of putrefaction.

We have seen that some organisms are destroyed by quinine. Krukenberg[1] found some which were killed by one in one hundred thousand. Others, however, require one in twenty thousand, and others may resist a stronger solution. Then we must remember that what one parasiticide fails to effect may be accomplished by another, and so when organisms resist one influence, we must seek for something that will be noxious to them. Perhaps the *bacterium subtile* is the most difficult to destroy. It will resist boiling for an hour. Yet M. Schnetzler has this year reported to the Paris Academy that it is easily killed by formic acid. To a drop of water teeming with these bacteria he adds a drop of liquid containing one-thousandth part of formic acid, and says that the effect is such that the liquid may be introduced into the digestive tract with impunity. If this be confirmed, the fact may probably be turned to practical account in preventing, if not in curing, disease. Experiments with tubercular sputa seem to show that it possesses great power of resistance to our ordinary antiseptics. Falk announced that its virulence was destroyed by putrefaction, but this has not been confirmed, and the bacilli have been found in putrefying sputa. Baumgarten thought the virulence of tubercular material was diminished by putrefaction. Parrot and Martin (*Rev. de Méd.*, 1883) found that the infective power was not destroyed by corrosive sublimate solution of one in one thousand; bromine, one in one thousand; salicylic acid, one in five hundred; carbolic acid, one in twenty. But a temperature of 100° to 120° C. was sufficient. Recently Dr. Niepel (*La France Méd.*, 1884) has stated that sulphuretted hydrogen destroys the infective property. Professor J. Sormani, of Pavia, brought before the late Hygienic Congress at the Hague (August,

---

[1] Krukenberg: Vergleichend. Physiologische Studien. 1880.

1884), the results of numerous experiments as to the possibility of destroying the tubercle bacillus. The following are his conclusions: 1. The bacilli of tuberculosis were generally very difficult to destroy; dryness, exposure to oxygen, putrefaction, and most disinfectants failed to produce any effect. 2. A temperature of 100° C. only killed the bacilli after at least five minutes of ebullition. 3. The artificial digestion of bacilli showed that they were the last of all living organisms to be destroyed by the gastric juices or hydrochloric acid. A very active digestion is necessary to kill this microbe. A healthy man may destroy the bacilli in his stomach, but an infant or an adult with his digestive faculties impaired would easily allow the germ to pass the stomach intact, and retain its virulence in the intestinal tube. This determined enteric ulcerations, etc. 4. The bacillus of tuberculosis can be preserved intact for a whole year when mixed with water. It is probable, though not proved, that it has retained its virulence during that time. Thus drinking-water may become the means of propagating tuberculosis. It is probable that contaminated linen retains its virulence for five or six months. 5. Alcohol does not destroy the germ; hard drinkers often suffer from tuberculosis. 6. Cod-liver oil, ozone, oxygenated preparations, and other similar remedies, have no effect in killing the bacillus, nor are benzoate of soda, salicylate of soda, sulphate of zinc and carbolic acid, iodide of silver, bromide, camphor, etc., of much greater use. They injure, perhaps, but do not absolutely destroy the bacillus, at least not in the doses that can be taken without danger. 7. A more decisive action may be attributed to creasote, eucalyptol, pure carbolic acid, the naphthols, and bichloride of mercury. 8. For disinfecting spittoons, a carbolic acid solution of five per cent. is thought sufficient, and Dr. Sormani asserts that the breath never contains any bacillus. He also suggested that oil of turpentine or eucalyptus should be diffused in houses as an agent for the destruction of this bacillus.

Although it may seem that if we could saturate the system without inconvenience with a substance fatal to microbes, we should thus be able to effect an immediate cure of diseases dependent upon them, it does not necessarily follow that less active parasiticides are useless. It is easy to imagine that a small quantity of an antiseptic, though insufficient to kill, may cause inconvenience to parasites, especially if such antiseptic should be excreted through an organ in which they have effected a lodgement. Moreover, it may well be that such an antiseptic might improve the condition of that organ, and thereby render it a less suitable nidus for the microbe. And again, it is possible that a substance in the process of elimination from the body, either by decomposition or otherwise, may be more noxious to micro-organisms than when employed in a culture-fluid. The conditions, in fact, in the living body differ greatly from those in the laboratory, otherwise we should only have to collect the organisms and destroy them. It must, too, be freely admitted that the reactions of the system, provoked by the parasites, form no inconsiderable proportion of the phenomenon. Nevertheless, they scarcely account for the whole, and while freely admitting all that is to be said in favor of the nervous theory, we are scarcely in a position to deny the reasonableness of the antizymotic explanation.

Antiseptic pneumatics belong to several groups. Some of them have an expectorant action and may be called properly enough antiseptic expectorants. But there are others of which the *modus operandi* is not so clear, and there are some general antiseptics which are useful in respiratory diseases, but are not expectorant at all, and perhaps possess even opposite properties, *e.g.*, tonic antiseptics, such as quinia, etc. That the antiseptic or disinfectant action is exercised on the sputa is reasonable enough to suppose when the remedy is excreted by the bronchial mucous membrane, and we have seen that several such true expectorants are calculated to act in this manner, but there is no doubt also an action on the mucous membrane itself. Thus an alterative expectorant may be antiseptic and quite a number of stimulant expectorants are antiseptics, such as essential oils, the balsams, the camphors, and the turpenes. Eucalyptus, which has been lately so much employed, may be classed among these. Cubeb contains an essential oil, and as this is perhaps eliminated partly by the bronchial membrane, it would seem more adapted for diseases of this tissue than the crude piper. But the value of this medicine has been much exaggerated of late years. Other essential oils are nicer and better.

Tar and its derivatives may be mentioned here, though in the present day they are more used as local antiseptics. Tar is the πίττα of Theophrastus, the κῶνος or πίσσα ὑγρά of Dioscorides, the *pix liquida* of Pliny. In the middle of the last century it was suddenly brought into vogue by Bishop Berkeley,[1] the great philosopher, who seems to have thought that he had discovered a panacea in tar-water, and whose book on the subject gave rise to a great number of pamphlets for and against the use of the remedy, one of them quaintly entitled "Cure for the Epidemical Madness of Drinking Tar-water." Nevertheless the good bishop and learned philosopher returned to the charge, professed unbounded belief in his panacea, and no doubt thought that he was serving his fellow-creatures in making known his favorite remedy. It soon, however, fell into disuse, though it has at times been revived. It is still occasionally prescribed in catarrhal affections and in phthisis, and indeed of late years an attempt has been made to introduce special preparations of tar into use. Dr.

---

[1] Berkeley, G. (Bishop of Cloyne): Siris: A Chain of Philosophical Reflections and Inquiries respecting the Virtues of Tar Water, etc. 1744. Also, Two Letters on the Usefulness of Tar Water in the Plague. 1747. Also, Further Thoughts on Tar Water. 1752.

Dunglison[1] reported considerable benefit in chronic bronchitis; he directed one ounce of tar to be digested in two pints of water for a week and strained; of this he gave from eight to twelve ounces daily mixed with milk. The inhalation of tar vapors is more rational and often of service.

The discovery of *Creasote* by Reichenbach,[2] who named it from κρέας, flesh, and σώζω, I preserve, led to the disuse of the crude remedy and the substitution of the supposed active principle, both for internal use and for inhalations. It seems, however, that creasote is not a simple body but a rather valuable compound containing a good deal of creasol, and there is little doubt that very impure preparations have found their way into the market.

As tar was displaced by creasote so this in its turn has been largely superseded by carbolic acid and its compounds, as have also some old preparations of soot, pyroligneous acid, rag oil, paper oil, animal oil, mummy, and other obsolete medicaments. Creasote, however, has been largely employed internally, and Elliotson,[3] Miquet,[4] and Sir John Cormack[5] have written in its favor, while it still maintains a place as a valuable inhalation. Carbolic acid, though chiefly obtained from coal-tar, is produced in small quantities during the distillation of benzoin and some other gum-resins, and is also said to be found in the urine of man, and some animals. In the Italian war of liberation, in 1859, some of the French surgeons used a powder containing coal tar and lime as an application to wounds. In 1863 Dr. Lemaire discussed Pasteur's germ theory, in a volume on carbolic acid.[6] He had previously published a work on coal-tar.[7] He recommended the new product as an antiseptic application in wounds, injuries, and diseases, and advised its internal use in diseases due to infective poisons. A couple of years later, M. Bobœuf[8] addressed the Academy of Paris on the value of phenol and the history of its use.

---

[1] Dunglison: Practice of Medicine. 1844.

[2] Reichenbach, C. von.: Das Kreosot, in chemischer, physischer, und medicinischer Beziehung, etc. 1833.

[3] Elliotson, J.: Med. Chir. Trans. 1835.

[4] Miquet, E.: Recherches chimiques et médicales de créosote. 1834.

[5] Cormack, J. Rose: A Treatise on the Chemical, Medical, and Physiological Properties of Creosote, illustrated by Experiments on the Lower Animals, etc. Harveian Prize. 1836.

[6] Lemaire, J.: De l'Acide Phénique, de son action sur les végétaux, les animaux, les ferments, les venins, les virus, les miasmes, et de ses applications à l'hygiène, etc. 1863.

[7] Lemaire, J.: Du coal-tar saponifié, d'sinfectant énergique arrêtant les fermentations; de ses applications à l'hygiène, à la thérapeutique, à l'histoire naturelle. 1860.

[8] Bobœuf, P. A. F.: Mémoire adressé à l'Acad. des Sciences sur l'acide Phénique, etc.; Propriétés du Phénol Sodique, etc. 1865.

About the same time M. Déclat[1] wrote on the subject and he has since contributed a number of papers respecting it. In England, Dr. Turner, of Manchester, communicated a paper to the British Association in 1863, in which he recommended carbolic acid as antiseptic and astringent in mucous discharges, etc. Dr. Grace Calvert about this time took up the study of the acid as a disinfectant. In 1867, Lister began that systematic use of antiseptic dressings which has been so fully developed.

Antispasmodic Pneumatics.

The ancients were well acquainted with a number of our antispasmodics and formed a tolerably clear estimate of their virtues. Early in the last century Bauer[2] turned his attention to the general antispasmodics, and soon after the middle the Dijon Academy awarded a prize to Godart[3] for an essay on the subject, and perhaps to the interest thus excited we owe Nonne's[4] work, which appeared soon after. At the beginning of this century a brief account of antispasmodics was published by Sproede,[5] and nearly thirty years later a short treatise by Salinger.[6] Of course, during this period, as previously, these medicines continued to be largely employed. The word was applied in a wide sense and many remedies were included in it which we now classify in a different manner.

Those antispasmodics which are employed in respiratory diseases, whether alone or in combination with pneumatics, furnish an interesting and important group, and although they might be otherwise classed, bringing them together under this head affords an opportunity of comparing their actions with advantage. In considering expectorants we have seen that many of them possess antispasmodic properties. They may therefore be called, as indeed they frequently are, antispasmodic expectorants. On the other hand, some antispasmodics, though given in respiratory diseases, are not expectorant at all, but restrain rather than promote the secretion of the bronchial membrane. It is therefore better to use the word pneumatics, in grouping them, rather than expectorant, which has too often been erroneously applied.

Whatever produces relaxation of muscular fibre may seem to be entitled to the name antispasmodic, whether the action be local or general. The inhalation of steam, therefore, might be included, or the use of vapor baths, or other baths, or any method of applying heat; but at present

---

[1] Déclat, G. : Nouvelles applications de l'acide phénique en médicine et en chirurgie aux affections occasionnées par les microphytes, les microzoaires, etc. 1865.
[2] Bauer, C. H. : De specificis antispasmodicis. 1704.
[3] Godart, G. L. : Sur les Antispasmodiques proprement dits. 1765.
[4] Nonne, J. P. : De Antispasmodicorum modo agendi et usu. 1769.
[5] Sproede, J. G. L. : De medicamentis antispasmodicis. 1800.
[6] Salinger, L. : De Antispasmodicorum differentia. 1829.

we are rather concerned with those remedies which when introduced into the system produce their effect. Among the expectorants which are also antispasmodic, we have seen that emetics and the nauseants possess this property in a high degree; though they are not often used on this account. As the act of vomiting produces intense muscular relaxation, it may sometimes be provoked for this purpose when the need is urgent. The sensation of nausea is also attended by relaxation, so that the nauseants are antispasmodic, but the influence in this case takes more time to produce and generally lasts longer. The anæsthetics, which are very powerful antispasmodics, have largely superseded the use of nauseants for the purpose of relaxing muscles.

The majority of our group of stimulant expectorants are to some extent antispasmodic, thus the essential oils, the oleo-resins, and the gum-resins may be so termed. Assafœtida, galbanum, and other odoriferous remedies have been known from very early times to possess the properties we call antispasmodic, and their effect on the bronchial membrane has already been explained. Directly opposed to these are some of the depressant expectorants, of which lobelia and tobacco are examples. They have both been used to relax spasm, though neither perhaps is entitled to much confidence, and should only be used with caution. Next come a number of neurotics which have been considered antispasmodic: opium, belladonna, hyoscyamus, stramonium, datura, etc., are used as antispasmodic-pneumatics. Their special uses depend chiefly upon their action on the nervous system; conium acts chiefly on the motor nerves. Then we have seen that nitrate of potash and iodide may sometimes serve as antispasmodics, and so indirectly may anything which causes a free secretion of mucus and so relieves the vessels, the effect being to remove the cause of the spasm. Ipecacuanha and senega sometimes effect this.

All depressants of the respiratory centre may serve as antispasmodics. Opium takes a chief place here, though it also acts over a more extended area; so alcohol, ether, and chloroform depress this centre and under certain circumstances are antispasmodic pneumatics, though we can obtain from them the stimulant action, giving them only in small quantities. The bromides also depress the centre, and when the system becomes affected by them have considerable power over spasm. This is well seen in some cases of asthma and of laryngismus. Chloral, again, is a depressant of the centre which may also produce similar effect. Then we have other nerve-depressants, which become antispasmodic by an action on the nervo-muscular apparatus rather than on the centre. It will thus be seen that the antispasmodic pneumatics differ greatly among each other. A few words may be added respecting one or two members of the group, which have not been previously considered.

Amyl nitrite, $C_5H_{11}NO_2$, is mostly used as an inhalation, being only seldom given internally; but we may employ it so in doses of one-half a minim to one minim. Discovered by Balard in 1844 (*Annales de Chimie et de Phys.* xii.), the attention of physiologists was directed to it by Guthrie in 1859, and in 1865 Dr. B. W. Richardson reported on its physiological action to the British Association for the Advancement of Science. Dr. Lauder Brunton then took up the subject, and in 1867 suggested to the Clinical Society the use of amyl in angina pectoris, basing his recommendation on its physiological properties. This valuable deduction has been fully confirmed by clinical experience, and affords us a brilliant illustration of the success of theory applied to therapeutics. In 1868 Dr. Arthur Gamgee communicated to the Royal Society the result of an elaborate investigation into the effect of the nitrite on the blood ("Philosophical Trans.," 1868). In 1871 the Warren Prize was awarded to Dr. H. C. Wood for his memoir on the subject, which was published the same year (*American Journ. Med. Sciences*, 1871).

Amyl nitrite is remarkable for the rapidity of its effect on the circulation. A short inhalation will bring on immediately palpitation with fulness and throbbing of the head, flushing of the face and neck, quickly extending to the trunk, tingling of the surface, and perhaps giddiness. Another inspiration or two, and pulsation of the carotids with great restlessness, disturbance of vision, depression, and cold sweats follow, with generally cold extremities, headache, and some confusion, but no loss of consciousness. It will be seen that the chief effect is on the circulation, two very distinct actions being observed; one acceleration of the cardiac beat, the other dilatation of the peripheral vessels. The pulse is immediately increased in frequency, and the blood-pressure is simultaneously greatly diminished. Different explanations have been offered of these remarkable effects, but the great point is the sudden fall of blood-pressure, so that the resistance to the left ventricle is taken off, its contractions being more frequent, though they can scarcely be more powerful; thus the heart has less work to do and there is more power to do it—that is to say, in a given time the number of contractions is increased. The vascular relaxation has been referred to an action on the vaso-motor nerves and muscular coat of the arterioles, or, on the other hand, to an effect on the centre in the medulla. It really appears to be peripheral, not central, for both Brunton and Wood found that it occurred after the vessels had been separated from the centres by section of the cord, showing that the sudden fall is produced by a direct paralyzing action of the remedy exercised upon the coats of the arterioles. The local action of amyl upon muscular tissue confirms this view. The quick beat may perhaps be partly due to depression of the cardiac centre or of the peripheral cardiac vagus.

The nitrite, however exhibited, greatly reduces the temperature, both in health and in a febrile condition; this effect seems to be due to a direct check to the tissue metamorphoses. Wood has shown that it is associated with diminished excretion of carbonic acid. It is independent of the nerve-centres, for it occurs after section of the cord, and even after death in those cases in which the temperature remains high or rises. Outside the body the nitrite has a remarkable influence over oxidation—for example, a few drops introduced into a jar of glowing phosphorus will extinguish it. Possibly within the body a similar effect is produced, though not so complete, or instant death would ensue. Diminution, not total arrest of oxidation probably occurs. All highly-organized tissues lose power in the presence of amyl. Muscles, nerves, and nerve-centres all have their functions restrained or suppressed by contact with it; but if the contact be only brief they recover their power, so that the poison does not destroy the vitality of the tissues.

Nitrite of amyl is very rapidly taken up by the blood, on which it exercises a remarkable influence, greatly interfering with the function of the red corpuscles. Under its influence, both the arterial and venous blood assume a chocolate hue, the cause of which has been carefully investigated by Dr. Gamgee, who finds that the nitrite unites with oxyhæmoglobin to form a compound which is in its turn easily broken up by ammonia, and by reducing agents. He found, too, that blood to which the nitrite had been added failed to take up an appreciable quantity of oxygen. Still the corpuscles retain in some degree their power of giving up ozone to substances having an affinity for it. Thus their respiratory function is not abolished, though it is greatly interfered with. The poison is eliminated by the kidneys and in poisonous doses gives rise to glycosuria, a phenomenon first observed by J. A. Koffmann (*Reichert's Archives*, 1872) when experimenting on rabbits. At the same time the amount of urine is increased. Perhaps disturbances of pressure in the kidneys and liver may account for these symptoms.

The most important use of nitrite of amyl is in angina pectoris, but in other cases of cardiac failure, as for instance in chloroform narcosis, it has sometimes been successful. Whenever it is essential to lower the blood-pressure rapidly it may be effected by amyl. Its action also naturally indicates it as likely to relieve asthma as well as other forms of spasm. As soon as Dr. Brunton brought it forward for angina, I began to use it to relieve spasm of the bronchial tubes, and obtained excellent results. I have seen a single whiff from a bottle containing a few minims put an end to a severe paroxysm of asthma, but sometimes the dyspnœa returns rather soon. We may then repeat the inhalation with due care, but it is not desirable for the patient to become habituated to the remedy. It seems almost necessary to intrust him with it, and the relief is so rapid and sometimes so complete and lasts so long that there is some danger of his resorting to it too freely. It is when the dyspnœa is accompanied with pallor of the face that this remedy may be tried; when the superficial vessels are already relaxed it is inappropriate. The patient should also be instructed to leave off inhaling the moment he feels the flushing begin, as the effect will continue and even increase for a short time after this. Dr. Kitchin (*Amer. Jour. Med. Sci.*, 1873) has used it in acute bronchitis as well as asthma, but I should scarcely resort to it unless distressing spasm were present.

In emphysema and cardiac dyspnœa it is not so successful, and, indeed, in heart disease it should only be resorted to with considerable circumspection. It is, as already stated, to take off blood-pressure and arrest spasm that it is chiefly indicated, though it has naturally been tried in many diseases, and a good epitome of experience concerning it will be found in Dr. Pick's pamphlet.[1]

Nitro-glycerine has been revived by Dr. Murrell[2] as a substitute for amylnitrite in the treatment of angina pectoris, and it may also be used as an antispasmodic in asthma, whooping-cough, etc. Its effects come on more slowly but last longer. Many years ago it was introduced by Mr. Field (*Med. Times*, 1858) as a very powerful remedy, giving rise to a sensation of fulness in the neck and head, some confusion, noises in the ears, etc., followed by headache. And he strongly recommended it in neuralgia and spasmodic diseases. His statements were confirmed by my late brother, Mr. Augustus James, then a student at University College, by Drs. Thorowgood, Brady, and others. But Drs. G. Harley and Fuller took much larger quantities without any effect, and their statements no doubt largely contributed to the remedy falling into neglect. It seemed obvious to me at the time that there were considerable differences in the susceptibility of individuals to the influence of nitro-glycerine. My brother was affected severely for a considerable time by a dose taken in my presence from the same bottle out of which I took an equal quantity, but in my case the effects passed off in a few minutes, while he was affected in the manner he has described (*Medical Times*, 1858). The average dose is one minim of a one per cent. solution, which may be taken in water, or on a lump of sugar, or in tablets as made by Martindale. It acts on the blood like nitrite of amyl, paralyzes muscle, and destroys the reflex function of the cord, causing death by asphyxia.

Nitrite of sodium has also been found to produce effects similar to those caused by nitrite of amyl. On the heart and vessels, as well as on the blood, the action seems to be precisely similar, but it is less sudden and less energetic; on the other hand, it lasts longer. Dr. Ralfe brought it before the Medical Chirurgical Society, and described the toxic symptoms to which doses then thought small had given rise. Dr. Ramskill corroborated this, and it has been shown that when it began to be used impure specimens were employed, and the dose therefore of the pure

---

[1] Pick, R.: Ueber das Amylnitrit und seine therapeutische Anwendung. 1874.
[2] Murrell, William: Nitro-glycerine as a remedy for angina pectoris. 1882.

nitrite, now readily obtained, is much smaller than was supposed. Drs. Reichert and Matthew Hay have shown that pure nitrite of sodium or of potassium will act like the nitrite of amyl (*Practitioner*, 1883). Gamgee has shown that the nitrites act on the blood in the same way as the amyl compound, and Drs. Ringer and Murrell have experimented (*Lancet*, 1883) on the relative activity of nitrites and nitrates on animals, and employed full doses, producing unpleasant symptoms on patients. When used it would be desirable to begin with smaller doses than have been recommended, say two to three grains. Of course, the alkaline nitrites are less diffusible than the amyl, take longer time to be absorbed, and it would appear that they are more depressant to the central nervous system, while they also act on the peripheral nerves and the muscles, not only through the centre but by local access through the circulation.

Ethyl-iodide or *hydriodic-ether*, C_2H_5I, is sometimes inhaled as an antispasmodic in asthma when the expectoration is scanty and tenacious. It must be regarded as a stimulant to the air-passages, and of course it introduces iodine into the system very rapidly. It can scarcely be regarded as anæsthetic. Occasionally the relief it gives in spasmodic asthma is almost instantaneous.

Ether and Chloroform.—Without producing any anæsthesia small quantities of ether or chloroform when inhaled often act at once on the respiratory passages and arrest spasm. Even spasm of the glottis may be stayed by a whiff of chloroform, so may the asthmatic paroxysm; so again may a useless, dry, persistent spasmodic cough, whether arising from irritation in the respiratory passages or in the nervous system. Yet these vapors unless largely diluted are very irritant. The use of chloroform liniment is sufficient to show this, or a small quantity on cotton wool with a covering to confine the vapor will be found an active rubifacient and even vesicant. The irritating effect of the vapor of ether on the respiratory membrane is manifest when it is given as an anæsthetic; cough is set up almost always and very often some blood-stained mucus is brought up. The reason we do not see this in using chloroform is probably because so much smaller a proportion of the vapor is employed, from three to four per cent. being sufficient, and we give ether in as concentrated a form as we can. Dr. Snow[1] stated that air at 80° F. saturated with ether contained about seventy-one per cent. of the vapor. In practice we scarcely reach this degree of concentration, but the nearer it is approached the more rapidly anæsthesia is produced.

How is the antispasmodic action of chloroform and ether produced? In larger doses we know that the respiration and circulation are both more frequent at first; but very soon the cardiac and respiratory centres

---

[1] Snow, J.: On Chloroform and other Anæsthetics, their Action and Administration. 1858.

become depressed, the pulse falls in frequency and power, and the breathing becomes slow, heavy, and stertorous. Although, therefore, these vapors may act upon the respiratory nerves, it is to the centre we must look for the chief explanation. Ether depresses the circulation less than chloroform; the heart beats after respiration has been arrested by ether, which is one reason of its greater safety; at the same time the respiratory centre is less depressed, though we must not forget that it is a depressant eventually. The stage of stimulation is, however, much longer with ether than chloroform and the anæsthesia is briefer and not so deep. Comparatively, therefore, ether is sometimes spoken of as a stimulant and chloroform as a narcotic. Each possesses both properties, as we have explained in regard to other narcotics, but the exciting stage of chloroform quickly passes into narcosis, while the primary effect of ether is more protracted and more decided.

Both ether and chloroform are antispasmodics when administered internally. The former is the more generally used for this purpose, being more exciting and commonly regarded as a powerful diffusible stimulant. They are both carminatives. As soon as they are taken into the stomach they stimulate the circulation and the nerve-supply of that organ, and act reflexly on the heart and respiratory system. At the same time a portion is at once absorbed and taken by the circulation to the nervous centres, where the first effect is excitant. By careful dosage we may contrive to confine the action to this stage, giving only enough to produce excitement and taking care not to repeat the dose until time has been given for the effect to pass off, so as to prevent accumulation. This is what we want when we give these medicines for their antispasmodic effect. This is obtained in the stage of stimulus without any perceptible degree of narcotism being necessary. Chloroform is much more pleasant to take, but more decidedly narcotic. Hence, ether, the more stimulant and the more diffusible and so the more evanescent in its action, maintains the first place as an antispasmodic, but by careful dosage—and this is all in all—spirit or tincture of chloroform may very often take its place and in some cases seems to be preferable. It is perhaps better suited to gastro-intestinal affections, while ether is far superior in spasm of the respiratory system. Ether may be given in capsules or pearls, or the spirit of ether may be used, but the compound spirit, the old Hoffmann's anodyne, is more decidedly antispasmodic. Acetic ether has been introduced as more agreeable in taste and odor and therefore more appropriate as a carminative.

Other anæsthetics naturally possess antispasmodic properties, but as they are not used for this purpose need not detain us.

Quebracho.—This bark has been lately introduced as a remedy for dyspnœa, and is said to be especially useful in emphysema; it may, therefore, perhaps, be called antispasmodic; but both the bark and the alkaloid, aspidospermia, reduce the frequency of the respiration as well as the heart's action and apparently the temperature, probably by operating through the centre. It has, however, sometimes been said that quebracho stimulates the respiratory system. Manifestly further information is needed as to its precise action, which is probably rather complex, as quebrachin and four other alkaloidal substances have been described as contained in it besides aspidospermin. The dose of this last is given as a quarter to half a grain. A tincture of the bark, one in five of proof spirit, is also used in doses of five to sixty minims; the drug appears, however, to be uncertain in its action, and when given should be carefully watched. Probably it will be found that its proper place is among the central depressants, though it may possess a preliminary stimulating action.

Sedative and Anodyne Pneumatics.

These have naturally been named in the other groups, but may be considered together for a moment as substances the use of which is indicated for the purpose of removing pain or uneasy sensations in the respiratory organs, and for restraining their excessive action, whether that be represented by spasm, cough, or other symptom. It is obvious that whatever dulls the perceptive faculties will seem to relieve pain or uneasiness. The sedative or anodyne action may therefore be cerebral, annulling consciousness, as in the case of anæsthetics, or diminishing it, as by narcotics. So it is also clear that whatever interrupts the conveyance of sensation from the periphery to the centre will appear to be sedative or anodyne. Consequently the depressants of the respiratory branches of the vagi, which, as we have seen, are antispasmodic, are naturally anodynes to the respiratory organs. Further, whatever relieves the mucous membrane or other tissue, and restores healthy breathing, may claim to be an anodyne pneumatic; therefore, moisture and warmth, when the membrane is swollen and dry, are remedies of this kind, so that fomentations and hot applications acting reflexly may give relief; but inhalations of warm vapor, bringing the remedy in contact with the surface are much more effectual. But we are concerned here with general rather than topical remedies. Sometimes expectorants alone are effectual, inasmuch as they cure the disease; but very often they are not required, or their action needs modification; therefore, combinations with opiates or other neurotics may be resorted to, or a dose of the neurotic may be required at long intervals, while small quantities of the expectorant are given more frequently. We may use for this purpose the stimulant depressant or alterant expectorants, according to circumstances, combining them with various neurotics, of which opium is perhaps the most frequently employed. But belladonna deserves very often the preference. As these neurotics both restrain secretion, they may be thought to be therapeutically incompatible with expectorants, but practically this is not the case, and often the greatest advantages are obtained by the judicious combination of differently acting medicines. Nowhere is this more decidedly the case than in diseases of the respiratory organs. And the combinations of the various pneumatics appropriate to individual cases afford ample scope for the exercise of the therapeutical knowledge and skill of the physician. In other cases our anodynes may require to be contra-expectorant, or we may turn from this class entirely and employ agents which act directly on the centre.

Contra-Expectorants.

These are medicines which diminish the sputa by restraining the secernent function: they therefore antagonize expectorants. Most of them have been considered among other groups. Thus we have seen that some of the antispasmodics restrain secretion, though others produce a freer flow of mucus. So, too, some of the sedatives and anodyne pneumatics restrain secretion and antagonize the expectorants; besides these, neurotics diminish the bronchial mucus; thus, both opium and belladonna do this, though their action on the nerve-centre is precisely opposite.

We now come to quite a different class of remedies, which restrain secretion whether by a local action or administered through the system.

Astringents, among which may be included certain acids, are distinct contra-expectorants, as they tend to diminish the amount of secretion, and though their action is not so manifest on the bronchial membrane as on other surfaces, it must still be admitted to exist and may sometimes be utilized. Some of our most valued tonics possess astringent properties, and where an astringent seems to be required in bronchial affections, a tonic of this character, particularly an acid one, will very often accomplish all that we require. When the use of an astringent becomes urgent to restrain bleeding instead of secretion, it may be necessary to resort to the most powerful members of the class, but generally in such cases astringents alone are not to be relied upon, although they often assist such powerful agents as complete rest and other appropriate measures.

Central Pneumatics.

These are such remedies as produce their effects by a direct action on the respiratory centre. Such action may be of two kinds, either stimulant or depressant; the first giving rise to more active respiratory movements, the other retarding them. This may seem to be a sufficiently broad line of distinction between the two groups, but if we try to tabulate the central pneumatics we shall find that some of them almost claim a place in each. Thus most of the depressants of the centre cause at first some excitement, if only very brief, as if the first impression on a nerve or organ excited a reaction in the part affected. We have seen such an effect in chloroform, ether, alcohol, and the narcotics generally, which appear to excite before they depress, the duration of the exciting stage rather than its intensity furnishing the most important distinction between them. Camphor has often been spoken of as both a stimulant and a depressant to the nervous system. So far as the respiratory centre is concerned, it seems at first to excite, but the depressant action soon comes on. Then stimulants of the cardiac centre, many of which act also on the respiratory, may be employed for the latter effect, and we must remember that excitement of the circulation would usually be accompanied by more rapid breathing, whether the stimulus directly affected the respiratory centre or not. Moreover, cardiac stimulants, as we have seen, are mostly followed by a stage of depression, and we have to vary our mode of using them according to the effect we are desirous of producing.

Opium—which, as we have seen, distinctly depresses the respiratory centre—has a full preliminary excitant action on the circulation, or possibly at first a slight stimulating effect on the respiratory centre, so slight as to be disguised by the circulatory excitement, and so brief that it is to be regarded as a characteristic depressant pneumatic. So it has been said that such potent depressants as prussic acid, tobacco, and calabar bean exhibit a slight but evanescent stimulating effect, which is soon quenched in the profound depression. A similar course of events may be observed in respect to the action on mucous membrane, those which increase secretion and even depress the circulation being found to produce a preliminary ephemeral stimulation. Even antimony and ipecacuanha are said to act in this way, both on the centre and the periphery, but the excitement is so slight that it can seldom be detected, and can never be reckoned on, unless it is to be maintained that the promotion of secretion is to be considered as necessarily stimulating.

In the same way it is to be observed that our most valued stimulants of the respiratory centre tend to terminate in a stage of depression, or if the expression be preferred, of exhaustion. Thus, belladonna in too large quantities and its allies, hyoscyamus, stramonium, and datura, may finally depress, though in therapeutic doses the real value is the powerful stimulant action on the centre, which sometimes enables us to maintain the failing respiration.

The difficulty of strictly classifying these central pneumatics will now be apparent, but we shall take the ordinary chief therapeutical action, and disregard so far the minor or subsidiary effects, and first of all we will consider

1. STIMULANTS OF THE RESPIRATORY CENTRE.—*Ammonia* is perhaps the most important of these. It is the most commonly used medicine for this purpose, and may be confidently relied upon to produce a rapid effect on the centre, the circulation being at the same time powerfully stimulated. As we have seen, it is also an expectorant, promoting the secretion within the tubes, its action on the centre increasing the coughing power and sustaining the breathing, while, of course, the general stimulant action is also produced.

Belladonna and Atropia.—This remedy, too, we have previously considered at length, and need now say but a few words respecting its action on the centre. It is a most valuable stimulant, the next perhaps to ammonia, if not its equal, for sustaining the failing respiration; the action is rapid, beginning almost immediately, and soon passing off, so that small doses can be repeated. The pupil is not a sufficient guide, nor is dryness of the throat, nor dimness of vision; these are symptoms caused by full doses and are somewhat uncertain. The patient should be watched and a small dose will generally cause flushing, which is a sufficient indication of the effect. The breathing will then become deeper, and when it again fails a dose may be repeated. It will be remembered that atropia also restrains the secretion and tends to dry the membrane, in this respect being contra-expectorant and opposing ammonia; at the same time it reinforces the drying action of opium, though it antagonizes that medicine in other directions, as we have already shown at length in our chapter on neurotics.

Nux Vomica and Strychnia.—The medulla is very powerfully stimulated by this remedy. The vaso-motor, the cardiac, and the respiratory centres all experience its influence; some of the effects are partially masked by the powerful stimulation of the cord, which in toxic doses exalts the reflex excitability of the motor centres to so great an extent. The frequency and depth of the respirations are increased by therapeutical doses, as shown both by experiment and observation. Prokop Rokitanski found (*Oesterreich Med. Jahrb.*, 1874) that strychnia caused the respiratory movements to reappear after they had been abolished by section of the cord. Dr. Milner Fothergill found that strychnia successfully antagonized lethal doses of aconitia, and he brought it forward at the International Medical Congress in London, in 1881, as an expectorant acting through the centre. Dr. Lauder Brunton has employed it to check the night-sweats of phthisis, thinking that when the centre becomes exhausted the accumulation of carbonic acid, being no longer sufficient to rouse the centre, excites the sweat-glands. He therefore gave strychnia to increase the excitability of the centre and so lead to more perfect respiration. His clinical results ("St. Bartholomew's Hospital Reports") seem to support his view. Dr. Thorowgood has also used this remedy in embarrassed respiration, and I can fully confirm its value as a stimulant of the centre. We must remember, however, that it excites the other centres. Dr. Fothergill also suggested to the Congress that strychnia should be employed to meet the disturbance in the respiration sometimes caused by digitalis; and when disease in the lungs or air-passages co-exists with impairment of the heart's power, he would combine strychnia with digitalis in their treatment. For this purpose I use tincture of digitalis with tincture of nux vomica. In other instances, where a decided expectorant is required, the tincture may be added to the mixture of senega and ammonia; when the senega is not tolerated serpentary is a useful adjunct. Quinine and other tonics I also often employ in conjunction with nux vomica. As a respiratory stimulant in bronchitis, emphysema, asthma, and phthisis, whenever it is desirable to increase the action of the centre, I regard strychnia as a most valuable remedy, and inasmuch as it also acts on the cardiac centre, it is at the same time a tonic to the heart, the ganglia of which it also rouses, while it favorably influences the vessels through the vaso-motor centre, by which the arterial pressure is raised. It seems, therefore, appropriate in cardiac dilatation accompanied by diminished vascular tension.

I usually select strychnia in preference to atropine when a more prolonged effect is required and when the need for sustaining the centre is not so urgent. The remedy takes a longer time to produce its effect. It is, indeed, absorbed quickly enough, but we give it in doses which have to be repeated several times before a full action is established; it is eliminated rather slowly. It seems, therefore, suited for keeping up a moderate degree of stimulation for a considerable period. It also affects the circulation less than belladonna. When the necessity for a respiratory stimulant is urgent, I use atropia or belladonna; the effect of this is much more rapid, almost immediate; but it is also less permanent. I have also used the two alkaloids alternately, giving doses of atropia from time to time to maintain the failing movements while the less frequent doses of strychnia were gradually accumulating. I would not hesitate to produce rather quickly the physiological effect of strychnia, if it seemed to be needed; but this is seldom the case in my experience, for atropia may be confidently resorted to in urgent cases, and may be alternated with ammonia. I have therefore come to regard strychnia as better in less urgent or more chronic cases, and would plead for its more frequent use in chronic bronchitis and emphysema.

The use of strychnia in the paralysis that sometimes follows diphtheria in some cases of laryngeal paralysis, and for other purposes, need scarcely detain us.

2. DEPRESSANTS OF THE RESPIRATORY CENTRE.—Some of these stimulants, as already pointed out, may at a late stage act as depressants, but we may pass them by.

The nauseants, especially antimony and ipecacuanha, are decided depressants of the centre, so much so as at times to be contra-indicated when otherwise they might be useful.

Lobelia and tobacco are both powerful depressants, but have been fully considered in a previous article.

Opium is the most constantly useful of the respiratory depressants; we have seen how it enters into a considerable number of our groups, and although it affects the entire nervous system—we might almost say the entire organism—we must never forget the special depressant action on the respiratory centre as a most important element in its use as a pneumatic. Paralysis of this centre is the cause of death in opium-poisoning, and its depressing influence in therapeutical doses comes on at an early period. We have seen, too, how it restrains bronchial secretion, blunts the sensations, impairs the action of the vagi, and lowers the pulmonary circulation as well as the general blood-pressure. Thus the entire effect on the respiratory organs is very depressing, though its immense value must be recognized in restraining cough, spasm, dyspnœa, expectoration, and even vascular excitement and hæmoptysis. Some of the undesirable or dangerous effects may also be modified or prevented by the simultaneous administration of other remedies acting as antagonists to it over a portion of its area of operation. Combinations, too, with synergists instead of with antagonists, as well as with various other pneumatics, may also render the greatest service. Further, the administration of other medicines, even such as scarcely claim to be considered pneumatics, may in their turn advantageously affect the narcotic; and again, an occasional opiate may be ventured upon under circumstances when its action on the centre might suggest that it would be scarcely appropriate. The action of opium itself is very complex, and he will prove himself the most skilful therapeutist who, whether by judicious combinations or alternations with other medicines, or by the most careful dosage, is able to obtain the greatest benefit for his patient.

Chloral.—Chloral acts first on the brain, but larger doses soon involve the medullary centres, which are much depressed. The heart is weakened partly through the centre, but a full dose is a direct cardiac poison, retarding and enfeebling the beat by diminishing the excitability of the intrinsic ganglia, and in toxic doses the ventricle is arrested in diastole. The blood-pressure likewise falls, apparently from a direct action on the vascular walls as well as on the vaso-motor centre, which is depressed. Thus there is altogether a very considerable weakening of the circulation, tending toward its arrest, while at the same time the respiratory centre is deeply depressed and the breathing rendered slower and feebler, a little later irregular and shallow, and ultimately it is completely arrested. These effects have been observed after section of the vagi, and so must be attributed to the influence on the centre. There is a remarkable fall in the temperature.

In spite of its powerful effect chloral has been used to restrain spasm, cough, and dyspnœa, but it very frequently fails to do so, and altogether in respiratory diseases it is much less useful than opium, though it is probably more risky, and when there is any fear of the centre becoming exhausted it is very dangerous. Moreover, large doses seem to favor congestion of the lungs and air-passages. In acute inflammation of these parts it is, therefore, to be avoided. It is useless as an anodyne and not much use as an antispasmodic; though it has sometimes relieved asthma, it more frequently fails. In whooping-cough and laryngismus it is perhaps more successful when given in small doses two or three times a day, but even in these cases it is inferior to the bromides, and should for the most part be reserved as a hypnotic, for which purpose it is invaluable.

Bromides depress the respiratory centre, and to this influence is perhaps due their effect in relieving spasmodic diseases of the respiratory system. But the depression is by no means confined to this part; the other centres in the medulla are also affected and probably all parts of the nervous system, though much of the influence on the brain is to be traced to the effect on the cerebral circulation. The depression of the cardiac centre is not very marked; true, the heart is weakened and retarded by the medicine, but this appears to be largely due to direct action on the nervo-muscular substance rather than on the centre, which explains the value of the remedy in certain nervous disorders of the heart. The tension seems to be reduced, but the effect on the vessels is hardly settled. The temperature is usually somewhat lowered.

The bromides are absorbed readily and elimination begins at once through the saliva and urine. The salt usually passes away unchanged; the largest portion leaves the system during the first day, but as Clarke and Amory[1] have shown, elimination goes on longer; on the second day, after a single large dose, the amount removed is much less than on the first, and the reduction goes on until elimination is complete. Namias reported (*Gazette Hebdom.*, 1868) that when bromide had been taken for a considerable period its excretion might not be complete for fourteen days. Disease of the kidney renders the process slower, and Dr. Lees mentioned to the Pathological Society ("Trans.," 1877) a case in which it was found four weeks after the medicine was discontinued. Rabuteau, operating upon large quantities of urine, found a salt of bromine at such a distance of time that he came to regard it as a natural constituent (*Gazette Hebdom.*, 1868) of the body. Armory found that a single dose was removed almost entirely in one day, but some of it often remained until the second day. After continuous doses it lingered longer

---

[1] Clarke, Edward H., Amory, Robert: The Physiological and Therapeutical Action of Bromide of Potassium and Bromide of Ammonium. 1872.

in the system. Namias found (*Comptes Rendus*, tome lxx.) bromide of potassium in all the fluids as well as in the brain, lungs, liver, and other viscera of a man who had died while taking a course of that medicine. Some elimination has been supposed to go on upon the mucous membranes, and Voison (*Bull. gén. de Thérap.*, lxxi.) has stated that the breath has a strong odor of bromine after the continued use of the salt, so that he supposed it might be decomposed to some extent in the respiratory passages, and the volatile element escape, and he and others observed that hoarseness, cough, and laryngeal and bronchial irritation sometimes followed the use of the drug. But Amory and Clarke conclude (*op. cit.*) that it is not eliminated by the breath, and regard the odor observed by M. Voisin as one which is produced equally by other salts of potassium, after the use of which they say a fetid smell is often given off and a disagreeable taste experienced by the patient. Echeverria [1] noticed that this odor occurred earlier and to a greater degree in persons who did not attend rigidly to the cleanliness of the teeth; a similar effect is observed in persons who have a habit of biting and chewing up portions of the lining of the mouth and lips. The laryngeal symptoms observed are generally attributed to impurities in the medicine; a very small contamination with iodate will bring on catarrhal symptoms. The skin assists in the elimination on the second day, and the eruptions so well known to occasionally follow the use of bromides may perhaps be due to a local action. The effect on the circulation is less marked than on the nervous system, but must not be forgotten as an important element in the action.

It will be observed that the action of the bromides is rather slow, compared with the iodides; or again, compared with chloral the effect is slower but more protracted, so that in so far as the respiratory system is concerned, chloral and the bromides are to each other as depressants of the centre much the same as belladonna and strychnia are to each other as stimulants; and here is one danger of chloral and the great recommendation of the bromides. We do not really want to suddenly depress the centre in the manner in which it is so often most urgent to suddenly stimulate it. Therefore, the quick-acting stimulant is often called for, but the quick-acting depressant is rather to be avoided when possible, though the slower and less intense depressant action of the bromides may not be injurious; in some instances it is manifestly desirable, and constantly it is allowable in order to secure the other remarkably valuable effects on the nervous system.

The bromides are useful in spasmodic respiratory disorders; they sometimes relieve laryngismus and prevent the recurrence of the paroxysms, and thus enable the tonic and hygienic measures so often neces-

---

[1] Echeverria, Gonzalez: On Epilepsy.

sary to be brought into play. In whooping-cough the value of the bromides was pointed out by Gibb[1] and corroborated by Dr. G. Harley (*Lancet*, 1863). In asthma the bromides have been freely employed, both during the paroxysm and the interval; more successfully in the latter case. Professor Sée (*Bull. de Thérap.*, 1865) found that it delayed the attacks, as did Dr. Warburton Begbie (*Edin. Med. Journal*, 1866), and M. Saison published a remarkable case in which, after full doses for a fortnight, no further paroxysms occurred ("Du Bromaure," Thèse, 1868). The same author recommended this remedy in recurrent tonsillitis, but it appears to have little influence, and would probably only be of use to persons in whom it was otherwise indicated. ' In phthisis it is sometimes useful in relieving spasmodic laryngeal cough, though it very often fails to do so, and is not appropriate where there is much anæmia or depression. In this disease, also, it will restrain dysphagia when that assumes a spasmodic character and there is a good deal of irritation but no severe disease in the larynx. Of course it would only aggravate this symptom if it arose from any degree of paresis. In diphtheria these salts have been taken internally and used locally, with the addition occasionally of some pure bromine. The records are not very encouraging.

In resorting to the bromides in respiratory diseases, they may often be advantageously given in combination, or alternately with other remedies, partly for the purpose of reinforcing their action, as by chloral, cannabis, opium, aconite, etc.; but more frequently with a view of modifying the effect or preventing disagreeable consequences. The antagonism of strychnia to the bromides was observed by Saison, and has since been confirmed. He examined post mortem the spinal centres, after using the two medicines, and found that after bromides the capillaries were scarcely visible, but after strychnia they were intensely congested. As the effect of strychnia on the brain is so much less than on the cords, we may partially neutralize the spinal action of bromides without interfering much with their cerebral influence, and this is occasionally an advantage. Combinations with opium are also sometimes of service. Da Costa (*Amer. Jour. Med. Sciences*, 1871) has shown that it is easy to correct the disagreeable action of opium by preceding it with a dose of bromide, and that with no loss, but rather with a gain in the hypnotic effect. So arsenic is sometimes given simultaneously to prevent or remove the cutaneous complication set up by bromides. Dr. Bill in a very interesting article holds that chloride of sodium (*Amer. Jour. Med. Sciences*, 1868) is in some degree antagonistic to bromide of potassium. Atropia and ergot are both partially antagonistic, but both also may at times be advantageously combined with bromide. In some cases expec-

---

[1] Gibb, G. D.: Treatise on Whooping-cough. 1854.

torants and other pneumatics are advantageously given, while the system is kept more or less under the influence of bromide.

The potassium salt is most commonly used. The ammonium salt has been freely employed, and seems of late to grow in favor, but it is the more nauseous of the two, and seems also to be more irritating to mucous membranes, for which reason it is perhaps less appropriate when there is much bronchial irritation. It has also been thought to be more evanescent. Brown-Séquard [1] found that the union of the two salts increased the sedative action of the dose. The *bromide of sodium* is less depressing, less likely to irritate mucous membranes, and rather less disagreeable in taste. As in other instances, I have been led to prefer the sodium salt; the dose is a little less, as, weight for weight, it contains more bromine. *Bromide of lithium* is still more powerful, containing about ninety-two per cent. of bromine against seventy-eight of the sodium salt and sixty-six of the potassium. I have found a smaller dose, that is to say, a dose proportionate to these figures, thoroughly effectual, less disagreeable to take, and sometimes successful when the other salts fail. Probably absorption, and perhaps elimination, may be a little more rapid, for sleep comes on more quickly after the lithium salt. It may be preferred in gouty and rheumatic constitutions. Dr. Weir Mitchell speaks as favorably (*Amer. Jour. Med. Sci.*, 1870) of the hypnotic qualities of this preparation. *Bromide of calcium* is another active salt. It has not come much into favor, perhaps because it is rather unstable, but it has a value of its own. Hammond has found it succeed after the potassium has failed, and I would once more urge the more frequent use of this and other calcium salts. Charcot has recommended (*Brit. Med. Jour.*, 1877) *bromide of zinc*, and some other metallic bromides have been employed, so have combinations with quinine and other *organic* bases; while *hydrobromic acid* has been supposed to produce many of the effects of the alkaline bromides without causing so much depression, but it is more frequently used, merely to prevent quininism than to produce the effects of bromine.

Prussic Acid.—The dilute hydrocyanic acid of the pharmacopœias, United States and British, contains only two per cent. of this powerful poison, which is absorbed with great rapidity and produces its deadly effects most swiftly. It changes the red corpuscles and converts the venous blood to a bright arterial color which speedily turns black. The first change seems to check the function of the corpuscles, the second to destroy them, reduction of oxyhæmoglobin being effected and cyanohæmoglobin formed. This body, discovered by Hoppe-Seyler, has no ozonizing power, at the same time some cyano-oxyhæmoglobin seems also to

---

[1] Brown-Séquard: Lectures on the Diagnosis and Treatment of Functional Nervous Affections.

be formed, but these changes which have been well studied out of the body, and which no doubt occur in cases of poisoning, do not altogether account for the therapeutical effects which appear to be largely produced by an action on the nervous tissues, all of which are greatly depressed. As soon as ever the poison is taken the respiratory centre feels the impulse, and after a momentary excitement is greatly depressed, so that the respiration falls, dyspnœa ensues, and asphyxia follows. It has been held that at the same time the respiratory nerves are depressed and reflex respiratory acts arrested, especially when the poison is inhaled, but Preyer[1] says that after division of the vagi lethal doses did not kill, though Boehm and Knie (*Archiv f. exper. Path. u. Therap.*) found that section of the vagus had no influence on the respiratory action of the poison.

The vaso-motor centre is affected like the respiratory; so also is the cardiac, but this in a much less degree, so that the heart continues to beat for some time after respiration has ceased.

A full medicinal dose may cause giddiness, disturbed breathing, and syncope; this effect may be so rapid as to cause great alarm. No doubt great differences in the preparation have occurred, and it is unfortunate that in dealing with so powerful a medicine more than one preparation should have been in common use. The so-called Scheele's acid was much stronger than that of the Pharmacopœia, and also somewhat uncertain in its strength. It is, therefore, a source of regret that a few physicians should still insist on using it. The officinal acid is strong enough for all therapeutical purposes, and a dose of one to three minims need not be exceeded. The dose given by the British Pharmacopœia is much too high; indeed, it may be said to be quite double what it ought to be. The British Pharmacopœia gives two to eight minims, but very few people happily will prescribe this maximum. A case has been communicated to me in which a patient, after four minims, fell down in a state of breathlessness, which was at first thought by friends to be fainting, but the great dyspnœa, constriction of the chest, confusion and giddiness without loss of consciousness, convinced them that the medicine just swallowed was the cause of the symptoms, and they sent for the gentleman who had prescribed it in the greatest alarm. As he happened to be near he arrived in time to see the effect, which left a severe headache and sense of prostration for several hours. Admitting that there may have been some unusual susceptibility in this case, considering that many other persons who have taken a less dose have experienced headache and other unpleasant symptoms, my contention seems reasonable that the officinal dose is too large. Moreover, as the effect of the medicine in small doses is very evanescent, it can be repeated every two or three

---

[1] Preyer, W.: Die Blausäure. 1870.

hours, and in most of the cases in which it is likely to prove useful small and frequent doses will be found most desirable. In my experience two or three minims always suffice, and more frequently one to two.

The chief use of hydrocyanic acid is to allay troublesome spasmodic cough in phthisis, asthma, and whooping-cough. It probably acts both on the centre and on the peripheral nerves; by this latter action it also arrests vomiting, for which it is often employed, and this also makes it specially useful in some cases of phthisis. We know that locally it allays irritation, as we sometimes see on the skin, and perhaps this is why the inhalation of the vapor is occasionally so effectual. The high value placed on this remedy by Majendie [1] is not easy to account for, in view of the disappointment to which it constantly gives rise. The still more extravagant praises of Dr. Granville,[2] who seemed to think it could cure advanced phthisis, did not perhaps exercise the influence of the eminent physiologist's confidence. Dr. Elliotson [3] defined its sphere in dyspepsia. Dr. Roe [4] had much confidence in its value in simple, uncomplicated cases of whooping-cough, but gave it in doses which Sir T. Watson in his lectures pronounced "gigantic." Still Dr. West [5] admits it sometimes "exerts a magical influence," and Dr. Atlee (*Am. J. Med. S.*) gives a favorable report based on two hundred cases. Dr. Lonsdale made some interesting experiments (*Edin. M. and S. J.*, 1838) and Dr. Nunneley contributed others ("Trans. Prov. Med. and S. Ass.," 1847). The more recent researches have already been cited, and a review of the whole evidence will, I think, confirm my estimate of its therapeutical value.

Cherry-laurel is only of use for the prussic acid it contains, and being of uncertain strength is best avoided. Its introduction into the British Pharmacopœia is greatly to be regretted.

Cyanide of potassium produces the same effects as hydrocyanic acid. Five grains have several times proved fatal, and it is not improbable that half that quantity will cause death. The medicinal dose would be a tenth to an eighth of a grain. Cyanide of zinc has also been employed, and is included in the French Codex; dose, one-fourth of a grain. Some other cyanides have been used, but not as substitutes for the acid.

Physostigma, after a very brief stimulation, powerfully depresses the respiratory centre, death being due to failure of the respiration. It has

---

[1] Majendie, F.: Recherches physio'ogiques et cliniques sur l'emploie de l'acide Prussique, etc. 1819.

[2] Granville, A. B.: Internal Use of Hydrocyanic Acid in Pulmonary Complaints, etc. 1819.

[3] Elliotson, J.: Cases Illustrative of the Efficacy of Hydrocyanic or Prussic Acid, etc. 1820.

[4] Roe, G. Hamilton: Treatise on Whooping-cough. 1838.

[5] West, Charles: Lectures on Diseases of Infancy and Childhood. 1854.

been used in several spasmodic affections. Subbotin speaks well of it (*Archiv f. klin. Med.*, 1869) in chronic bronchial catarrh, attended with dyspnœa. We have already mentioned its antagonism to atropia. It can scarcely be said at present to have obtained a position as a respiratory remedy.

Aconite and *veratria*, although they depress the respiratory centre, are employed rather for their effects on the circulation, which have already been considered. A similar observation may be made as to gelsemium.

Conium may be mentioned here, and has been used more or less from the time when Socrates drained the fatal cup presented by the Athenians. But although the ancients employed it as a medicine, it fell into disuse until Baron Stoerck[1] revived it in a treatise of nearly three hundred and fifty pages, which he followed by a second smaller work and besides that a "supplement," all three of which were in a short time translated into English. No one can doubt that Stoerck's statements are highly colored, as indeed his contemporaries asserted, but the interest he excited secured a long trial for the drug, and a very considerable literature accumulated concerning it, of which Bayle has furnished us ("Bibliothèque de Thérapeutique," 1835) an excellent summary. Sir Charles Scudamore added conium to his iodine inhalations; and in our own time Dr. J. Harley[2] has carefully studied its action and shown how it may be intensified by opium. A further important study of the subject has been made by Martin-Damourette and Pelvette.[3]

Conia or *conine* seems to have been discovered by Brandes and Giesecke in 1826, and was isolated by Geiger in 1831, but the first important examination of its properties was by Professor R. Christison ("Trans. Royal Soc. Edinburgh," 1836). It appears that the herb contains also an essential oil which is not poisonous, and a crystalline base, conhydrin, which is less poisonous than conia. This last is a yellowish, strong-smelling, oily liquid, more soluble in cold than hot water, very unstable, very poisonous, and when locally applied an irritant. It may be dissolved in alcohol, and it forms salts which are more manageable than the alkaloid. Conia is readily taken into the system, where it has been supposed to be destroyed, but Zaleski and Draggendorff have detected it in the urine, and Orfila found some in the spleen, kidneys, and lungs of animals poisoned by it. Perhaps on account of its volatility it may easily escape

---

[1] Stoerck, A.: Libellus quo demonstratur Cicutam non solum uso interno tutissimo exhiberi sed et esse simul remedium valde utile in multis morbis. 1760–1. Also, Libellus secundus quo confirmatur, Cicutam usu interno tutissime adhiberi, etc. 1761. Also, Supplementum necessarium de Cicuta. 1761.

[2] Harley, John: The Old Vegetable Neurotics. 1869.

[3] Martin-Damourette et Pelvette: Étude de physiologie expérimentale et thérapeutique sur la Ciguë et son Alcaloïde. 1871.

in the breath. Its chief action is on the motor nerves, paralysis taking place first in the extremities and proceeding upward. In the end the respiratory centre as well as the nerves are paralyzed, and death ensues from asphyxia. But it is not so entirely without effect on sensory nerves as has often been said. Gubler (*Bull. de Thérap.*, 1875) found that it numbed cutaneous sensibility, and Lautenbach (*Phil. Med. Times*, vol. v.) declares that it greatly impairs the functions of the peripheral afferent nerves. It is not a hypnotic or a narcotic, as was once supposed, and its great use is to restrain excitement of the motor system.

Conia locally applied is extremely irritant, and probably fatal to the more highly organized tissues. Christison (*loc. cit.*) proved this in regard to muscular fibre, and yet we use it as a soothing inhalation. When mixed with hot water it rises with the vapor, but curiously enough its own boiling point, unmixed, is much higher than that of water. It would appear that in ordinary inhalations the vapor is so diluted that it no longer irritates the mucous membrane, and some local effect is exercised on the nerves; at the same time, by its central action it tends to relieve spasm. The British Pharmacopœia vapor is a bad preparation; in fact, we have been very unfortunate in our officinal preparations of hemlock; the extract is very often useless, and the mixture of it with liquor potassæ a method of dissipating the small amount of alkaloid that may be present in a superior specimen. The best way is to employ the succus, the only useful preparation in the British Pharmacopœia, and the alkali ought to be added at the time of use, so that what conia separates may be inhaled. Moreover, an alkaline carbonate is best. This vapor, carefully prepared, I have found very soothing in laryngeal phthisis and some other diseases. Internally the succus may be used in whooping-cough and other spasmodic affections; it must be continued until the physiological effects are manifest. The new United States Pharmacopœia has an abstract and a fluid extract, but I have not yet fully tried them. Perhaps a salt of the alkaloid will in time become more frequently used.

CHAPTER XIX.

TOPICAL PNEUMATICS.

The attempt to introduce vapors or fumes into the air-passages dates from the most remote antiquity, and the attempt to vary the conditions of the atmosphere must be quite as ancient. It could not but be that the pernicious effects of fogs should be noticed at the earliest period, and that men should not only avoid as far as they could what seemed injurious, but should when ill attempt to subject themselves to opposite conditions. The breathing of various emanations from the earth, especially in the neighborhood of volcanoes, was probably also resorted to in the earliest ages, just as other natural agents were employed, either accidentally or from superstition; and plants which were not found to be edible have been used as medicines by most uncivilized tribes.

Fumigations with cyphi (κῦφι) entered into the practice of the ancient Egyptians, these cyphi being, according to Dioscorides, a mixture of various drugs, and as the Egyptians had made great advances in the use of spices, balms, and other odorous medicines, it is probable that these entered largely into their cyphi. As soon as men began to use warm baths, indeed, as soon as they made water hot, they would become acquainted with its vapor, and probably notice the soothing effect of breathing steam, and endeavor to turn it to useful account.

The early Greeks were well acquainted with the use of inhalations and fumigations, and handed their knowledge down to the Romans, from whom the Arabian school obtained it. Then it may, perhaps, be said that wherever incense was burned in religious ceremonies the effect of breathing it must have been noticed, and to it would probably be ascribed the greatest benefit both curative and preventive. Homer mentions fumigations with sulphur ("Iliad," xvi., 228, and "Odyssey," xxii., 481). Hippocrates frequently recommends the inhalation of vapors and fumes of various balsamic and resinous substances, and on his authority these remedies long held a high place in the practice of his successors. He even describes an apparatus for the purpose, consisting of a saucepan with a hole in the lid through which a reed was passed. As the vapor escaped it was inhaled through the open mouth, wet sponges being employed to prevent scalding the lips. The works of Hippocrates further show his knowledge of the effects of air and the use of change of climate. Galen's genius and erudition are only equalled by his practical sense and observation, and we know the reliance he placed on this class of remedy. A little later flourished Cælius Aurelianus, the only Roman methodist whose works have descended to us, and though this sect rejected previous opinions, and ridiculed the Hippocratic system, terming it a "meditation on death," we find this author[1] recommended inhalations and fumigations. Pliny had already recorded ("Hist. Nat.") that in his time diseases of the lungs were believed to be benefited by a residence among pine-trees, so that the inhalations could pass into the air-passages, and Celsus ("De Medicina") recommended sea voyages to consumptives. As the Arabians took their inspiration from Hippocrates and Galen, they handed down this method of treatment, extending it by the numerous medicines and compounds which they added to the recognized materia medica.

During the dark ages, when the incubus of an apostate church weighed down the nations of Europe, medicine largely passed into the hands of a corrupt priesthood, many of whom were vicious, many ignorant, and not a few both. But while they lost many other valuable remedies and replaced them by superstitious observances, they recognized the value of fumigation, though they may have attributed its use to religious observances. On the revival of learning, men turned back with impatience to the writings of the Greeks, and soon the spirit of inquiry in various departments added to what had been previously known. In the sixteenth, seventeenth, and eighteenth centuries a large number of writers recommended frankincense, myrrh, amber, camphor, styrax, assafœtida, cloves, sulphur, the balsams, and, indeed, all strongly odorous fumes and vapors as a method of applying remedies directly to the bronchial membrane. Benedict in his "Theatrum Tabidorum" recommended this method of treating consumption, and Boerhaave in his "Materia Medica" furnished a number of formulæ for such fumigations. Dr. Mead recommended fumes produced by throwing the medicament on hot coals to be inhaled through a tube, and regretted the undeserved neglect into which this plan had fallen in his day, and bore testimony to the signal service he had obtained from balsamic fumes thus employed. Alberti[2] described the use of various inhalations and Buxtorf[3] and Caccialupi[4] wrote treatises on the subject.

Toward the close of the last century a series of brilliant discoveries directed attention to the elastic gases. It is true that hydrogen had

---

[1] Aurelianus, Cælius: De morbis Chron., lib. iii., c. 4.

[2] Alberti, M.: Dissertatio de spirandi difficultate. 1726.

[3] Buxtorf, J. R.: De inhalatione. 1758.

[4] Caccialupi, A.: De usu et præstantia Halituum, Vaporum, Suffituumque in morbis respirationis organa obsidentibus. 1795.

been discovered a little earlier, namely, in 1766, by Cavendish ("Phil. Trans.," 1766), but it was not till 1774 that Priestley[1] announced the discovery of oxygen, and Scheele that of chlorine, while two years later Priestley discovered nitrous oxide. This gas received its name of nitrous oxide from Sir Humphry Davy,[2] whose remarkable researches respecting it have a special interest, inasmuch as he foreshadowed the use of anæsthetics in surgery. Dr. Beddoes[3] took up the subject of elastic gases and their use in medicine with great energy, and had the advantage of Sir H. Davy's assistance. He tried the various gases in a great number of diseases, and recorded their effects with care. He was followed by Hill,[4] Cavallo,[5] and others, and in another work he[6] himself recommended the inhalations of hydrogen mixed with air as a remedy for consumption, on the assumption that in that disease there was excessive oxygenation. He prescribed inhalations for about fifteen minutes at a time, to be used several times a day. About this time the value of acids, chlorate of potash, and other substances containing a considerable proportion of oxygen was often attributed to that element being liberated in the system, as it was not till some time later that Wöhler and Stehberger (*Zeitschrift f. Phys.*, 1824) showed that these medicines were eliminated unchanged.

Under the influence of the impetus thus given to the inhalation of gases, even carbonic acid mixed with air was freely tried again; for it had been previously employed by Percival,[7] who considered that it ameliorated the febrile symptoms of phthisis, in which he was confirmed by Hulme,[8] Beddoes,[9] Cavallo,[10] and some others. The later observers thought that it lessened the stimulation of oxygen, which they considered excessive in consumption, and that it also relieved hectic and expectoration. Brera[11] recorded some important observations respecting this gas, which was also tried by Priestley,[12] who, however, turned his attention to

---

[1] Priestley, J.: Experiments and Observations on Different Kinds of Air. 3 vols. 1775-77.

[2] Davy, Sir Humphry: Researches Chemical and Philosophical chiefly concerning Nitrous Oxide. 1800.

[3] Beddoes, T., and Watts, J.: On the Medicinal Use and Production of Factitious Airs. 1796.

[4] Hill, D.: Practical Observations on the Use of Oxygen or Vital Air in the Cure of Diseases. 1800.

[5] Cavallo, T.: On the Medicinal Properties of Factitious Airs; with Appendix on Blood. 1798.

[6] Beddoes, T.: On a New Method of Treating Pulmonary Consumption.

[7] Percival, T.: Essays, Medical and Experimental. 3 vols. 1773-76.

[8] Hulme, N.: Easy Remedy for the Relief of Stone, Gravel, Scurvy, Gout, etc. 1778. [9] Beddoes: Op. cit. [10] Cavallo: Op. cit.

[11] Brera, V. L.: Osservazioni e sperienze sull' uso delle Aere Mefitiche inspirate nella Tisi Pulmonare. 1796.

[12] Priestley, J.: Directions for Impregnating Water with Fixed Air. 1772.

its use in solution in water. In the hands of Mühry[1] inhalations of carbonic acid entirely failed to relieve consumption. Although carbonic acid continued to be used in a number of diseases, we cannot affect surprise that it could not maintain the position which these writers assign to it. It may be stated, however, that it can be used when mixed with air more easily than might be supposed, and it is sometimes resorted to at Ems, St. Moritz, and other places in the present day. The good effect believed to arise from sleeping in cow-houses and stables was also supposed to depend on the carbonic acid in the air, but perhaps, if it existed, it was rather due to the warmth and moisture present. It has also often been remarked that asthmatics sometimes breathe easier in the close air of large towns, than in the country, and this also has been attributed to the carbonic acid ; but it may quite as likely depend on some other condition, and it is by no means a general rule. The relief sometimes afforded to asthmatics by the fumes of nitre papers has also been attributed to the same agency.

At the opening of the present century Ward[2] collected a number of observations on inhalations, and Zallony[3] employed them freely, while Dr. Paris[4] advised the evaporation of water in the rooms of his patients during the prevalence of dry east winds, as well as other modes of inhalation. A little later than this Sir Alexander Crichton[5] turned back to tar vapor, as an inhalation in phthisis, in which he was followed by Lazzaretto,[6] and Hufeland.[7] About this time, also, Dr. Forbes (*Med. and Phys. Journ.*, 1822) and Pagenstecher (*Hufeland's Journal*, 1827) advocated the use of inhalations of tar, which continued to hold their ground until quite recently, Sales-Girons[8] recommending them, in addition to the pulverized liquids with which his name is so honorably connected, and some others, especially in France, employing them at the present time. But this use of the fumes of tar was largely superseded by creasote, as recommended by its discoverer, Reichenbach, whose researches excited no little attention. Elliotson at once employed creasote inhalations in phthisis, as did Miquet and Cormack. The writings of these four authors have already been cited in the article on creasote, and

---

[1] Mühry, G. F.: De Aeris fixis inspirati usu in Phthisi Pulmonali. 1796.

[2] Ward, G.: De Medicina Pneumatica. 1800.

[3] Zallony : Traité de l'asthme. 1800.

[4] Paris, J. A. : Pharmacologia. 1812. Ninth edition, 1843.

[5] Crichton, Sir A. : Account of some Experiments made with the Vapor of Boiling Tar in the Cure of Pulmonary Consumption. 1817.

[6] Lazzaretto, E. : Practical Hints on the Nature and Cure of Consumption, with Cases in Proof of the Beneficial Effects of Inhaling the Fumes of Pitch, as a Powerful Auxiliary to other Treatment. 1818.

[7] Hufeland, C. W.: System der Pract. Heilkunde. 3 vols. 1818-28.

[8] Sales-Girons: Traitement de la Phthisie pulmonaire par l'inhalation des liquides pulvérisés et par les fumigations de gondron. 1860.

may be supplemented by Frueh's[1] work. These inhalations are still used, and often render good service in various affections of the respiratory passages. Creasote has, however, fallen into comparative neglect since the introduction of carbolic acid, which is now so extensively employed.

To return. Murray[2] and Scudamore[3] turned to iodine in the hope that its vapor might prove to be an effectual remedy in phthisis and other respiratory diseases. They were gratified with the results obtained, and it is not surprising that their expressions of confidence are somewhat exaggerated and scarcely supported by their cases. Those of Dr. Bardsley[4] certainly inspire more confidence. The use of iodine inhalations became gradually extended, and their value in some conditions of the air-passages is now generally acknowledged; though we no longer entertain the hope that this or any other agent will prove a panacea.

While so much attention was being directed to the inhalation of vapors and gases it was not likely that chlorine would be overlooked, and accordingly we find that this powerful irritating and suffocating gas, when adequately diluted, was tried as an inhalation. Gannal[5] gives an account of eight cases of consumption in which diluted chlorine appeared to him to alleviate the symptoms and prolong life. In one case recovery was said to have taken place, but then the only proof of the existence of tubercle was the presence of general symptoms. Elliotson[6] and other physicians also employed these inhalations in various diseases, and with somewhat conflicting results. At the present time we employ the vapor chlori as a stimulant and disinfectant inhalation in bronchiectasis, advanced phthisis, and in gangrene of the lungs. The air of bleaching-works is believed by many residing near them to be effectual in the cure of chronic coughs, and they sometimes resort to these works for that purpose; and it has been reported that the bleachers, although they may suffer from the emanations of chlorine, are particularly free from consumption.

We have already seen that sulphur fumigations have been employed from time immemorial; we need not, therefore, dwell on them specifically, but pass on to the more general subject of the various kinds of inhalations. Many of these are described by Maddock,[7] and about the same

---

[1] Frueh, G.: Ueber die Anwendung des Kreosots in der Pneumophthisis. 1836.

[2] Murray, James: On the Influence of Heat and Humidity; with Practical Observations on the Inhalation of Iodine and Various Vapors in Consumption, Catarrh, Croup, Asthma, and other Diseases. 1829.

[3] Scudamore, Sir Charles: Op. cit.

[4] Bardsley, J. L.: Hospital Facts and Observations. On Iodine, etc. 1830.

[5] Gannal, J. N.: Du Chlore employé comme rem'de contra la Phthisie Pulmonaire. 1830.   [6] Elliotson: Op. cit.

[7] Maddock, A. B.: Cases of Pulmonary Consumption, Bronchitis, Asthma, Chronic Cough, and Various Diseases of the Lungs, Air-passages, Throat, and Heart Successfully Treated by Medicated Inhalations. 1837.

time Dr. Coxe [1] urged their curative power. A little later than this a very remarkable impulse to the study of inhalation was given by its application for the purpose of producing anæsthesia during the performance of surgical operations, the year 1847 being remarkable for the appearance of between thirty and forty pamphlets respecting the new discovery, and the next year being almost as fruitful. Somewhat later Sylvestri [2] gave a full account of the subject of inhalation and Desruez [3] published his practical guide to this method of treatment.

At many continental spas inhalation-rooms were provided in which the steam of the mineral waters was employed, but of course it was felt that none of the mineral constituents were present, and ordinary water might have been employed for the same purpose. The idea of atomizing the mineral waters seems to have originated, or at any rate to have been first carried into effect by Auphan at Euzet-les-Bains in 1849 ; he projected a jet of the mineral water on the wall of the inhaling-room with sufficient force to break it up into a spray, which was inhaled by his patients. This method was adopted at several spas. But at length Sales-Girons constructed a portable apparatus for atomizing fluids, and brought it before the Academy of Medicine of Paris in 1858. This was undoubtedly an epoch in the history of inhalations, and the greatest interest was excited. It was not, however, till 1862 that the committee of the Academy appointed to investigate the new method brought in its report, and during the interval prolonged discussion had taken place as to whether the spray penetrated deeply into the air-passages. The report stated that it was proved that both the water and the mineral constituents employed penetrated not only to the bronchi but even to the air-cells ; and this report was founded on extensive independent experiments and was almost unanimously adopted by the Academy. The conclusions and practice of Sales-Girons,[4] who has been called the Father of Atomization, thus received the highest authority and rapidly spread over the civilized world. Demarquay [5] was one of the earliest to adopt it, and to prove independently that the atomized liquids rapidly pass into the respiratory passages.

---

[1] Coxe, E. J. : Practical Treatise on Medical Inhalation and its Curative Powers in Bronchitis, Consumption, and other Diseases of the Respiratory Organs. 1841.

[2] Sylvestri, G. : La Pneumojatria ossia l'arte di curare le malattie coi medicamenti sotto forma gazosa, giusta le expe ienze di rinomati medici antichi e moderni. 1851.

[3] Desruez, J. M. R. : Inspirations Pulmonaires ou Fumigations internes ; guide pratique de la méthode thérapeutique. 1854.

[4] Sales-Girons : Thérapeutique Respiratoire : Traité th'orique et pratique des Salles de Respiration nouvelles (à l'eau minérale pulvérisée) dans les établissements thermaux pour le traitement des maladies de poitrine. 1858.

[5] Demarquay : Mémoire sur la pénétration des Liquides Pulvérisés dans les voies, respiratoires et leur application au traitement des maladies des yeux, du pharynx, et du larynx. 1862.

It seems surprising that this should have been doubted, considering that solid particles are constantly introduced in the cases of workmen engaged in certain trades and give rise to serious diseases, e.g., miner's lung, and grinder's disease. Dr. Bergson simplified the apparatus by employing a pair of tubes fixed at right angles, one descending vertically into the liquid to be atomized, the other for driving the air through. Siegle substituted steam for the bellows, and Dr. Beigel added a screen, for which a tube is sometimes substituted, to direct the spray into the mouth.

Methods of Inhaling.

When we wish only to change the conditions under which respiration is carried on, we may to some extent do this by change of air or change of climate, and instances have already been mentioned in which patients have been sent to localities where they could breathe certain emanations from the earth or air laden with artificial impurities. The atmosphere of gas-works, bleaching-works, stables, and other places has often been resorted to by the public.

Attempts have also been made to change the air in a patient's room. The most familiar instances is when we add moisture and raise the temperature. In addition to this, the liberation of elastic gases has also been tried; thus oxygen has been set free in the room in the hope thereby of providing an atmosphere in which respiration should be carried on more efficiently. On the other hand, hydrogen and even carbonic acid have been set free with a view of restraining oxidation. This method is difficult to manage, ventilation is essential, the diffusion of gases rapid, and we have no mode of measuring in what degree such a plan can operate in an ordinary room. Of course, in specially constructed chambers this process might be carried out for short periods, but for a longer time the necessity of ventilation defeats our efforts.

The liberation of irritating gases, such as chlorine or sulphurous acid, is easy enough, as very small quantities speedily make their presence known, and only such small quantities can be tolerated. The fumes obtained from burning nitre papers are rapidly diffused through the apartment, and can be approximately measured by using definite sizes of paper saturated in solutions of known strength. The amount of sulphurous acid liberated may be regulated by burning a known weight of sulphur. Pastilles of definite strength can be made for this purpose. The fumes of tar may be obtained by boiling, or by simply stirring with a hot poker.

Turpentine, by reason of its volatility only requires to be sprinkled about a room to secure a portion of the vapor being perceptible in the air for a considerable period, and it can easily be removed. A solution of carbolic acid may be employed in the same way, but it is less volatile, and when we wish to secure its full effect it is necessary either to vaporize it or to distribute the solution over a considerable surface, which may be accomplished by dipping towels or sheets in a solution of known strength and suspending them as curtains by the doors, and if necessary by the windows or any other openings, so that the entering air must traverse them. This has been done lately at the Victoria Park Hospital, and Dr. Heron calculates (*Lancet*, September 6, 1884) that patients in the inhalation-room pass seven cubic feet of carbolic vapor through their lungs in twenty-four hours. But for a room to be permanently used in this manner, special provision ought to be made for ventilation. Even when a room is only used for a short time, it is not an easy matter to keep up a definite supply of the substance with which the air is to be laden. Dr. Hassall has constructed a special inhalation-chamber at San Remo, in which he has endeavored to accomplish this (*Lancet*, 1884). But it is often inconvenient, or even impossible to resort to a special chamber,

Fig. 1—Bumstead's Vaporizer.

and therefore various modes of evaporating carbolic acid and other substances are employed. Such a vaporizer as is here shown will very quickly fill an ordinary room with the vapor of carbolic acid, eucalyptus, or other volatile antiseptics, and it will keep on evolving small quantities for about three hours without needing a fresh supply. For this purpose the little lamp has only to be trimmed with oil, taking care that the wick is short, as a very small flame is sufficient. Indeed, in the absence of a lamp an ordinary night-light will give off sufficient heat, and too much will be produced if the lamp be filled with spirit or petroleum.

Iodine gives off its vapor at a low temperature, and a little of the element may be placed in a watch-glass, and this floated on a saucer of hot water. Balsamic substances may be thrown upon live coals or heated over a lamp.

In order to diffuse an increased amount of moisture through the air a saucer containing water is often placed on a stove when that method of warming is employed. For an open fire or gas-stove an ordinary croup kettle, which has a long tube to project into the room, may be used.

For the same purpose, as well as to permeate the apartment either with aqueous vapor alone or combined with carbolic acid, a steam-draught inhaler may be employed.

In these cases a supply of steam is diffused through the air; but sometimes instead of vaporizing the water we may simply atomize it, and in this case it carries with it particles of such solids as it may contain in solution. In this way attempts have been made by employing salt to

Fig. 2.—Croup Kettle.

Fig. 3.—Steam-draught Inhaler.

fill the patient's room with a kind of artificial sea air. Any of the larger sized spray-producers will answer for this purpose.

When we do not desire to charge the air with the remedy, but only to supply such an amount as the patient can use for a short time, we have to resort to inhalers.

For the elastic gases a very simple apparatus will suffice; but Barth's oxygen inhaler is the most convenient, and this gas may be produced condensed in strong metallic bottles, from which a given quantity can be let into the inhaler as required. Limosin had an apparatus made to liberate the oxygen at the time of use, and Robbins also contrived an inhaler for this purpose, but I have always used the condensed oxygen. For nitrous oxide Barth's apparatus is also excellent.

For iodine Nelson's inhaler can be used, but it is better to have one of glass, such as is here figured, which will also serve for chlorine.

For steam inhalations a very simple plan is to breathe through a large

FIG. 4.—Iodine Inhaler.

cup sponge which has been dipped in hot water and rapidly squeezed. Children willingly do this. A less efficacious method is to lean over a jug or basin, half filled with hot water. In this case the patient should not envelop the head in a towel, as many often do, for this is likely to produce flushing and headache; it is easy to arrange a towel around the brim of a jug so as to enclose only the mouth and nose. A simple dou-

FIG. 5.—Double-valved Inhaler.

ble-valved inhaler is more convenient than the domestic jug and very often necessary.

When the patient cannot sit up a long elastic tube is necessary, as in

Hunter's inhaler (see Fig. 6), an improvement on which is Maw's double-valved, which provides a separate channel for the expired air.

In Hunter's inhaler the air is made to traverse the water. Bullock's hospital inhaler, and Martindale's portable one are both cheap and effective.

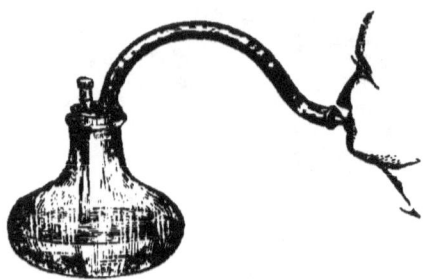

FIG. 6.—Hunter's Inhaler.

The eclectic, which professes to combine the excellencies of several inhalers, is much more expensive, and some patients consider it cumbrous, but it is very complete.

For any patient who can sit up Lee's steam-draught inhaler (Fig. 3) is the least troublesome. It brings a full stream of vapor out at the end of the tube at a temperature fit for breathing, so that no suction is required, and no effort on the part of the patient, who simply sits with his

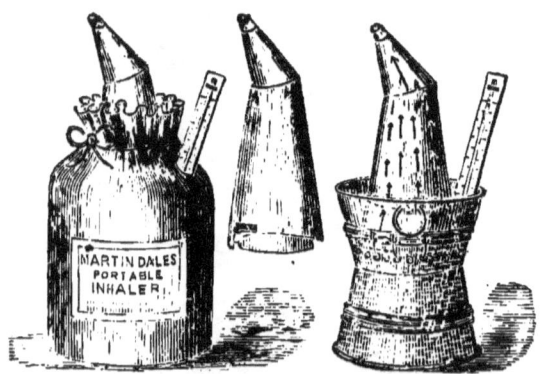

FIG. 7.—Martindale's Portable Inhaler.

mouth near the opening and breathes in the ordinary way. This inhaler is also well adapted for carbolic acid, as Dr. Lee has shown that this vapor is given off with the steam in a constant proportion.

For the inhalation of atomized fluids or sprays a different method is adopted. A pair of Bergson's tubes constitutes the most important part of the atomizer. They may be made of metal, and with a hinge, so as to be easily carried about, as seen in Fig. 9, but they are more fre-

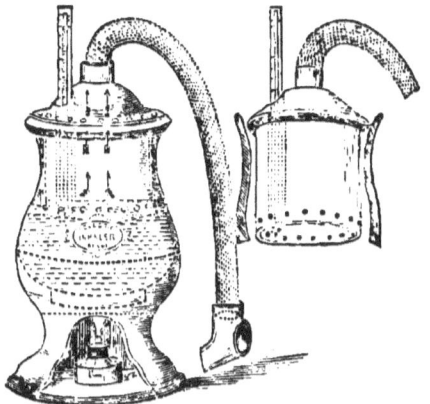

FIG. 8.—The Eclectic Inhaler.

quently made of glass. One of the tubes descends vertically into the bottle containing the liquid for atomization, and a rush of air or steam

FIG 9.—Bergson's Portable Tubes.

through the horizontal tube exhausts its fellow, when the fluid rises and is sent forward in a fine spray, as seen in the following engraving of Bergson's instrument.

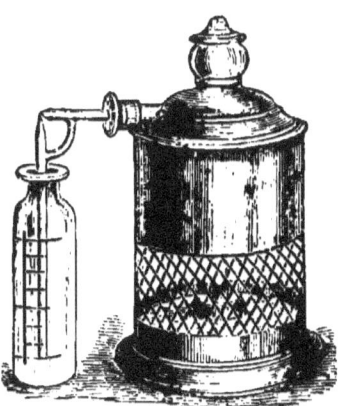

FIG. 10.—Bergson's Atomizer.

288 THERAPEUTICS OF THE RESPIRATORY PASSAGES.

The bottle being held in one hand, the bellows can be worked with the other, or an assistant may do this. Wintrich modified the form of the tubes so that they might be introduced into the mouth, and the spray thus prevented from being projected upon the face.

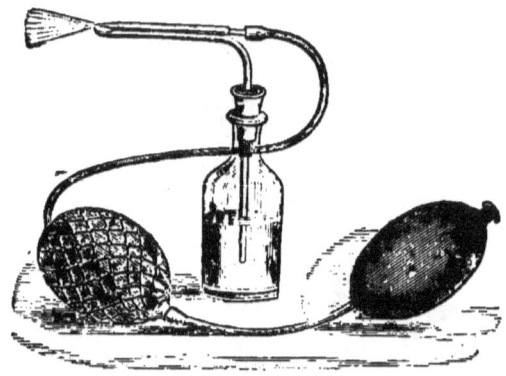

FIG. 11.—Wintrich's Atomizer.

Schnitzler's apparatus is constructed on the same plan with a vulcanite tube protecting the fine points.

Newman's instrument with glass tubes is very much cheaper and quite as effectual. But the cheapest of all, which I have used for many years, and which I have named the "Simple Atomizer," is constructed somewhat differently. The tubes are placed parallel to each other, and the distal end of the upper one is formed into a cup which holds enough liquid for each occasion; the fluid finds its way by gravity to the point, and the

FIG. 12.—Schnitzler's Atomizer.

air is blown through the lower tube by the ordinary double bellows; the bottle is thus dispensed with, and the apparatus simplified. The tubes can be held far enough in the mouth to prevent the spray spreading over the face. This simple atomizer I introduced to the London Medical

Society many years ago, but a very similar form seems also to have been used by Dr. Rumbold, and also by Dr. Solis-Cohen. A somewhat modified form with a larger cup has also been made.

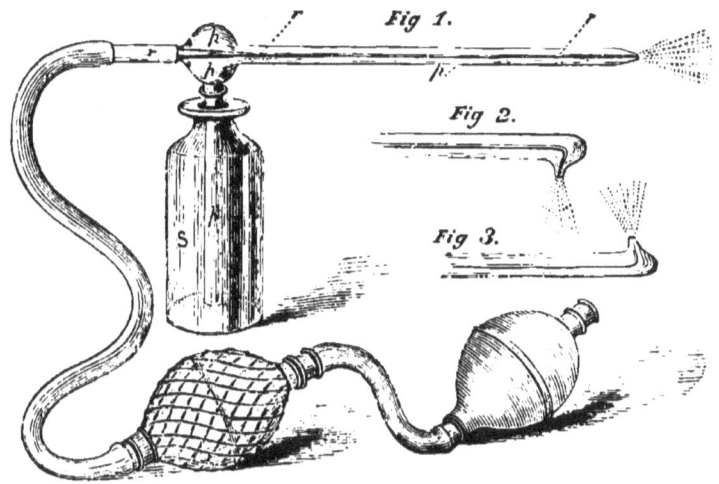

FIG. 13.—Newman's Atomizer.

Instead of the hand-bellows, Dr. Siegle employed steam as the motive power, and the patent which he took out having now expired, steam atomizers are made of various shapes, and at greatly reduced prices. Fig. 16 shows his instrument.

FIG. 14.—Simple Atomizer.

Another apparatus constructed on the same principle, with the addition of a tube to convey the spray to the mouth, is represented in Fig. 17.

To these may be added, when price is not an object, Codman & Shurtleff's Complete Steam Atomizer (Fig. 18), which I have used with satisfaction.

All these methods of employing inhalations only permit us to carry our remedies into the air-passages for a short period, at a sitting. In

Fig. 15.—Simple Atomizer, Large Size.

order to prolong the application, or rather to extend it over a long period, we have to resort to respirators; but here another principle is involved. In using them we dispense with the steam, but we filter the air which the patient breathes, and not only so, but we warm it as the stream of expired air imparts its heat to the respirator, which gives it up again to the inspired current. A simple woollen respirator may easily be knitted of some such pattern as shown in Fig. 19; and if the upper end be left open, cotton-wool soaked in antiseptics can be introduced. Or any

Fig. 16.—Siegle's Atomizer.

of the small respirators sold in shops may be employed, but by far the best is Jeffreys', in which the air traverses carefully arranged strands of gold wire, which takes and gives back the heat from the breath most readily, and maintains a current of air of uniform temperature. In order to raise the temperature still more, Dr. Seaton has proposed (*Lancet*,

1884) to utilize the body heat, by compelling the air to pass through a tube applied to the surface, but this would seem almost necessarily to in-

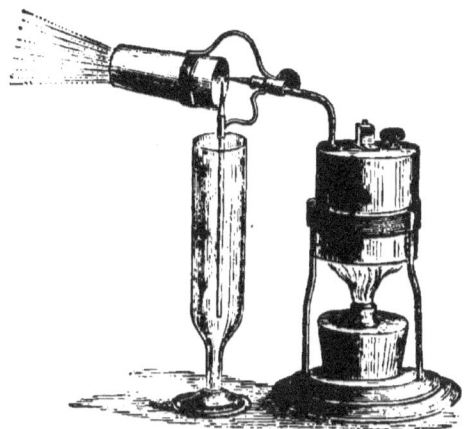

Fig. 17.—Atomizer with Tube.

volve greater labor in breathing, and Jeffreys' respirators have been found to regulate the temperature without much inconvenience. Nevertheless,

Fig. 18.—The Complete Atomizer.

the objection to all respirators is that they may to some extent impede the breathing. Dr. Ramadge, however, deliberately impeded the expira-

Fig. 19.—Knitted Respirator.

tion by inserting a graduated valve for the purpose, and he attached more importance to the alteration of pressure thus brought about, than to the inhalations which he also prescribed. The following engraving shows a combined inhaler-respirator, but quite as frequently, a small tube only is used, with the necessary valves, so as to compel the patient to expire slowly, leaving the respiration free.

Respirators are also used in order to charge the air inspired with antiseptic vapors, and this method of treatment has received a great impetus from Koch's discovery of bacilli in tubercle. If we can secure a portion of the medicament passing into the air-passages, it is easy to suppose that it may produce a favorable influence, even although it is unable

FIG 20.—Inhaler-Respirator.

to destroy bacilli or to repair organic mischief in the lungs. We are not, therefore, shut up to the theory that only germicides can be useful in this way. Dr. Coghill's antiseptic respirator has been largely tried, and he reports (*Brit. Med. Journ.*, 1881) very favorable results from its use. Some modifications of his original form have been made by Dr. Hunter Mackenzie, whose instrument has the disadvantage of being much heavier; and by Dr. Roberts, but they all very much resemble Fig. 21.

Dr. Hassall has investigated the question how far the antiseptic substances usually employed can be recovered from the sponge or cotton-wool with which these inhalers are charged. He found (*Lancet*, 1883) that after one or two hours' inhalation he was able to recover more than four-fifths of the creasote, carbolic acid, or thymol which had been placed in the respirator, from which he concludes that most of the oral and oro-nasal inhalers are of little use, and he has devised others in which the antiseptic substance is distributed over a much larger extent of **surface**.

Our view of the value of antiseptic respirators must depend somewhat on the intention with which we use them. If we wish merely to introduce a certain amount of vapor into the respiratory passages this can be accomplished by very simple means. A glass tube containing some of the substance on cotton-wool or sponge will answer the purpose, and such a simple tube packed with blotting-paper has lately been sold as the "patent pocket inhaler," though why it should be patented must be a mystery to those who have used it many years before; perhaps the patentee relies upon the shape of his blotting-paper. A vapor may be introduced either through the nose or mouth, and a Frankfort physician a few years ago contrived the simplest possible nasal inhaler, the advantage of which he held to be its invisibility. It was, in fact, a small celluloid tube, which was to be passed into the nostril and then a piece of

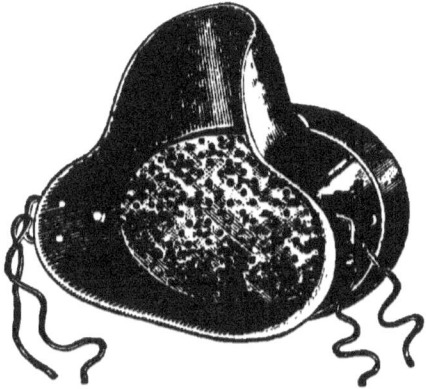

FIG. 21.—The Antiseptic Respirator.

cotton-wool charged with the antiseptic introduced, the tube merely serving to protect the lining of the nostril. Dr. Ward Cousins has lately proposed (*Lancet*, 1884) a somewhat larger tube, the end of which can be placed either in the nostril or mouth and supported by a wire which suspends the inhaler and compresses the nose. He considers it useful not only for introducing antiseptic vapor but as a mechanical method of sustaining forced inspiration. Dr. Burney Yeo recommends a small respirator of perforated zinc (*Brit. Med. Journ.*, 1884).

If we wish to compel a patient to breathe a pure atmosphere and attempt to attain this end by means of a respirator, it is obvious that an oro-nasal one would be preferable and that it must be such an one as does not impede respiration, since it would have to be constantly worn. On the other hand, if the object were only to kill bacilli believed to be present, it would be necessary for the antiseptic to be sufficiently noxious to these organisms and yet not poisonous to the patient, and if such a germicide could be found it need only be breathed for a sufficient time to accomplish this object. Again, if we only wish to act upon the respiratory membrane or the lungs in the same way as when we employ steam inhalations, then it would suffice for the respirator to be charged with a volatile medicament of such a strength as should be found sufficient, and for the instrument to be worn for a short time at suitable intervals. Further, if we abandon these efforts and revert to the use of respirators for the purpose of partially filtering and warming the air, then the instrument must be regarded rather in the light of protective clothing and the times at which it is worn regulated accordingly. Even in this case it is most important that the respirator should not impede inspiration, or the patient will soon give up its use. Heavy instruments are apt to cause a feeling of oppression, even when sufficient provision is made for the passage of air, a point often overlooked. Those which press closely on the face are found to be very irksome, and it is essential that a patient should be able to take sufficient exercise without experiencing any difficulty or oppression of his breathing. It seems obvious, therefore, that the simpler and lighter the instrument the better. Those which are complex and cumbrous are sure to be discarded.

Uses of Inhalations and Other Topical Pneumatics.

Though we have considered inhalations as topical remedies for the entire respiratory passages, it must not be forgotten that the absorptive power of the membrane is such that many of them become the most potent and rapid of general remedies. It is, therefore, only as a matter of convenience that they have been included in this chapter. The anæsthetics and nitrite of amyl are examples of this. Then, again, the inhalation of oxygen can only be regarded as a general, not a local remedy. But in most other cases the topical effect is so important as to obtain the first consideration.

Steam inhalations may be considered first. The contact of simple watery vapor with inflamed mucous membrane is so grateful as to deserve to be called anodyne. It cuts short the congestive stage of catarrh by supplying moisture, and so relieving the dryness, heat, and oppression, and at a later stage it dilutes and assists the removal of the secretion. In acute bronchitis the inhalation of steam is often of the greatest service, and the vapor may be made the purveyor of other volatile remedies, either anodyne or stimulant. Much of the good effect of many inhalations may be traced to the influence of the steam. In croup and diphtheria, great good arises from keeping the atmosphere of the sickroom saturated with steam. In the first edition of my "Sore Throat" I insisted on the value of a more thorough and systematic use of steam in these diseases, and have been advocating it ever since. During the last few years it has also been recommended by Oertel and other German writers.

Any volatile substance can be easily employed in the form of inhalation, and those herbs, the therapeutic value of which depends on an aromatic volatile principle, are often thus used; or this principle is previously extracted, as in the case of essential oils. The soothing properties of steam are often increased by employing hops—the vapor of the freshly-made infusion being charged with the sedative properties of the plant; the oil of hops is too irritating to substitute for the crude drug. Chamomile flowers may be used in the same way. Another aromatic sedative is obtained by putting a teaspoonful of tinct. benz. comp. into the inhaler, with a pint of hot water. Vapor coniæ is more distinctly sedative—its efficacy depending on the conia being set free by the alkali, for which reason it should be added at the time. The succus conii is to be preferred to the extract, as previously pointed out. The vapor acidi hydrocyanici is employed with cold water, but may be ranked among sedatives. The volatile parts of opium can be utilized by putting the tincture or the solid drug in the inhaler with hot water. Ether and chloroform may be used with water at a low temperature. Conium or opium, in combination with chloroform and similar compound inhalations will often prove of service when a simple one fails. It is, however, obvious that very volatile substances may be as well administered without the medium of water. A very simple inhaler, or a little cotton-wool in a cone of paper, will suffice for chloroform, ether, and nitrite of amyl.

Besides anodyne inhalations, those possessing stimulant properties are most important. In the British Pharmacopœia there are only three —vapor chlori, creasoti, and iodi. The mode of using differs in each case, while each vapor is a special stimulant. For vapor chlori cold water is used; for vapor iodi the water is heated after the addition of the tincture, but the other plan is quite effectual. Camphor is a good stimulant for inhalation. Ten drops of the spirit may be put into the inhaler to begin with: it is better to dilute it with more spirit. Camphor is also a very useful addition to many other inhalations. In like manner most of the essential oils can be used as stimulants. They can be dissolved in spirit or diffused through water by means of magnesia, or powdered silex, in the manner often resorted to for making medicinal waters. The oils of aniseed, cajeput, cloves, cinnamon, marjoram, myrtle, rosemary, and others are adapted for this purpose. Ammonia is a general stimulant often resorted to, and may be utilized for its local effect.

When the effect of steam is not needed, we may try the plan recommended by the late Dr. J. A. Symonds,[1] viz.: inhaling solutions of balsams in ether, or pyro-acetic spirit from an ordinary wide-mouthed

---

[1] British Medical Journal, 1868.

bottle, the warmth of the hand holding it being quite sufficient to volatilize the liquid. I prefer ether to the pyro-acetic spirit. Spirit of chloroform may also be used by this method. Half an ounce of benzoic acid in an ounce of ether forms a standard solution, to which two drachms of balsam of Peru, or of Tolu, or of any similar substance, may be added. Turpentine has also been used in this way, but might then be made the menstruum. Other rather volatile drugs, such as creasote, carbolic acid, iodine, the essential oils, etc., can, if desired, be employed in this manner without steam, but very often the warm vapor is an important element in the treatment.

Sometimes warm inhalations are not desirable. We can then use sprays or atomized fluids. Sprays are generally useful when it is an advantage to administer the inhalations at a low temperature; while for warm applications I mostly resort to the mode of inhaling steam impregnated with the remedy. It is usually advisable to administer astringent sprays cold, though, of course, they may be used warm. Anodynes are applicable either way, but more frequently should be taken warm. After warm inhalations it is often desirable for the patient not to go into the open air, or into a cold room; but the cold spray is the best possible preparation for such sudden changes of temperature.

The remedies most commonly used in the atomizer are solutions of metallic salts. As astringents, sulphate and chloride of zinc, alum, perchloride of iron, and sulphate of iron. The strength of the solutions of these salts may vary from two to ten grains in the ounce, or more, according to circumstances. Permanganate of potash, one to five grains per ounce, is disinfectant and stimulant, and in some cases exercises a most happy influence on mucous membranes. In other cases calx chlorata, two to five grains, is preferable; or we may use the liq. calcis chlorate, or the liq. sodæ chlor., ten to sixty minims, or liq. chlori., two to thirty minims per ounce. Carbolic acid, one to five grains, is a good stimulant, particularly when the membrane is unusually dry. This remedy is also much used as an antiseptic and disinfectant in various conditions of the membrane. A much weaker solution thus is required, for this also exercises a soothing influence in cases of great irritability. It is not to be forgotten that carbolic acid is readily absorbed by mucous membranes. Lactic acid has been strongly recommended as a solvent of the false membrane of diphtheria. About half a drachm in the ounce will be strong enough generally, but I have had occasion to use it much more concentrated. Nitrate of alumina (two to five grains) was tried by the late Dr. Beigel,[1] who also used many other substances, including acetate of lead, chloride of sodium, and even cod-liver oil. I have used chloride of aluminium rather freely as an astringent.

---

[1] On Inhalation. London, 1866.

As an anæsthetic bromide of potassium has been used, ten to fifteen grains; but weaker solutions are serviceable for other purposes. Solutions of the alkaline carbonates and their salts form a very useful series. Thus, the carbonates of soda, potash, or lithia may be tried, two to ten grains. Muriate of ammonia, eight to fifteen, is reputed to possess peculiar effects on the faucial membrane. The local influence of chlorate of potash, now fully recognized, may be obtained in this way, two to ten grains. A number of mineral waters are also used as sprays. Those containing sulphur, or chloride of sodium are most in repute. Corrosive sublimate is sometimes used in specific cases, but such remedies should only be used with circumspection.

Demarquay recommends glycerine to soothe an irritated pharynx, and the late Dr. Scott Alison employed this fluid in laryngitis and tracheitis. Laudanum and solution of the salts of morphia can be used in the form of spray, five minims to twenty at a time, properly diluted; so can the tinctures of hyoscyamus, conium, belladonna, etc. In asthma, success seems to have followed Fowler's solution, five minims at a time, administered in this manner. Recently Professor Sée has recommended sprays of iodide of potassium in this disease. Other uses for this drug will occur to the reader, and solutions of iodine have also been used. Sprays of vin. ipecac. have of late years come into use and sometimes seem very efficient, but to many patients they are extremely disagreeable. Sulphurous acid is generally prescribed too diluted. It may be employed pure, as advised by Dewar[1] and Pairman,[2] or it may be diluted with one, two, or three parts of water. It should be recently prepared. Tannin, as an astringent, varies much in the dose, and is often given too weak to be effectual. According to the effect required the strength may vary from a single grain per ounce to fifteen or more.

Iodoform has been used as a spray as well as in other modes. Professor Sormani contrived a special apparatus for the purpose of experimenting upon animals, and by means of compressed air he forced an ethereal solution of iodoform deep into the lungs, in order to determine its power as an antiparasitic in phthisis. That the remedy was thus carried into the air-cells he proved experimentally; he has also reported three cases of phthisis treated by iodoform sprays. The patients gained in weight and some of the symptoms were relieved but the expectoration continued to contain bacilli, though their number diminished. The expectoration injected into guinea-pigs was still infective, but seemed to be rather less virulent. Other attempts are being made to bring antiseptics to bear in the treatment of lung diseases by inhalation.

---

[1] On the Application of Sulphurous Acid to the Prevention and Cure of Contagious Diseases. 1867.

[2] The Great Sulphur Cure brought to the Test. 1868.

So far we have been considering those topical remedies which operate on the whole of the respiratory passages at once. It may, of course, be desirable, even for a disease which only involves a portion of the mucous membrane, to subject it all to the action of remedies; but there are other cases in which we may wish to restrict the extent of the application. And then, again, there are remedies which only affect a small portion of the surface, and some of which we must now notice.

Sprays may very easily be localized and for this reason may claim a place between vapors and liquids; but enough has been said about them and we pass on to the latter, beginning with

Gargles.—This form of local remedy is very ancient and was freely used by Hippocrates. It is still popular, though considerable discussion has at times taken place as to its value. While some physicians have almost excluded gargles from their practice, others have relied upon them to a great extent. The former have maintained that they never come in contact with more than the anterior surface of the velum and uvula, and perhaps a portion of the tonsils. The latter have endeavored to show that they penetrate much farther. Even were the first allegation correct there would still be a use for gargles, but it is now generally abandoned.

These diverse views have no doubt partly depended on the inclusion of several distinct acts in the term gargling. A mere mouth-wash may be so employed that the anterior surface of the velum is subjected to its influence. The word gargling, however, is generally understood to imply that the air is to be expelled through the liquid with sufficient force to make the bubbling[1] noise which many seem to consider so essential to the process. But a moment's thought will suffice to show that the liquid may be permitted to remain in the position it occupies as long as the breath can be held and, further, that expiration may be carried on so gently as to prevent any bubbling noise being heard. If now any attempt be made either to swallow or to inspire to a very slight degree, the liquid may pass farther and yet be arrested before it sets up any spasmodic action. The act of gargling may be well studied in conjunction with that of swallowing, and the reader will no doubt be aware of the diversities of opinion that have prevailed respecting the physiology of deglutition. We all know that soft bodies produce little irritation in the larynx compared with hard ones, though, as previously pointed out, a drop or two of liquid coming unexpectedly on certain parts may at once set up spasm.

It is believed by many that the use of the epiglottis in closing the air-

---

[1] Gargle and gurgle were originally the same word—both forms are derived from the Latin *gurgulio*, throat, and this doubtless from the older Greek, γαργαρεών. The reduplicated syllable *gargar* also appears with the same meaning in Hebrew and other oriental languages.

passage during deglutition has been somewhat exaggerated. Certainly I have met with many cases in which great destruction of the epiglottis had occurred—some in which it had been entirely destroyed—without the power of swallowing being greatly affected. On the other hand, we constantly see swelling and ulceration of the epiglottis associated with pain and difficulty in deglutition.

M. Krishaber having masticated and insalivated a little bread crumb, pushed it with his finger over the edge of the epiglottis, and then by an inspiration drew it into the air-passage, expelling it again by a sudden forcible expiration. M. Guinier, of Montpellier,[1] had previously observed on himself, by means of the mirror, that such a morsel of soft bread could come upon the closed glottis without causing any uneasiness. The last-named author subsequently attained such command over the parts as to let liquids enter the larynx, and founded upon his experience the plan of laryngeal gargling.[2] We must remember, however, that the presence of the mirror in the throat during an effort to swallow completely changes the conditions to be studied. In the natural act of deglutition there can be little doubt that the air-passage is generally closed. The impossibility of breathing during that act, and the spasm excited by the entrance of small quantities of the food or drink, seem to show this. Besides, if we eat any substance that will impart a distinct color to the membrane over which it passes, and then practise autolaryngoscopy, we find that it discolors the anterior surface of the epiglottis, but not the posterior, still less the mucous membrane of the vestibule of the larynx.

From what has preceded it will be readily understood that the act of gargling may vary very much with the individual. Accordingly, we find that some persons never learn to gargle properly, while others achieve what at first seems impossible. Singers generally acquire considerable control over the parts, while those who have never learned to gargle sometimes find it no easy task. The majority of patients, in point of fact, require some instruction respecting the end to be attained. In using a mouth-wash the velum and uvula descend so as to cut off all communication with the pharynx. In gargling, as frequently understood, the result may be almost the same, the liquid coming in contact with the anterior surface of the velum and uvula only. If, however, the patient should now raise the velum, as many can do, some of the liquid will flow into the pharynx. There it produces a desire to swallow, and unless the patient can control this some of it will pass into the œsophagus and stomach. Some persons, however, instead of yielding to the desire to swallow, suddenly jerk the head forward, and a quick, forcible expiration taking place

---

[1] Nouvelles Expériences sur la Déglutition faites au moyen 'de l'Auto-Laryngoscope (L'Union Méd., 1865).
[2] Étude du Gargarism Laryngien. Paris, 1868.

at the same moment the liquid is expelled through the nose. What they do involuntarily can be accomplished by others deliberately and without inconvenience. In persons possessing this control over the parts we may often obtain good results from a natural nasal douche thus employed.

Is is by no means so easy to let the liquid enter the larynx, and, in spite of persistent efforts, many will totally fail in the attempt. Of course, the glottis must be kept closed if the fluid is to rest upon it, and therefore the duration can only be while the patient can hold his breath. M. Guinier, who has demonstrated his method with the laryngoscope, says that the head should not be thrown back, as the less it is raised the less urgent is the desire to swallow. He also advises that the mouth should not be quite closed. With these precautions he takes the liquid into the mouth, brings forward the lower jaw, and closes the glottis by the uncompleted act of emitting a vowel sound. The volum in this disposition of the parts is raised, and the base of the tongue perhaps falls a little, so that the liquid finds its way into the larynx, where, if the patient can completely control the sensibility, it may remain as long as the breath is held. The slightest attempt to inspire will bring on spasmodic cough. Only the few can expect to attain success in this method, and its use is therefore very restricted, especially when we remember that there are other modes of applying liquids to the laryngeal mucous membrane.

In gargling it is more important to manage the respiration than deglutition. If compelled to swallow, the patient merely receives a little of the fluid in his stomach, and, unless the gargle should contain some noxious ingredient, there is an end of the incident. On the other hand, entrance of the fluid into the air-passages may give rise to severe laryngeal spasm.

Astringents, disinfectants, and antiseptics are the remedies most frequently required in this form, but anodynes may be utilized in the same way. One of the best gargles is a solution of alum, the strength of which may be varied according to the effect required. Chloride or nitrate of aluminium may be employed for the same purposes. For a powerful astringent tannin may also be used—one or two drachms in half a pint of water, to which a drachm of rectified spirit or an ounce of glycerine has been added. Borax and chlorate of potash are also useful as both gargles and mouth-washes. The former is slightly alkaline as well as astringent; the latter possesses special value in an aphthous condition of the buccal and faucial mucous membrane, and is often advantageously combined with decoction of bark. As a disinfecting wash and gargle Condy's fluid is the best, but chlorated gargles are also valuable. Carbolic acid is a good stimulant to the pharyngeal mucous membrane, but its flavor is very disagreeable to some patients. Weak gargles of carbolic acid are not merely slight stimulants, but exercise a soothing influence over irritable mucous membrane. The sulpho-carbolates are also useful. I have had excellent results from lactic acid, both as a gargle and in the form of spray. Three or four drachms or more may be diluted with eight or ten ounces of water. This remedy is of special value in diphtheria. The lactates and lime-water are also used in the same disease. Mineral acids ought not to be employed as gargles, as they destroy the patient's teeth, and less injurious substances are equally or more efficacious.

It is obvious that it is not desirable to order gargles for children who have not learned to use them, or for persons whose fauces are so inflamed as to make all movements of these parts painful.

Gargles are usually employed cold, but occasionally—especially when anodyne—they are ordered warm.

The *nasal douche* may be mentioned here, inasmuch as it is in one sense supplementary to gargling. By it the fluid is brought into contact with a portion of mucous membrane which, in the majority of persons, is otherwise inaccessible to local treatment such as the patient can employ.

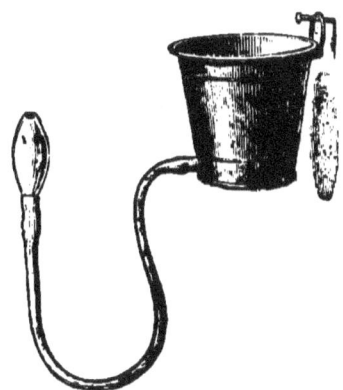

FIG. 22.—Siphon Douche.

I usually recommend the simple siphon douche, Fig. 22. The nasal douche is, of course, most serviceable when the disease is located in the posterior nares,[1] but is also of value in some laryngeal cases. The liquids most serviceable for the nasal douche are weak saline and alkaline solutions. Half a drachm of carbonate of soda in a pint of water is one of the best forms, being both safer and more agreeable than the chloride of sodium so often prescribed. By this the mucous membrane may always be thoroughly freed from the discharges, and thus prepared for other remedies. Astringents may be afterward resorted to in many cases, the chlorides of aluminium, zinc, and iron being employed *in very small quantity*. The sulpho-carbolate of zinc is a good astringent. The permanganates are still more useful, possessing local action and being deodorizers. Chloride of lime I have also found very valuable. Carbolic acid has also most valuable properties and is a recognized disinfectant. The nasal douche should be used tepid at first in all cases. Occasionally it may be desirable to decrease the temperature, especially when astringents are employed. A temperature of 85° Fahr. may then be employed, and gradually reduced to 75°, or even 70°. Very rarely do we descend to the ordinary cold water. The patient should breathe

---

[1] See the author's paper on Ozæna, read before the Medical Society of London, October 20, 1871 (Medical Press and Circular, December 6, 1871). Reference may also be made to the author's contribution to the same journal on Syphilitic Diseases of the Throat (January 10 and February 28, 1872).

quietly and restrain any movement of deglutition, in order that the Eustachian tubes may remain closed. It can no longer be doubted that in using a nasal douche the accidental entrance of fluid into the ear has given rise to serious results. It is not necessary to have much pressure—just enough to cause the fluid to flow through the nares will suffice, and this will be produced with the siphon douche, when the vessel is not raised much above the patient's head. It is a good plan for the patient at first to hold it himself—he will scarcely then be able to lift it too high. The soda solution above named is the best for removing hardened secretions. I have also found it less irritant than the salt solution so constantly used, and Weber-Liel[1] has shown that it is less likely to injure the ear—a conclusion quite in accordance with my experience. A weak saline produces less stuffiness than pure water. In using astringents care should be observed to use them weak enough. Half a grain per ounce of sulphate or acetate of zinc is quite enough. More is apt to give rise to headache, local irritation, and other unpleasantness. Alum is said to have impaired the function of smell. The douche should be used once or twice a day—seldom oftener, but it is necessary in many cases to continue it for a long time.

Irrigations.—I have designed tubes with perforated extremities, by means of which the nasal passage can be carefully irrigated by the physician, both from the front and back. In this way we may apply fluid remedies which should not be entrusted to the patient.

Syringes and atomizers constructed for the purpose can also be used to convey medicated fluids to the nasal passages. Besides these methods we may sometimes avail ourselves of the more primitive plan of snuffing up the liquid by a series of sudden short inspirations. Fluids may thus be taken up from the hollow of the hand or from a convenient vessel. Some persons acquire a trick of this kind, and call it "drinking through the nose."

The Pharyngeal Douche and Irrigator.—It is easy to direct a stream of liquid on the posterior wall of the pharynx. For this purpose I have arranged mouth-pieces which can be used in place of the nose-piece of the nasal douche. Some patients learn to employ this method themselves. The liquids are such as have been recommended for the nasal douche, and for gargles, but their strength should be between these two.

My pharyngeal *irrigator*, like its nasal fellow, consists of a properly shaped silver tube perforated at the extremity by numerous minute holes, so that the liquid can be projected in a fine shower by means of a small Higginson's syringe attached to it. The *atomizer* may be resorted to when it is desirable to break up the fluid into a fine spray, but both the pharyngeal and nasal irrigators will often be found of service.

---

[1] Berliner klin. Woch., April 1, 1878.

Linctus.—In order to somewhat prolong the local action of medicines which are to be swallowed they may be rendered thicker and more glutinous by sugar and other additions. We then have the linctus, loch or lohoch, called also eclegma, eclectos, ecleitos, elegma, and illinctus—various terms derived from ἐκλείχω. Various syrups and mucilages may be used for the same qualities, or may enter into the composition of the linctus. Glycerine, from its slowness to evaporate, may be used to attain the same end, and is, indeed, rather too popular, for since the glycerine of tannin and borax were introduced to the Pharmacopœia their use has become quite an abuse.

Solids.

As sprays take an intermediate place between vapors and liquids so between these and solids we might place *confections* or *electuaries*, in which we have a tenacious semi-solid substance, which may be slowly dissolved in the mouth, and thus the local action of its ingredients prolonged; though these preparations are also used as vehicles for systemic remedies. In confectio opii the galena, mithridate, philonium, and theriaca of antique pharmacy survive.

Lozenges, being completely solid, are slower of solution, and therefore enable the medicament to exercise a much more prolonged influence in the mouth and fauces, and have always been popular remedies. The Greek hypoglottides are represented in our lozenges. Although the London and Dublin Pharmacopœias omitted them, the Edinburgh retained them, and the British has restored them. In doing so, however, the lozenge has been made as much use of as a vehicle as for local effects. Thus some officinal lozenges (as morphia and iron) are chiefly to be regarded as dosed, general remedies, though others (as tannin and chlorate of potash) are valuable for their topical influence. It is obvious that the two qualities may often be combined.

For special topical use lozenges should possess the following qualities: 1. They should dissolve slowly in the mouth, so that the resulting solution of the medicament may remain as long as possible in contact with the mucous membrane. 2. They should possess a certain degree of softness, so as not to hurt the diseased surface mechanically. 3. For the same reason their shape should be without corners. 4. Their flavor should be agreable, or as little distasteful as possible. 5. They should keep without change for an indefinite period, as they cannot be advantageously made in small quantities.

The lozenges of the British Pharmacopœia are most defective, on account of their hardness. They irritate the mucous surface; and the sharp corners of some shapes in common use, or of the broken pieces of others, may enlarge ulcers, tear congested membrane, or do other injury. Of course, when used for their constitutional effect these objections may scarcely apply. A softer consistence has been attained by the employment of fruit-paste—as in the favorite black-currant lozenges; and this substance has been more extensively used of late years. Extract of liquorice, as in "Pontefract cakes" and gelatine, have also been utilized. A more recent innovation is the effervescent base introduced by Mr. Cooper, which, for some purposes, is of special value. The French, so famous for all kinds of confectionery, have given us the *pâte de Guimauve;* but the defect of this is that it does not keep well. We owe to them also our best jujubes, a sweetmeat first made with the juice of the *Rhamnus zizyphus,* but now never containing that agreeable fruit. Experimental experience, extending over more than a quarter of a century, leads me to conclude that a *pâte de jujube* of the best French method of manufacture will be found most generally useful as a base. It fulfils all the indications required; it can be variously flavored and colored, divided into lozenges of any size or shape, and medicated with the most suitable remedies. It does not excite nausea or cause indigestion, and does not change too much after months of exposure. It is, therefore, adapted for lozenges prescribed for their topical influence, and is equally available for those given for their effects on the system.

Lozenges are more extensively used than could have been supposed when the London and Dublin Pharmacopœias rejected them. Every one who remembers that time will know that, in spite of that discouragement, every large pharmacy was obliged to keep a considerable number. A list of upward of one hundred and fifty tried formulæ in use at that date lies before me. It comprises nearly all in common use now. Rhatany, an excellent astringent, still extensively prescribed for local purposes, is in that list, and was known long before. So with cubeb lozenges, which have lately been forced into extensive sale by a vendor who vaunts them as "bronchial troches." But this is according to the common practice of quacks, who take some of our tried formulæ and advertise them as their own discoveries. We have, in fact, few new lozenges. Red gum has been introduced; so, too, has carbolic acid; chlorodyne can scarcely be counted, being only morphia disguised; superior glycerine jujubes may be had at any leading pharmacy, or of inferior quality as an advertised panacea. The lozenges comprised in this long list may be classified according to their therapeutical uses, *e.g.,* astringents, demulcents, sedatives, special stimulants, etc. The lozenges made from my formulæ by Messrs. Allen & Hanburys, have been generally pronounced a distinct advance in medicated lozenges, since after a long trial they were exhibited at the Cambridge meeting of the British Medical Association a few years ago.

As the words *trochisci* and *tabellæ* have become associated with the harder lozenges, as jujubes seem to savor too much of sweetmeats, and as these are distinctly medicinal agents, they were named pastils, an old English word more familiar in the French *pastilles*, and derived from the Latin *pastillus*, which was used by Celsus for such a purpose; *pastilli* will, therefore, be an appropriate name in prescriptions.

With regard to dosage, those pastilles which are intended to replace the British Pharmacopœia lozenges have been made of similar strength, as it was considered advisable not to burden the prescriber's memory too much. This is specially the case with the pastilles of morphia, of morphia and ipecacuanha, and of opium; in each of these the pastille may be regarded as an agreeable substitute for the lozenge. So, too, with the simple ipecacuanha pastille, which will be found much more popular with children than the lozenge. The same remark applies to pastilli ferri. Each pastillus aconiti may be considered equivalent to half a minim of British Pharmacopœia tincture, and prescribed accordingly. The pastillus expectorans, or morphiæ et ipecacuanhæ compositus, is a combination of the simple one with other expectorants, and will be found most serviceable in bronchitis, chronic coughs, etc. The chlorate of potash pastilles are not so strong as the lozenges, and may be taken in twice the usual doses; they are, however, very efficacious, and the disagreeable flavor is so successfully concealed that few can detect it. If large quantities are needed, other modes of administration may be tried, such, *e.g.*, as the compressed tablets. The pastillus sodæ chloratis I introduced as an efficacious and pleasant substitute for the potash salt. The lithia pastille contains a grain of the carbonate, and is valuable for both its local and remote effects. The benzoated pastille will be found the most agreeable of all mild voice-lozenges, and may be taken shortly before speaking, reading, singing, preaching, etc., to give tone to the vocal apparatus. In obstinate or chronic cases the camphorated pastille is a still more powerful voice-lozenge, but, unfortunately, its flavor is not nearly so agreeable. This is, in fact, the only one of the series that can be considered unpalatable.

Powders.—Solids reduced to fine powder may be applied to various parts of the upper portion of the respiratory passages by means of insufflators, or by still simpler forms of apparatus. A tube of convenient shape, with a common puff-ball attached, is an efficient instrument, only requiring a little management. Such a tube may be of silver, vulcanite, or glass, according to its destined use. A little more complicated contrivance is the wash-bottle we have all used in the laboratory. Any one who can bend a glass tube over a lamp can himself make such an instrument of any size he likes. A one-ounce or two-ounce phial will be convenient, and the distal end of the tube should be drawn to a smooth point, as in (Fig. 23). A puff-ball, or the bellows of the atomizer completes the apparatus, and when a little powder is put into the bottle it affords a simple means of projecting some of the drug on any easily accessible point of the surface.

The therapeutic action of powders thus used is somewhat complex. An insoluble substance applied in this manner to a healthy portion of mucous membrane provokes some irritation and increased secretion; the mucus thrown out envelops the powder and it is soon removed. In morbid conditions the result will be modified ; soluble powders will be to some extent dissolved in the secretion, and thus we have superadded the effects of a solution of the substance. Besides the shock of the impact, which in some parts is particularly to be considered, and the irritation

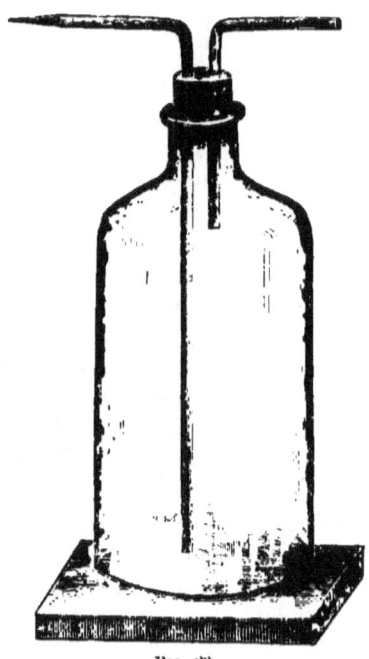

Fig. 23.

occasioned by a cloud of powder when that passes into the larynx, or nares, we have to consider the therapeutical qualities of the remedy—astringent, absorbent, anodyne, stimulant, caustic, special stimulant, etc. This is not a suitable way of applying caustics, but the other classes are well represented. Tannin and gallic acid are favorite astringents ; bismuth and zinc oxide, absorbents ; morphia is perhaps the most frequently used anodyne. Calomel, chlorate of potash, borax, alum, and many other powders may be thus employed. Excellent results are obtainable from the use of iodoform by this method. This and other drugs are often mixed with inert powders with advantage. Starch answers well, as it adheres easily to the moist surface. The author is desirous of again urging the value of iodoform by this method. He believes he was the first to employ it thus, and his experience proves it to be most valuable. Those who have watched the remarkable influence of this remedy when dusted on other surfaces will not be surprised at its effect here. In regard to powders used in this way it must not be forgotten that the remedy finds its way to the stomach, and therefore its action on that organ and on the system must not be forgotten. When it is desirable to apply powders to the interior of the larynx a well-made insufflator should be used, and the application is to be made with the aid of the laryngoscope.

Caustics.—Sometimes it is desirable to use solid caustics in the mouth or throat. Here aluminium wire or glass rod, with a little of the caustic fused on it, may be used. An instrument of this kind is much safer than ordinary caustic-holders, which have given rise to accidents, from the caustic breaking off and being swallowed. It should be added that caustics are often resorted to unnecessarily, and therefore not seldom injuriously.

INDEX.

Acids which best serve in artificial digestion, 34
Aconite, accidental poisoning, 178
  and veratria as depressants of the respiratory centre, 274
  contrasted with veratrum in effects, 182
  effect of medicinal doses, 179, 183
  effect on dogs, 178
  effect on secretion, 182
  effect on the alimentary system, 183
  effect on the circulation, 180
  effect on the muscular system, 181
  effect on the nervous system, 181
  effect on the respiratory system, 181
  history of its use, 177
  in acute bronchitis, 184
  in acute pneumonia, 185
  in asthma, 185
  in catarrh of the upper air-passages, 184
  in pleurisy, 185
  in tonsillitis, 184
  overdose of, 180
  tincture of the root, 185
  various methods of administering, 184
Aids to digestion, 74
  ancient use of liquids taken from the stomachs of animals, 74
  extract of malt, 74
  pepsin and pancreatin, 75
Albuminate of mercury fatal to bacteria, 250
Albuminates, 22
Alcohol, 97
  administered with food, 106
  amount a strong man can oxidize, 99
  as an antipyretic, 100-104
  as a diaphoretic, 103-134
  as a nutrient, 97
  effect in the stomach, 101
  effect on temperature, 103
  effect on the blood, 101
  effect on the cutaneous system, 103
  effect on the general system, 98
  effect on the heart and vessels, 101
  effect on the liver, 102
  effect on the mucous membrane, 100

Alcohol, effect on the nervous system, 102
  effect on the respiratory system, 102
  in excessive quantity, 100
  in phthisis, 105
  in pneumonia, 105
  therapeutic uses, 103
  time required for elimination, 100
Aliment unsuitable for the tissues, 9
Aliments as remedies, 44
Alkalies to promote bronchial secretion, 239
Alkaloid aconitia, 186
Alterant expectorants, 239
Ammonia and its carbonate as stimulant expectorants, 224
  a stimulant of the respiratory centre, 265
  as a volatile stimulant, 224
Ammoniacum, as a stimulant expectorant, 234
  effect on the respiratory membrane, 234
  history of, 234.
Ammonium chloride, effect on the system of, 242
  history of, 241
Amounts of digestive fluids secreted daily, 33
Amyl nitrite, as an inhalation, 256
  effect on the circulation, 257
Anæmia, 60
Analeptics, 48
Animal and vegetable diet, 26
Anodyne inhalations, 295
Antagonism of neurotics, 205
Antimony in respiratory diseases, 129-132
  in pneumonia, bronchitis, and croup, 132
  its effect on the system, 131
Antiphlogistic depletion as a remedy, 109
  diet and regimen, 108
  use of mercury, 125
Antiphlogistics, 107
Antipyretic, quinine, 156
  treatment in pneumonia, 150-153
Antipyretics, 142
  cold affusions, cold packs, cold sponging, 154

Antipyretics, cold first of the direct, 142
Antiseptic action of quinine, 159
and disinfectant pneumatics, 249
expectorants, 252
remedies, uses in respiratory diseases, 250
Antiseptics, history of, 249
through inhalation, 297
Antispasmodic pneumatics, 255
pneumatics contrasted with expectorants, 255
Apnœa, 20
Apomorphia, as an expectorant, 220-223
as a spray in phthisis, 221
injected into the veins of animals, 129
in respiratory diseases, 129
Arsenic, as an anti-malarial, 160
Arteriotomy, 110
Assafœtida, as an expectorant, 235
effect on the bronchial secretions, 236
effect on the system, 235
Assimilation of food, 4
Asthma, relief from nitrites, ethyl-iodide, etc., 259
Astringents, as contra-expectorants, 263
used in atomizers, 296
Atomizer with tube, 291
Atomizers, illustrated, 288
solutions of metallic salts in, 296
Atropia and aconite, antagonism of, 204
and bromal, antagonism of, 204
administered for opium-poisoning, 193
counter-poison to morphia, 193
and physostigma, antagonism of, 203
and pilocarpine, antagonism of, 203
as an expectorant, 221
and prussic acid, antagonism of, 204
effect in phthisis, 200
effect on circulatory system, 195
effect on the gland-cells, 198
elimination of, 196
with iodine, 246

BACTERIUM subtile, 251
Balance between food supply and power of the system, 44
Balm of Gilead, 234
Balsam of Tolu in chronic coughs, 232
Balsams as expectorants, 231
Barth's oxygen inhaler, 284
Baths for typhoid, 148
quinine and salicylate of soda as antipyretics, 143
Bdellium, 234
Beef-tea as a nutrient, 46
Belladonna and aconite in laryngeal catarrh, 199
and atropia, 193
and atropia as stimulants of the respiratory centre, 265
as an antispasmodic, 197
as a respiratory and cardiac stimulant, 198
effect in asthma, 199
effect in diphtheria, 200

Belladonna, effect in influenza or catarrhal fever, 199
effect in laryngismus, 200
effect of excessive doses, 194
effect on the eye, 196
effect on the glandular system, 196
effect on the muscular system, 196
effect on the respiration, 195
effect on temperature, 196
moderate doses of, 194
or atropia, effect on the nervous system, 195
or atropine to relieve pain, 197
with opium, 201
Benzoin, effect on the mucous membrane, 232
Bergson's atomizer, 287
Beverages, 89
fermented, tea and its allies, 90
simple diluents, mucilaginous, saccharine, liquid foods, aërated drinks, acidulous drinks, salines, medicinal beverages, 89
Bile, 37
as an aid to digestion, 76
as a medicinal agent, 76
Bleeding carried to excess, 113
history of, 109
for pneumonia, 115
for undue blood-pressure in acute diseases, 116
in distention of the right heart, 116
in respiratory diseases, 116
present view of it as a remedy, 115
reaction to stimulation, 113
Blisters in chronic synovitis and acute rheumatism, 119
in thoracic diseases, 118
pustulants, issues and moxa, 118
Blood too oxygenated, 20
too venous, 20
Bromides as antispasmodics, 269
depress the respiratory centre, 268
effect on the respiratory system, 269
elimination of, 268
in combination with other medicines, 270
value of the different compounds, 271
Bronchial secretion hindered, 214
troches, 304
Brown-Séquard and others on nerve-centres, 19
Bumstead's vaporizer, 283

CAFFEINE and morphia, antagonism of, 205
Camphor and iodine inhaled, 243
as an expectorant, 207
effect on the system, especially the mucous membrane, 231
effect in bronchitis, 207
effect on the system, 207
Canada balsam as an expectorant, 231
Calcium phosphate, 64
amount in the body, 64

INDEX. 311

Calcium phosphate, effect of excluding it, 65
Calomel and opium, 123
Carbo-hydrates or amyloids, 23
Carbolic acid, history of, 254
Carbonate of ammonia as an expectorant, 225
Cases of antipyretic treatment in the German hospitals, 153
Catarrhal pneumonia, 165
Caustic, safest method of applying it, 307
Central pneumatics, 263
Chemical daily needs of the tissues, 25
Cherry-laurel, its use, 273
Chinolin or quinolin, 174
Chloral as a depressant of the respiratory centre, 267
Chlorate of potash as an alterant expectorant, 240
    effect on the mucous membrane, 241
    effect on the system, 240
Chloride of sodium in food, 24
Chloroform, effect on the respiratory membrane, 260
Chlorosis, use of iron in, 61
Codein, 193
Cod-liver oil analyzed, 49
    best methods of administering it, 51
    history of its use, 48
    in digestion, 50
    in phthisis, 50
    with ether, 52
Coffee, its effects, 90
Cold affusions as an antipyretic, 154
    as an antipyretic, 145
    bath treatment to check heat-production, 143
    baths, in diphtheria, 152; in pneumonia, 149; in scarlet-fever, 152; in typhoid, contra indications, 149
    by sprays, as an antipyretic, 155
    drinks as antipyretics, 155
    enemata, as antipyretics, 156
    history of its internal and external use, 145
    in acute rheumatism, 148
    in typhoid, 147
    locally applied as an antipyretic, 155
    packs as an antipyretic, 154
    sponging as an antipyretic, 154
    treatment, various opinions of its value, 156
Comparative value of different diets, 30
Conia as an inhalation, 275
    best preparation to use of, 275
    or conine, effect on the respiratory centre, 274
Conium, 274
Contra-expectorants, 263
Copaiba, 237
Corrosive sublimate fatal to bacteria, 250
Counter-irritation, 117,
    blisters, pustulants, issues, and moxa, 118
    cupping, foot-baths, etc., 117

Counter-irritation, hot water, 117
    its effects, 118
    poultices and fomentations, 117
Cough, a natural relief, 214
    with no secretion, 214
Creasote, 254
Croup-kettle, 284
Croupous pneumonia treated with cold baths, 147
    use of quinine, 164
Cubeb, 253
    lozenges, 304
Cuca analyzed, 93
    as a beverage, 92
    as a stimulant, 93
Cucaine as a nerve food, 93
Cyanide of potassium, safe dose, 273
Cyanosis, 18

DENUTRIENTS, 107
Depressant expectorants, 222
    antimony and ipecacuanha, 222
Depressants of the respiratory centre, 266
Deprivation of amyloids, 41
    of proteids, 41
Destruction of tissue, excessive rapidity of, 7
Dextrin, 23
Diaphoretics, jaborandi, 138
    opium, jaborandi and its alkaloid, pilocarpine, 134
    spirits of nitre and liquor ammoniæ acetatis, 138
    their action, 133–135
    Turkish bath, 135
Digestion disturbed through deficiency of food, 41; through excess of food 40; through the quality of the food, 41
    of iron, 56
Digestive process disturbed by the body, 40
Diminution of iron in the system, 60
Digitalis as an antipyretic, 175; in phthisis, 165
    in bronchitis, 176
    in hæmoptysis, 176
    in phthisis, 176
    in pneumonia, 175
Diphtheria, cold bath treatment, 152
Disinfectants, history of, 249
Dried blood as a nutrient, 77
    foods, 24
Double-valved inhaler, 285
Dover's powder, 190, 191
Dyspnœa, 20

EFFECTS of change of external temperature, 142
    of cura on fatigue, 92
    of counter-irritation, 118
Eggs as nutrients, 47
Emetics, 128
    antimony, 128
    effects on the system, 128

312   INDEX.

Emetics, in croup, diphtheria, and bronchitis, 130
Elastic gases set free in a patient's room, 282
Elemi as an ointment, 237
Errhines, effect on the respiratory membranes, 213
Essential oils analyzed, 230
  as expectorants, 230
  elimination of, 231
  effect on the alimentary membrane, 230
Ether and chloroform as antispasmodics, 260; compared in their effects, 261; effect on the circulation, 260
  effect on the respiratory membrane, 260
  effects on the system, 54
  in digestion, 52
  methods of administering in dose, 54
  ordinary dose, 55
Ethyl-iodide, 245
  or hydriodic-ether as an antispasmodic, 260
Evacuants, 110
Excessive amount of liquid drunk, 86
Excretions from cold bath treatment, 143
Exercise and rest, 94
  its effect on the general system, 96
Expectorants, 213
  action of different remedies as, 215
  which affect the quantity of sputa secreted, 215
  which diminish the secretion, 216
Expectoration in health, 214
Experiments on the secretion of mucus in animals, 217
External heat as a diaphoretic, 134
Extract of malt as an aid to digestion, 74

FARINACEOUS and saccharine nutrients, 47
Fats in digestion, 23, 37, 39
  or hydrocarbons, 23
Fatty nutrients, 47
Feeding by the nose or rectum, 77
Fetid gum resins, 235
Fomentations, 117
Food and diet, 21
  introduced through a fistulous wound, 38
  of an invalid, 42
Forced feeding by the stomach, 77
Formic acid fatal to bacterium subtile, 251
Free phosphorus, 68
  as a nutrient, 70
Function of respiration, 14

GALBANUM as an expectorant, 236
Gargles, alum, chloride of aluminium, tannin, borax, as astringents, 300
  astringents, disinfectants, and antiseptics as, 300
  Condy's fluid, carbolic acid, as disinfectants, 301
  of lactic acid, 301

Gargles, their use, 298
Gargling, the act of, 298
Gas in the stomach, 35, 40
Gastric juice, 34
  ancient artificial use to aid digestion, 74
Gelatinoids, 22
General expectorants, 212
Glandular system of the respiratory tract, 211
Gout from habitual excess of food, 41
Grape cure, 47
Gum resins, history of, 233

HABITUAL excess of food, 41
Hæmatinic effects of iron, 59
Hæmoglobin as the purveyor of oxygen, 15
Hemorrhages, iron used as a spray, 62
Hot water as a medicine, 88
Hunter's inhaler, 286
Hydrocyanic acid, chief use of, 273
  in whooping-cough, 273
Hydropathy, 88-145.
Hyoscyamus and stramonium, 206
  effect on the respiratory organs, 206
Hypodermic injection of blood and food, 83
  of milk, 84
Hypophosphites, 72
  use in North London Consumption Hospital, 73

IDIOPATHIC anæmia, 60
Indol, 36
Ingredients of milk, 21
Inhaling solutions from a bottle, 296
Inhalation chamber at San Remo, 283
  of camphor, 295
  of carbonic acid, 278
  of chlorine, 280
  of fumes of sulphur, 280
  of fumes of tar, 279
  of iodine, 280
  of steam, 294
  of steam with hops, and many aromatic sedatives, 295
  of stimulants, 295
  of vapor chlori, vapor of iodi, 295
  rooms at mineral spas, 281
Inhaler-respirator, 292
Inhalers illustrated, 283
Injections into serous cavities, 82
  into the peritoneum, 82
Inorganic substances in food, 24
Inula helenium, effect on the system, 237
Ipecacuanha, its effect on the system, 132
  in respiratory diseases, 129-132
Iron, 56
  and manganese in the blood, 25
  as a nutrient, 57
  best given with food, 62
  effect on the pulse, 58
  in bronchial affections, 63
  increasing oxidation, 58

INDEX. 313

Iron, quantity assimilated and eliminated, 60
    supplied to the blood, 12
    used in hemorrhages, 62
    use of, in chlorosis, 61
    with the red corpuscles, 57
Irrigation of the nasal passage, 302
Irregular secretion of bile, 37
Iodide of iron in phthisis, 63
Iodide of potassium as an expectorant, 219
    sprays for asthma, 245
Iodine, antispasmodic expectorant, 245
    as an alterative expectorant, 243
    as an expectorant, 244
    effect on the system, 243
    excreted, 244
    inhaler, 285
    in pneumonia, 247
    in phthisis, 246
    in syphilis, 246
    its use in respiratory diseases, 246
    stimulates the respiratory membrane, 243
    susceptibility of patients to, 245
    with belladonna, 246
    with carbolic acid inhaled, 243
    with morphia, 245
Iodism, symptoms of, 244
Iodoform, applied to the throat as a powder, 307
    as a spray, 297
    effects of, 248
    with croton-chloral-hydrate, in phthisis, 248

JABORANDI and pilocarpine, their effect, 138
    antagonized by belladonna, 140
    effect on the temperature, 139
    effect on the sight, 140
    in renal dropsy, asthma, pleuritic effusion, and diphtheria, 140
Jurgensen's method of baths in pneumonia, 149

KAIRIN, an antipyretic, 171
    history and uses of, 171
    illustrations of its use, 171
Koumiss, 46

LARIX, effect on the system, 238
Leeches, cupping, etc., 117
Lee's steam-draught inhaler, 286
Leucin and tyro-in, 36
Liebig, Pettenkofer and Voit on food converted into tissue, 10
Linctus, 303
Lobelia and tobacco as depressant expectorants, 224
    effect on the system, 224
Lozenges, their use, 303
Lungs, amount of air after a full inspiration, 15

Lungs, amount of air at every respiration, 15
    labored respiration, 15
    vital capacity under different circumstances, 15

MAGNESIUM phosphate, 66
    amount in the system, 66
Martindale's portable inhaler, 286
Mastich, 237
Meal-times, 42
Medicine, influenced by the digestive organs, 11
    general effect in passing through the system, 12
    regarded as food, 11
Mercury as an evacuant, 121
    effect on the system, 124
    eliminated by the excretions, 124
    history of its use, 121
    in acute bronchitis, 126
    in broncho-pneumonia, 126
    in diphtheria, 126
    in pleurisy, 128
    in pneumonia, 125-127
    in tonsillitis and in parotitis, 126
    its extravagant use, 122-124
    perchloride and sublimate in diphtheria, 127
Methods of inhaling, 282
    of vaporizing a room, 283
Micro-organisms, experiments, 251
Milk as a nutrient, 46
    for invalids, 42
Morphia and atropia, 201
    and atrophia, neutralizing effects of, 202
    salts, 192
Mucous-ferment in digestion, 35
Myrrh in phthisis and as a gargle, 234
    stimulant to the mucous membrane, 233

NARCOTICS, general use of, 187
Nasal douche, 301
    best astringents in, 301
    mode of employing it, 301
Nauseant diaphoretics, 136
    Dover's powder, 136
Nauseants, 130
    antimony, 131
    as depressants of the respiratory centres, 266
    in pneumonia, bronchitis, croup, 132
Nerve centres of perspiration, 19
Nervous control of nutrition, 10
Neurotics, 187
Newman's atomizer, 289
Nitrate of silver, application of, 220
Nitrite of amyl, in angina pectoris, 258
    in asthma, 258
    its effect, 137
    of sodium, effect on the circulation, 259
Nitrogen, 22

Nitrogenous diet, 28
   plastic or albuminous foods, 22
Nitro-glycerine, dose of, 259
   in angina pectoris, asthma, whooping-cough, etc., 259
Nœud vital, 19
Non-nitrogenous foods, 23
Nutrients as promoters of construction, 45
   beef tea, 46
   buttermilk, 46
   milk, 46
Nutrition in relation to respiratory diseases, 11
   in relation to therapeutics, 3
   from atmospheric oxygen, 3
   too abundant supply of, 8
Nutritive value of foods, 26
Nux vomica and strychnia stimulants of the respiratory centre, 265

Oil of anise, effect on the mucous membrane, 231
   of eucalyptus as a stimulant expectorant, 231
Opium and aconite, 192
   and bromide, 192
   and chloral hydrate, 192
   death from, 189
   effect of a small dose, 188
   effect on children and extreme age, 192
   effect on the respiratory centre, 264-267
   in coryza, influenza, asthma, and bronchitis, 190
   toxic dose of, 188
Opoponax, as an expectorant, 237
Ordinary condiments as stimulants to digestion, 78
Oxidation explained, 16
   in life and death, 6
   in the tissues, not in the blood, 16
Oxygen, effects of, absence of, 18
   inhalers, 284
Oxygenation impeded or arrested, 17

Pancreatic juice, 36
Pancreatin, the best method of administering it, 76
Pancreatized foods, 76
Parasite, origin of disease, 250
Pastilles, 305
   of morphia, etc., 305
Pâte de jujube, 304
Pepsin, 34
   action of, 35
   and pancreatin as aids to digestion, 75
   prepared with salt, 75
Peptonized foods, 76
Perchloride in blood-poisoning, 62
   in diphtheria, 62
   in phthisis, 63
   in scarlet fever, 63
   of iron, 62

Pilocarpine and emetia as expectorants, 220
   in diphtheria and phthisis, 140
Phosphate of lime, 65
   best method of administering, 66
   in anæmia, 66
   used in rickets, 65
Phosphates eliminated in bronchitis and phthisis, 67
   in food, 25
Phosphorus and its compounds, 64
   as a stimulant in the exhaustion of fever, 71
   best method of administering it, 72
   free, its use and elimination, 68
   in diseases of the nervous system, 71
   in phthisis, 72
   in pneumonia, 71
   its action on the system, 68-70
   its use in diseases of the osseous system, 71
Phthisis, quinine as an antipyretic, 165
Physostigma, 274
Pneumatics, 207
Pneumonia, cardiac failure and its treatment, 151
   cold effusions, 154
   treated with cold baths and stimulants, 150
   use of stimulants, 151
   various treatments, 165
Potash, as an expectorant, 219
Potassium phosphate, action on the system, 68
Powders applied to the respiratory passages, 305
   how best applied to the respiratory organs, 306
Preparation of food stuffs, 32
Prepared foods to be introduced per rectum, 77
Prussic acid, effect of an officinal dose, 272
   effect on the blood and respiration, 271
Ptyalin, 33
Purgation, its use, 119
Purgatives, 119
   drastics, 120
   mild aperients, 120
   saline laxatives, 120
Pyrexia from septic conditions, 144
   its causes, 144

Quebracho as an antispasmodic, 261
   dose given of, 262
Quinia, effect on the excretions, 143
Quinine and cold baths, 152-163
   as an antipyretic, 157
   effect on the blood, 157
   effect on the heart, 161
   effect on the nervous system, 161
   effect on the system, 157
   elimination of, 161
   in acute rheumatism, 163
   in catarrhal pneumonia, 165
   in croupous pneumonia, 164

INDEX. 315

Quinine in hay-fever, 164
   in pneumonia, 152
   in phthisis, 165
   in whooping-cough, 164
   its action, 144
   its antiseptic action, 159
   large doses, 161, 162
   power of destroying micro-organisms, 160-164
   power over fever heat, 162

RAPID rubefacients and vesicants, 119
Relative proportions of food at different meals, 42
Remedies in accumulated secretion of the bronchial tubes, 215
Rennet in digestion, 35
Resorcin, 174
Respiration, 14
   controlled by the nervous centres, 211
   during balloon ascension, 17
   affected by the quality of the air, 209
   effects from varied altitudes, 17
   in mines, diving-bells, etc., 17
   labored, 15
   more active when arterialization is defective, 19
   number required to renew the air in the lungs, 16
   performed by the skin, 18
Respirators, 292
   their use, 293
Respiratory centre of nerve action, 19
Rest as an antiphlogistic agent, 108
   as a first remedy in respiratory diseases, 95
   promotes nutrition, 95
Rickets, 65
Rossbach's experiments on secretion of mucus, 217

SAGAPENUM, effect on the respiratory and alimentary mucous membrane, 236
Salicin, antipyretic and antiseptic, 168
   as an antipyretic, its history, 166
   as a substitute for quinine, 167
Salicyl compounds, 166
Salicylate of soda, its use, 169
Salicylates, effect on the stomach, 168
Salicylic acid an antiseptic, 168
   best method of administering it, 170
   its action, 164
Salicyls as antiseptics and disinfectants, 169
   effect on the blood, 169
   effect on temperature, 169
   in acute rheumatism, 170
   in diphtheria and scarlet fever, 170
   in pneumonia, 170
   process of elimination, 169
Saline diaphoretic, mindererus' spirit, 136
   diaphoretics, 133
Saliva, 33
Salts, 24
   as nutrients, 48

Salts of lithia as an expectorant, 219
Scarlet fever, cold bath treatment, 152
Scheele's acid, 272
Schnitzler's atomizer, 288
Scilla as an expectorant, 225
   its effect on the mucous membrane, 226
Secretion of mucus, effect of soda on, 217
   effects of temperature on, 217
   experimental application of medicinal substances, 217
   in the normal state, 217
   Petronne's experiments on animals, 218
Sedative and anodyne pneumatics, 262
Senega as an expectorant, 227
   effect on the system, 227
   history of, 227
Siegle's atomizer, 290
Simple atomizer, 289
   inhaler, 295
Siphon douche, 301
Soda, as an expectorant, 219, 239
Sodium and potassium phosphate, 66
   and potassium phosphates, action on the system, 67
   phosphate, action on the system, 67
Solids slowly dissolved in the mouth, 303
Spirits of nitre, its effect, 137
Spiritus ætheris nitrosi, 136
Sprays, cold and warm, 296
   remedies and amount required, 297
Starvation, 8
   death accelerated by a low temperature, 8
   death hastened by reduction of external air, 8
   death-point four-tenths the normal weight, 8
Steam-draught inhaler, 284
   inhalation, in acute bronchitis, 294; in diphtheria and croup, 294
   inhalations, their use, 294
   with hops inhaled, 295
Sternutatories, 213
Stimulant expectorants, 224
Stimulants of the respiratory centres, 265
   of the respiratory centres, in various combination, 266
   to digestion, 78; ordinary condiments, aromatics, carminatives, bitters, general tonics, sedatives, 78
Storax as an expectorant, 232
Stramonium, effect and use in asthma, 206
Strychnia, for night sweats of phthisis, 265
Succus entericus, 38
Sulphate of copper as an emetic in poisoning from vapor of phosphorus, 69
Sulphur, effect on the mucous membrane, 242
   history of, 242
Symptoms produced on respiration by varied altitudes, 17

TABLE of food required for rest and for work, 27

Table of the nutritive values of food, 27
Tannin, application to the larynx, 220
Tar and its derivatives, as antiseptics, 253
    as a stimulant expectorant, 239
    vapor, 254
Tea, coffee, etc., analyzed, 90
    its effects, 91
The antiseptic respirator, 293
    complete atomizer, 291
    eclectic inhaler, 287
    pharyngeal douche and irrigator, 362
Topical pneumatics, 276
    pneumatics, history of, 276
Transfusion, 79
    history of, 79
    of blood of animals, of blood-serum, of milk, 80
    of blood into the pleural cavity, 82
    of carbonate of ammonia, 81
    of defibrinated blood into arteries, 81
    of saline fluids, 80
    of saline fluids into arteries, 82
Tubercle bacillus, difficulty of destroying them, 252
Turpentine as an antidote for poisoning from vapor of phosphorus, 69
    as a stimulant to the mucous membrane, 228
    effect of an overdose, 228
    oil of, effect on kidneys, 229
Turpentines analyzed, 229

Uses of inhalations and other topical pneumatics, 294
Uric acid in relation to gout, 7

Vagi divided, 19
Valuable properties of quinine, 160
Vapor of phosphorus, its effects in large doses, 69
    of phosphorus, its use, 68
Variations in the digestive process, 39
Vegetable acids, 24
    diet, 28
Venesection, 110
Veratria as an antipyretic, 174
Vomiting to empty the air-passages, 129

Warm beverages in respiratory diseases, 87
Warm blood as a nutrient, 77
Waste and repair in health, 4
    products, accumulation of, within the tissues, 7
Water, amount in the body, 26
    as a dilutant in the system, 86
    as a regulator of temperature, 87
    as food, 26
    average amount taken in health, 85
    cold, to assist digestion, 87
    diluents, beverages, 85
    hot, as a medicine, 88
    necessary to digestion, 85
    quantity in the system, 86
    rapid introduction into the circulation, 86
    removed through the skin, 18
    temperature best suited for digestion, 87
    warm, in painful affections, 87
Willow, history of its use in medicine, 166
Wintrich's atomizer, 288

Ziemssen's method of baths, 148

www.ingramcontent.com/pod-product-compliance
Lightning Source LLC
Chambersburg PA
CBHW030743230426
43667CB00007B/820